To Erica Doreen Miller, 1911–2007

WHO DO YOU THINK YOU ARE?™

ENCYCLOPEDIA OF GENEALOGY

THE DEFINITIVE GUIDE TO TRACING YOUR FAMILY HISTORY

NICK BARRATT

HarperCollins*Publishers*

HarperCollins*Publishers*
77–85 Fulham Palace Road,
Hammersmith, London W6 8JB

www.harpercollins.co.uk

First published 2008

10 9 8 7 6 5 4 3 2

BBC and the BBC logo are trademarks of the
British Broadcasting Corporation and are used under licence

BBC © BBC 1996

Text © Nick Barratt 2008

ᐯᐱᐯ
wall to wall

Who Do You Think You Are?
Series Producer: Jamie Simpson
Executive Producers: Lucy Carter and Alex Graham

The right of Nick Barratt to be identified as the author of this work
has been asserted in accordance with the Copyright, Design and
Patents Act, 1988

A catalogue record of this book is available from the British Library

ISBN-13 978-0-00-726199-4
ISBN-10 0-00-726199-3

Printed and bound in China
by South China Printing Co. Ltd

Contents

Introduction

Congratulations. By picking up this book, you have just taken your first step on a unique journey into your past, one that will gradually reveal lost generations of your family that you never knew existed; their place in history; and the path that has led you to stand here today, reading this. Thousands of others have started a similar voyage of discovery in recent years, each on a personal mission to reveal who their ancestors were, and what their lives were like.

> *Every one of us is the living embodiment of the strands of personal history woven by our ancestors.*

One of the driving forces behind this phenomenon is the hit BBC TV show *Who Do You Think You Are?*, where every week a celebrity investigates their ancestral roots. Essentially a social history of Britain and its wider role in the world, the programme has stimulated millions of people to challenge their memories from school that history was a dull, academic subject and explore the past from a fresh perspective – that of their own relatives, rather than the politicians, generals and royals that tend to populate our textbooks. As a result, history becomes real, living and relevant; it's a personal journey into the past, with your own relatives as the tour guides. Events that you might once have read about in a textbook suddenly take on a new meaning once you realize that your ancestors were there as eye witnesses or even participants. The most exciting thing of all is that everyone can trace their family tree – it's not the exclusive preserve of those with privileged blood lines or aristocratic roots, but something that each and every one of us can do.

So why start looking into your family's history? Traditionally the press have given genealogy a rather negative image. Indeed, one commentator was moved to write that family history was 'self-indulgent navel gazing'. What utter nonsense! There are so many reasons why it is important to look into your family's background, leaving aside the sheer joy of discovery that makes it such an addictive pastime. Perhaps the most important reason for starting is that you are going to discover more about yourself and your family, and gain a real understanding about where you have come from, who the main people were that shaped the fortunes of your family, and how small decisions in the past have had a knock-on effect over the years. In essence, every one of us is the living embodiment of the strands of personal history woven by our ancestors, all of whom contributed in some way to making us who we are today. In turn, their struggles to survive in a variety of changing conditions allow us to gaze into wider British social history, and ask questions about how we fit in. What was our class or cultural background? How did we fare in some of the great social upheavals in the past, such as the Industrial Revolution? Each generation faced a new challenge, and you can revisit these moments in time through your investigations.

There is another important reason to set out on this voyage of discovery, namely to preserve these links with the past, which – once broken – are very difficult to repair. This is why, as you will see, you should always talk to your elderly relatives and record their stories, anecdotes and knowledge. It is a sad fact of life that we often take an interest in our past when it is no longer possible to talk to those who played such an important part in shaping it. Yet it is not just about preserving the past; by looking into your family's background, you will be creating a legacy that can be passed on to future generations – children, grandchildren and those still to come. This is especially important in the digital age, when we are no longer creating the treasured artefacts that we now look for and preserve as keepsakes from years gone by – photographs, letters and postcards. Our means of communication – email, text, mobile phone – are instantly disposable unless we take active steps to preserve them, so it is just as important to record our thoughts and feelings now, or future generations won't have the material to hand to understand us, or the people that made us who we are.

However, there are several myths about family history that you might have heard, and these need to be explained briefly. One misconception is that it is all about building a family tree as far back as possible. Whilst it is important to name ancestors and place them in history, the family tree is simply a map of your roots, showing you how your ancestors are related to one another. The real purpose of family history is to bring these names back to life by researching where they lived, what jobs they did, how their community changed over time, and the ways in which their lives were touched by local and national events. In many ways, the phrase 'family history' has rather outlived its use; instead, we should be thinking of 'personal heritage', as you'll be creating a far richer, brighter and more interesting picture of your ancestors than a list of names on a page.

The second myth that needs dispelling is that family history can be done quickly via the Internet, and that all you have to do to track down your distant ancestors is log on, subscribe to a few websites, and you'll have a family tree ready within hours. Sadly, this is completely untrue and misses the entire point of starting in the first place! As you will discover, there are certainly plenty of websites that will help you get started – and the datasets and databases that they contain continue to grow almost daily – and you will be able to achieve an amazing amount before you have to consider heading into an archive or museum. By putting the basic sources online, such as indexes to birth, marriage and death records, census returns and some wills, the process of constructing an initial family tree has been revolutionized. However, the Internet only provides a fraction of the resources you'll need to flesh out the bones of your family tree and – as indicated above – bring your ancestors to life as real people who faced real challenges. This is where this book takes over, and leads you into the world of Britain's archives, where there are original documents – often dating back centuries – that contain details of your ancestors' lives, or were even written by them.

The aim of this book is to provide you with the relevant guidance, advice, information and inspiration to start out on this journey with confidence and realistic expectations. For the first time, a practical step-by-step tutorial for the beginner is combined with an overview of the basic sources you'll need to get started, as well as the most comprehensive guide to the main family history topics that you'll encounter as you progress: the brave men and women who took up the call to arms and served their country in one of the branches of the military; those who left the land and found work in a factory, ironworks or coal mine and therefore became the lifeblood of Britain's industrial success on a world stage; the hopeful new arrivals coming to Britain to start a new life, at the same time as thousands more left these shores to emigrate to foreign lands that were brought into the orbit of the British Empire; or even those family secrets and hardships that our ancestors sought to hide from us. Furthermore, each section incorporates an overview of the social history behind the topic and surviving records that you'll need to consult to track down your ancestors. Every chapter is illustrated with case studies from celebrities in the show, who have unearthed some amazing stories during their investigations. There are also two reference sections that offer practical research tips for the most common lines of enquiry; the meaning behind the top surnames and occupations you are likely to encounter; and detailed guides to the main archives in the UK and Ireland that you will have to visit to dig for information.

I've been fortunate enough to have been involved with *Who Do You Think You Are?* from the start, as consultant genealogist for the first four series and appearing at the end of each episode in series 1 to help explain some of the research that underpinned the programme. However, I came to genealogy via a background in academic history – my PhD is in medieval history and I first joined The National Archives as a research advisor in 1996, although I've worked in the field of personal heritage for over ten years, ever since I became involved in the BBC's *House Detectives*. In 2000 I started my own research agency, Sticks Research Agency, to provide historical research and consultancy for the media.

The fascination with one's past still creeps into battle-hardened professionals, and even though I deal with other people's family history research most of the time, I still get drawn back to the mysteries in my own background. My mother's maternal lineage is completely unknown, thanks to my mysterious grandmother who has taken her secrets to the grave with her. All we are left with are some photographs, tantalizing stories of being spirited out of Europe on the eve of the First World War, and a connection with the American financier Arthur Chase. Nevertheless, it is still possible to make a surprising break-through, even when you think there's nothing left to find. One recent discovery has revealed a paternal great uncle who was caught spying for the Soviet Union in the 1920s, which only came to light via a random trawl of The National Archives' catalogue. His name appeared in some newly released intelligence files, and on inspection they contained a dossier over 100 pages long chronicling his activities, including photographs of him as a young man; his First World War service papers; surveillance notes and phone intercepts; and an account of a sting operation that was meant to incriminate him. The material also answered a long unsolved mystery surrounding his death, as it revealed that he had committed suicide at the point when he realized that he was about to be unmasked as a spy.

As you will find when you start work on your own family, each question that you answer will lead to a whole raft of new questions – one of the main reasons why seasoned genealogists have been working away for decades. At the end of the day, the history of your ancestors is your story to tell – so enjoy the detective process, the thrill of the chase, and happy hunting!

> *The history of your ancestors is your story to tell – so enjoy the detective process.*

How to Use This Book

The *Who Do You Think You Are? Encyclopedia of Genealogy* is the definitive, comprehensive guide to tracing your roots, and putting them into the correct historical context so that you fully understand not just who your ancestors were, but also the way they used to live. This book is split into five sections.

SECTION ONE

Section One concerns the preparatory stages you need to do before you even start logging on or heading to the nearest library, record office or archive – the sort of work you can do at home and with your family to hand. It includes gathering initial information from your family; organizing it into a family tree; setting your research goals; and working out which archives you'll need to visit first. These are crucial steps to take, and are often skipped over by enthusiasts straining at the leash – often with disastrous consequences later on. Remember, perfect planning prevents poor performance!

SECTION TWO

Section Two introduces you to the key resources you'll need to build and expand your family tree, in particular civil registration certificates of births, marriages and deaths; census returns; wills and probate documents; and parish registers. These will provide sufficient clues to bring your relatives back to life as real people who lived interesting lives. Many of these are now available online, and should be used as building blocks to construct a secure foundation for your research.

SECTION THREE

Section Three is where it all gets personal. By this stage you will have built your family tree, and this section provides more detailed information about the ways you can investigate the historical context surrounding the names you have uncovered. There are several sub-sections, each reflecting a major theme of British social history over the last few centuries, many of which are likely to have directly affected your ancestors. The main topics covered include military history, as it would be a great surprise if at least one ancestor wasn't involved in the forces at some point; occupations over the ages; migration into and out of Britain; family secrets, since we all have a skeleton or two lurking undiscovered in the closet; and wider aspects of social history, such as working further back in time and looking for blue-blooded ancestors. What makes the book unique is that there is a history of each theme as well as a description of the records you'll need to consult, where to find them and how to use them.

SECTIONS FOUR & FIVE

Sections Four and Five provide supplementary practical advice and support to help structure your work as it progresses. Section Four contains troubleshooting guides which take you along some of the most common lines of research step by step, such as searching for military ancestors, or those who entered or left Britain over the last few centuries; whilst Section Five provides profiles and meanings of some of the most frequently occurring surnames and occupations that you are likely to encounter during your research; information on genetic genealogy, where you trace distant relatives through your DNA; and information about the key archives, institutions and websites that you will visit or use during your work.

So, if you are a novice family historian and you want to get the best out of this book, don't skip over Sections One or Two. More experienced researchers might want to focus on Section Three. And if you get stuck, head for Sections Four and Five to kick-start your research in another direction.

Getting Started

The aim of this section is to encourage you to take those initial steps! Family history can be daunting and many people are put off because they simply don't know where to start. The chapters in this section will take you through the key preparatory stages, from talking to your family to building your family tree, setting your research goals (so that they are realistic!) and locating the best place to start your research.

CHAPTER 1

First Steps

Every journey starts with a single step, and in the case of tracking down your ancestors your first step should be little more than a small pace. Your key resource is the knowledge contained within your family – biographical data, anecdotes and stories, personal documentation, and treasured family heirlooms, objects and artefacts. All of these can be used to build up a picture of your family in terms of acquiring facts such as 'who is related to whom', as well as an idea of what your relatives were like.

There are a couple of questions you need to ask yourself before you get stuck in, as the answers will determine the direction your research will take. So ...

Why Do You Want to Start Your Research?

> The first step is to gather as much information as you can about your family, from your family.

If you've ever watched an episode of *Who Do You Think You Are?* you'll notice that the initial focus of attention is the celebrity researcher themselves – what they know, what they want to find out, and how they feel about their family. This process of self-reflection is exactly what you need do in real life, in the sense that it is your quest, and therefore you need to set your own research goals. You should take some time to reflect on why you want to find out about your family's heritage.

Everybody has a different reason. It may be to find out the truth about a long-standing family myth, such as the story passed down to Sue Johnston that her grandfather once drove the *Flying Scotsman*, or perhaps to find out more about yourself and why you have certain

character traits, which was the original motivation for Bill Oddie's investigation into the background of his mother – to find out why he felt abandoned by her at a young age, and understand the circumstances of her prolonged absences from the family home. Maybe your reason is to preserve the memory of the people who have shaped your destiny, in the way that Natasha Kaplinsky uncovered the truth behind what happened to her relatives who were killed in the Holocaust; or you may simply want to find out about your family out of personal interest, so that you have some stories to tell your children or grandchildren about their ancestors.

▲ Family photographs are a vivid reminder of our links to the past.

How Do You Want to Approach Your Research?

Your reason for starting out will largely determine what your initial research aims are. Although there are no 'right' or 'wrong' ways of tackling research, there are several common ways to start out. Do you want to trace back as many generations as you can as quickly as possible, or would you rather look at each generation in detail and work back gradually? Some people prefer to concentrate on one side of their tree first, either their mother's or father's branch, and then begin the other side once they feel they have found all they can about the first branch.

You may find it easier to build a skeleton tree as far back as you can to start with, because this does not need to take very long now that the Internet has brought genealogy into our living rooms, and then concentrate on putting some more flesh on the bones of the characters you found most interesting from your preliminary research by digging around in the archives. Even if your goal is to discover as much as possible about one particular ancestor, it can be helpful to investigate who came before and after them, as these are the people who would have influenced that ancestor's life and been a part of their world.

Reality Bites …

Who Do You Think You Are? at times can make family history appear to be quite easy – but in reality, it can be anything but! Months of research underpin each programme, and many of the actual steps taken to arrive at a pivotal moment in the storyline aren't filmed or shown, simply because there isn't enough time to squeeze them into the programme.

However, one thing that is reflected accurately on screen is the all too real sense of disappointment when a promising line of enquiry comes to an end. You have to be realistic with your initial aims; some families are going to be harder to trace than others, particularly the further back in time you progress. Similarly, if you have a very common surname in your tree, such as Jones or Smith, you will encounter difficulties tracing ancestors along that branch due to the sheer number of people who will share their name.

Equally, if you know very little about your family to start with, it will take that little bit longer to get the ball rolling and you will probably have to purchase more certificates until you can work your way back to the nineteenth century, where the availability of another important set of documents, the census records, helps to speed up the process. Both of these key sources are explained in more detail in Section Two.

It's important not to be disheartened when you encounter setbacks such as these. You simply have to keep persevering, and you will find that the reward when you do discover that missing link is worth all the additional work. Genealogy is a detective process, and just like any investigation there will be times when you hit a brick wall and can seemingly go no further. There are tips about how to seek help to overcome these obstacles in Chapter 4.

Starting Out

The first practical step you'll need to take is to gather as much information as you can about your family, from your family; and where better place to begin than with yourself? After all, it's your journey. Write down everything you know about your immediate family, from your date of birth to your parents' names, dates of birth and marriage, and see how far back you can go from there. Can you name all of your grandparents? Do you know their dates of birth, marriage and death if applicable? What about your eight great-grandparents? Can you name them, and provide similar details? It's not as easy as it seems, and many people simply can't give all of this data from memory. Nevertheless, even if you have doubts about what you think you know, it's important to write down as much as you can remember about everyone at this stage.

Apart from this important biographical data which, as you'll see in Section Two, you will eventually use to start tracking down original documents, you also need to focus on other aspects of their lives. In particular, you need to focus on where they were born, married, lived and died, as geography plays an important part in the detective process. Indeed, make a note of any scrap of information that you can find out about them, such as what jobs they did, whether they moved around and when they lived in certain places. It helps to write down the names and age differences of any siblings you know of as well, as these may enable you to narrow down a search in the archives later on.

Are there any family stories that were passed down to you that you want to find out more about? These will provide the colour in your family tree, and even trivial details can prove to be important in the next phase of your research, when you ask members of your family to comment on your memories. Did Uncle Albert serve with the Merchant Navy? What about great-grandpa, whose tales of valour in the Great War were retold regularly at Christmas? Memories of growing up in a foreign land, such as colonial India, can help you locate missing branches of your family when the time comes to search official records. Write down what you can remember about these snippets of information, and who told them to you. Research into these stories and family myths can run in parallel with your work constructing your family tree. You may find that as you build the basic tree the truth about some of these stories emerges, or it may become clear that more complicated research will be required to piece the jigsaw together, in which case Section Three of this book will be able to aid you.

Widening Your Search: Talking to Your Family

Having written down as much information as you can about your fore-bears, it's time to cast the net a little wider and draw upon the collective wisdom of your living relatives. Holidays such as Easter or Christmas, when the family tends to congregate, are often good times to begin your research, because festive gatherings tend to generate a sense of nostalgia, when folk naturally start to reminisce about happy times from the past, swapping anecdotes about relatives who may no longer be around to enjoy the festivities. If you can't wait for a natural opportunity to arise, you can always organize a family reunion, making sure to invite as many

SUMMARY

Recap of what to write down when making your initial notes:

• *Your full name, dates of birth and marriage, names and dates of birth and marriage of your children and grandchildren*

• *The names and vital details of your siblings*

• *Your parents' names, dates of birth, marriage and death if applicable*

• *The names and vital details of your aunts and uncles*

• *Your grandparents' names, dates of birth, marriage and death if applicable*

• *The names and vital details of your great-aunts and uncles (the siblings of your grandparents)*

• *Anything you know about your great-grandparents, their siblings, and anyone who came before them*

• *Family stories you have heard and who told them to you*

▲ Old family letters can hold many clues to family relationships.

of the older generations as you can. You will probably find that others are just as interested in your research as you are and will be eager to help you – story-telling is as much fun for the narrator as the audience.

Although you may have heard the same stories told year after year, there are probably plenty more that you haven't heard, mainly because it's easy to play down moments in one's life that we think are uninteresting, but are actually fascinating to someone who wasn't there. An 'everyday' childhood memory of growing up during the Blitz is still a powerful, unknown and chilling story to a later generation who have no concept of what it would have been like.

Aside from these colourful stories, it's important to focus – as before – on simple biographical details of names, dates and places. This is why it's important to talk to older members of the family; they can tell you about their parents and grandparents, folk that you are unlikely ever to have met other than in faded photographs. However, don't forget to record details of their lives as well – where they lived, what their jobs were and, most important of all, what they were like as people. It's all too easy to treat family history as an academic exercise, but these are the details you'll want to pass on to other members of the family. You'll be amazed at what you can uncover by spending several hours talking to a great-aunt – details of your grandparents as children, growing up in the countryside for instance, and working on a farm before moving to the city later in life. These conversations will peel back time and you'll see your family in an entirely new light – your grandparents as children; your uncles and aunts as brothers and sisters; and generations of your relatives at work and play, in love and in mourning.

Sadly, many people leave it too late to start this important process, or simply don't have any living relatives to help with this initial research. Whilst this makes things a bit harder, and removes the colour from the first stage of your research, it is still perfectly possible to start your family history from scratch, using the information on your birth certificate to find your parents' marriage and birth certificates and then work back from there. If this is the case, the information in Chapter 5 will help you get started.

Interview Techniques

It may seem like the most natural thing in the world to sit down with your relatives and extract information, but in reality a great deal of planning ought to go into this process, not only to focus your attention on what you need to find out, but also to put your family at ease. After

all, you don't want them to think they're about to face the Spanish Inquisition! It can be rather unnerving for both interviewer and interviewee at first, so you need to go out of your way to make the process as simple and fun as possible. For example, if you've set up a family gathering, you could even have a bit of fun and turn it into a game – initially asking the same few questions to everyone and comparing the answers afterwards to see who remembers the most, stimulating discussion and allowing you to focus on the most likely source of further information.

However, if you are spending time visiting members of the family individually, make sure you've compiled a clear set of questions, topics and people that you want to ask them about. Who was Great-aunt Alice? When was she born? Who was her husband, and when did they marry? Where was the ceremony? Did they live in the same area? So, Great-uncle Herbert was a farmer? Where was the farm? It's also important to focus on one family member at a time, so that neither you nor your relative becomes confused. In general, you should concentrate on obtaining initial information about:

- Names, including Christian and nicknames, surnames and maiden names
- Dates of birth, marriage and death
- Places of birth, marriage, death and abode
- Occupations

Once you've obtained as much biographical data about a person as you can, it's then time to ask about what the people were like. Having found out that Great-aunt Alice was born in London, but ended up marrying a farmer called Herbert in a remote part of Norfolk, the burning question is how did they meet? How did she adapt to life on a farm, having been brought up a Londoner? What was she like as a person?

This is where you'll have to exercise your diplomatic skills, as people can ramble on a bit, and memory will play tricks if the events being described took place a long time ago. You will need to balance the desire to learn about a particular subject with the ability to let someone talk about their past without too much interruption, because we all love telling anecdotes. However, your relatives may not want to talk about everything that's happened to them. Attitudes to illegitimacy have changed over time, and what to us is an interesting story might be a stigma that's caused pain and misery for decades. If you sense that someone is uncomfortable talking about certain matters then do not

If you are visiting family members individually, compile a clear set of questions, topics and people to ask them about.

force them to continue. It is better that you leave that topic of conversation so your interviewee does not feel pressurized. They may even decide to come back to talk to you about it at another time when they feel more comfortable.

Alternatively, they might want to talk to a third party or non-family member about what's happened to them. This is particularly true of painful memories that relate to war. You would be amazed how many former combatants don't tell their families about their experiences to shield them from what they went through, but will happily talk to a military historian who they believe has a greater understanding. As a final resort, you can always suggest that your relative writes down their secrets in a sealed envelope and leaves it to you in their will. Although this may appear frustrating, it does give them the opportunity to take their secrets to the grave with them, yet still reveal what it was that they thought too sensitive to talk about.

If there is a particular story or person that fascinates you, it's going to be important to talk to as many members of the family as possible, and compare different versions of the same tale – where accounts agree or overlap, there is likely to be a greater degree of truth. However, it's going to be your job as a family historian to verify everything you hear, which is why good note-taking is essential to this process.

Oral history is invaluable to genealogists and historians, creating a living link with the past. It is preferable if you can record your interviews, because as each generation gradually dies out there are fewer and fewer people to speak to who remember a way of life that will never return, and can tell stories about colourful characters that would otherwise be forgotten by time. There are many ways of recording interviews now, so, if your interviewee gives their consent, make the most of the opportunity to video-record the conversation using a camcorder or digital camera. This way their memories and stories are preserved for future generations to watch and enjoy, and you could find

▼ Weddings are a popular subject in the family photo album – but can you name all the people in yours?

a way of incorporating this material into a digital presentation of your research once you have finished. Ian Hislop made a very poignant observation when reviewing old cinefilm footage of his family on holiday: most of the time the camera was pointed not at his father, but at him – yet it was his father's thoughts, feelings and stories that he wished to hear in later life.

This is your chance to record your family, and you should urge them to be filmed – always respecting their decision to decline if they are really not comfortable with this process. At the very least, with permission try to keep a tape recorder handy to catch gems of information, and be sure to take clear, detailed notes that you will be able to go back to at a later date and decipher how each person is related, what details are relevant to which people, and who told you each scrap of information. It is important to keep a track of the sources of all your information so that you can return to that relation if you need to ask any further questions once you have begun your research.

> *Oral history is invaluable to genealogists and historians, creating a living link with the past.*

Family Secrets and Myths

You should be prepared that you might find out more than you bargained for, and your discoveries may even change your perception of some family members. It can be exciting to uncover a skeleton in the closet of a distant relation who died long before your time, but there are often secrets kept within families even now that come as a surprise to a generation that is far more open than its predecessors.

Issues such as illegitimacy, adoption, bigamy and even criminal activity may creep into your family tree at some point in time, all of which will be discussed in greater detail in Section Three. If you do discover a potentially revealing aspect of a close relative's past life, think carefully about how to deal with passing that information on to other family members or talking to the people it may have a direct effect on. It can take a long time before family secrets are accepted as being out in the open, and although moral values have changed so that what used to be considered scandalous behaviour is now not so, there are still many people who hold traditional values and find it difficult to discuss such behaviour. Furthermore, as the chronicler of your family's history, it will ultimately be your research that becomes public knowledge, and your decision as to how much you tell people. Once information is out there, it can't be taken back – so think long and hard before pursuing a family secret.

Verifying Information

While the information you glean from your extended family is vital to your research, be wary of believing everything you are told as fact. Oral historical accounts are invaluable, but are also subject to a certain amount of exaggeration and human error. Cross-reference the information you are given with lots of different family members, because details often get confused as time passes and people get older. Various individuals may have a contradictory account of the same event. The more information you are armed with, though, the easier your research should be. It is up to you to untangle the stories and find out who is right using primary sources in the archives.

Never assume that dates and places you are told by relatives are correct, even if your source is adamant they are right. These are simply guidelines for you to follow to speed up the research process. Everything should be verified using official documentation where possible, such as civil registration certificates for births, marriages and deaths (described in Chapter 5) which will carry the official date and place that an event occurred. This is often at odds with the supposed 'truth' you were told by your family.

When collecting information from relatives, be aware that the names they knew ancestors by may not have been the same names they were christened and registered with. For example, Granddad Liam's real name was actually Martin William, but he chose to use his second name, and then shortened that; Aunty Julie was actually born Mabel Julie. You will find that when looking for relatives' documents in the archives it is essential to know their official name because you will usually need this to locate their records in alphabetical indexes. Nevertheless, if a relative was known by more than one name it helps to be aware of the various options, so that you have an alternative to look for if you do not find them under their official name.

False relationships can also be planted in the information you receive from other family members. For example, you may need to do a fair bit of pruning of the family tree to remove lots of aunts and uncles who earned the title through family familiarity rather than blood ties, as well as tidy up the loose use of 'cousin'; Aunty Marie, who your Granny told you all about, was actually a close family friend rather than a blood relation, and Cousin Joyce may actually have been your Granny's aunt, but because there was such a close age difference between them they were brought up more like cousins rather than aunt and niece. Make sure you gather the specifics about exactly how each person is related so that you are not misled in your research before it even begins.

Looking for Clues

Now that you have as many names from living memory on your tree as you can gather, along with dates, places and occupations to work with, it's time to cast the net a little wider and start looking for physical clues. These can be tucked away in all sorts of unlikely places, such as in old boxes packed away in the attic or cellar; hidden in stuffed drawers; locked away in forgotten photo albums; or in safety deposit boxes in banks. You will be looking for a wide range of material, some of which may only take on a relevance once you've done a bit of initial research in archives. Given that you may not know precisely what you're looking for at this stage, it's important to try to get as many members of your family involved in the search for clues as possible, so that if anyone else stumbles across an interesting photo or family heirloom in the future they will let you know. Perhaps you will find some army medals or a wedding photo with names and a date on, giving you an immediate link to a military archive or the search for a marriage certificate. These forgotten objects can help immensely with your research, as they usually contain clues and spark up new lines of investigation.

Names and dates are often written on the back of old photos, regiment or ship names often inscribed on military and naval medals, or written on badges and uniforms; and all this memorabilia is evidence of your ancestors' existence and can fill in the gaps that are no longer within living memory. On the other hand, if you come across family heirlooms and are unsure where they have come from, be sure to ask other people in your family that might know. You are bound to find old photos and not be able to name one single person in them, so why not scan them and email a copy round to the rest of the family to see if they can help.

Make a special effort to locate birth, baptism, marriage and death certificates, as these can help verify the information you have been given from relations and can save you the money you would otherwise need to spend ordering duplicate copies from the General Register Office (see Chapter 5). Any copies of other official records you can find that may have been kept, such as wills, title deeds and legal documents, are a great stepping stone for your research, giving you a concrete foundation to work from and often supplying you with more names to add to your tree. Wills are particularly useful because they

▲ Part of the certificate of service for Adrian Veale, who enlisted into the Coldstream Guards in 1934 – it has been kept by his family all these years.

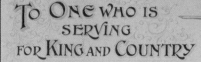

▲ A military memento found amongst personal papers within the family.

very often name members of the extended family and explain how they are related to the deceased person, and indicate where someone lived, who their dear friends were and what they did for a living – as well as possibly lifting the lid on a family secret or two, such as an illegitimate child given a sum of money.

If official documentation has not survived you may be lucky enough to find newspaper articles about your relatives that are often cut out, kept and treasured, perhaps if somebody did something that deserved special comment in the local paper. Wedding announcements and descriptions of the special day were very popular in the nineteenth century; or if one of your ancestors was well respected within the local community, an obituary may have been written about them shortly after their death. As well as newspaper reports, school reports can be just as enlightening, giving you an idea of what that person was like as a child.

Some families used to keep a family bible, handed down through generations, in which details of births, baptisms, marriages, spouses' names, deaths and special family events might be recorded. If you are lucky enough to have a surviving copy of this your workload will be instantly cut down. Family bibles can detail names and dates going back way to the early nineteenth and even the eighteenth century, and can often pre-date civil registration which, as you will see in Chapter 5, was first introduced into parts of Britain in 1837. As with all the sources you find, however, it is wise to double-check every bit of information that it contains against official records, because some family bibles may have been added to at a later date and could contain discrepancies.

Name patterns are usually a clue to the past. If an unusual first name or middle name has been passed down through a few generations, this can be an indicator that it was a maiden name of one of the women in your family tree that was passed down to her child and their subsequent descendants as a Christian name so that it was not lost after she took her husband's surname. For example, Basil Fanshawe Jagger was the father of Mick Jagger, lead singer of The Rolling Stones; a few simple searches revealed that Basil inherited his rather unusual middle name from his mother, Harriet Fanshawe. Keep an eye out for these distinctive names among the documents you uncover and see if you can locate the original source of the name when you start your research in the archives.

Until the late twentieth century, handwritten or typed letters were the main form of communication between family members who lived apart. The advent of email and mobile phone communication has changed all of that, so that correspondence with loved ones can be disposed of with the click of a button. Therefore, old letters that have survived can be of tremendous sentimental value to family historians, as well as being a great practical research aid, giving not only names and addresses, but also an idea of your ancestor's personality from their style of writing and sometimes giving an insight into their day-to-day lives. You should also look out for old postcards that can give you an idea of the kind of social standing your ancestor may have had. If they travelled abroad before cheap flights made this a common phenomenon, you will know that they probably lived quite well.

The First and Second World Wars produced an enormous amount of central government administration, a lot of which is stored safely in our national archives and is discussed in Section Three. But many of the by-products created by officialdom also ended up in people's homes. Some soldiers held onto their discharge papers after they had completed their military service, or would have received letters granting them exemption from compulsory conscription into the army, and many of those that did serve received medals or kept part of their uniform apparel as a souvenir of their contribution to the war effort. If you can find documents proving that your ancestor fought during either World War, or even that they were in the army, air force, navy or merchant navy before or after the wars, these will give you an indication of where you need to start looking to find any more documents that may be held in the archives for them. If you have an idea of the date they served and their rank, this is often enough information for you to start with, as will be explained in Section Three. Most medals were awarded for service at a particular time or for specific battles or events, so if you do find medals around the house you can examine the design to identify what they were awarded for.

In addition to the military paperwork generated by the wars, a wealth of civilian material also survives from that era, such as ration books, letters to and from loved ones separated by conflict, and telegrams from the army informing next of kin of the death of a soldier, all of which illustrate how difficult that time would have been for your ancestors. Civilian documents issued at other times are equally informative, like passports with a person's photo, vital details and stamps from the

Do not be scared to take a photo out of its frame to ensure that there are no names or other written details hidden on the back.

SUMMARY

Clues to look out for around the house:

• *Civil registration and religious certificates confirming births, marriages and deaths*

• *Wills, deeds and legal documents*

• *Newspaper articles and obituaries*

• *School reports*

• *Family bible and name patterns*

• *Letters and postcards*

• *Military, naval, air force and merchant navy documents, medals and uniform apparel*

• *Civilian wartime letters, ration books, identity cards*

• *Passports and citizenship documents*

• *Old receipts, magazines, tickets to the theatre or to football matches*

• *Photos*

places they visited, or identity papers and naturalization certificates if they settled in Britain from a foreign country.

Do not pass off as junk the general day-to-day items you might find when hunting in the attic or through drawers. Old receipts, tickets to the theatre, ballet, opera or to a football match, magazines that have been kept, all give an idea of what your ancestors enjoyed spending their money on and doing in their spare time. These are key indicators to what their lifestyle would have been like, and what they were like as people. If they believed that these bits and pieces were worth holding onto then that is an obvious clue as to what was important to them.

Photos are by far the most fascinating of our family artefacts. Even if we cannot name the majority of people in the frame it is always interesting to observe the different fashions, expressions and landscapes, and to try to work out when the picture was taken and what those people's lives would have been like. Photos in the Victorian and Edwardian periods were often very formal. Most people did not have a camera of their own and would have visited a photographer's studio or had their picture taken at a photographer's stand at a fair. The rarity of a photo opportunity during these eras meant that people wore their finery or would borrow clothes from the studio's wardrobe to dress up for the occasion. The clothes worn by the subjects can help you to identify a rough date for the photo, as specialists can establish when specific types of dress were fashionable. Your local archive or museum may be able to help you date the costume or background in an old photo. Photographic studios frequently printed their company name and address on photos, so you can trace this in trade directories to establish when that studio was in business, and to work out the rough geographical location where the person in the photo was living.

Do not be scared to take a photo out of its frame to ensure that there are no names or other written details hidden on the back. If the picture does not have any names or a date written on it, show it to as many elderly relatives as you can to see if anybody recognizes the faces or location. It might also be a good idea to make copies of photographs you find in relatives' houses, either by scanning them or taking digital photographs of the images – having obtained permission first. This way you can write on the back of your copies each time you identify a new face. Carry the pictures around with you so that you can keep adding to them as you show more relatives. You can also find out about how to preserve old photographs, or restore fading images, from local archives and specialist companies who now offer fairly cheap methods of storage and restoration techniques.

Preserve Your Past for the Future

Whilst talking to your extended family and delving into the family treasures they have hoarded, it will become apparent just how important it is to preserve your own family photos and mementos for future generations. Your children, nieces, nephews or grandchildren may not seem interested in their past while you are enthusiastically hunting away in the archives, but there will more than likely come a time when they will be curious to flick through old photo albums, read old family letters, and learn more about a past era that seems so different from today, but that their parents and grandparents were a part of.

The fantastic thing about genealogy is its educational element. Whilst finding out the names, dates and places of each person on your family tree, it is essential to put their lives into a social context, to find out what the major political and social events were that would have shaped their lives and affected their standard of living. Might they have visited the Great Exhibition when it opened in Hyde Park in 1851? Did they fight in the Boer War at the turn of the last century? Would they have been shocked to hear news of the *Titanic*'s sinking in April 1912? Placing your family history into a wider national and international historical context brings textbook history to life. These people you are related to really did exist, and while the name 'John Briggs' on a census return may not seem immediately exciting, when you look at the bigger picture and learn more about what life was like for him living in a Victorian slum, his existence gains meaning and our combined national past seems closer. As you find out more about each ancestor and can pin major historical events to their lives, you can help the children in your family to understand their history.

Genealogy is not just about the past; it is also about preserving the present for the future. So why not keep hold of a few items that may seem inconsequential today but will help to illustrate some of the defining events of your life in the future. Just as you write down the names of the faces you learn about on old photos you find, make the same effort with your own photos so that people will be able to identify you and your loved ones. The following chapter will explore ways of storing your research and organizing your findings so that the whole family can enjoy your hard work, but remember – it's all too easy to concentrate on the past at the expense of the present. Make sure that you are at the heart of your research, so don't forget to leave behind an impression of what *you* were like. After all, you have just become the chronicler for your family, and future generations will want to know all about you!

SUMMARY

• *What is the aim of my research?*

• *What do I know about my ancestors?*

• *What do my relatives know about our ancestry?*

• *Are there any family mysteries to clear up?*

• *Are there any family heirlooms to give me some clues?*

• *What information do I need to verify?*

CHAPTER 2

Building Your Family Tree

By now you will have spent many hours writing down what you know about your family, talking to relatives and looking for physical clues and objects that have accumulated over the years. The next stage is to organize this information into a family tree, and use this to choose which path then to follow – verifying information you are uncertain of; searching for new ancestors; or pursuing an interesting relative or family story in more detail.

Creating a Family Tree

A family tree is a diagram that shows at a glance how your relatives and ancestors are related to one another. This will become the foundation of your future work, a growing document that incorporates all the biographical information you uncover as you hunt for documentation in archives, libraries and museums. The importance of building a family tree from the instant you start your research at home has already been touched upon in Chapter 1, but you will learn here just how vital it is to keep updating your tree after every discovery so that you can see at a glance what your next research step should be.

People can get quite confused about drawing up a family tree, assuming it is a more complicated process than it really is. There are many software packages on the market that promise you an all-singing, all-dancing family tree with generational reports, photo uploads and print-outs. But if this is your first attempt to put a family tree together, it's probably best to go back to basics until you're more familiar with the procedure, and simply use a large piece of paper and a pencil.

> *A family tree shows how your ancestors are related to one another – and to you.*

This section will show you the various methods of writing family trees and the abbreviations and genealogical terminology used. Some of this may be familiar. If you've watched *Who Do You Think You Are?* regular graphics appear on screen to show you how, for example, John Hurt is related to Walter Lord Browne. Or you may have seen pedigrees published in books or newspapers that relate to the royal family or members of the aristocracy. Even though you may not have such distinguished roots, the principle behind a family tree's construction remains the same.

However, before attempting to build your first family tree, it's important to have a basic grasp of some of the terminology used, since you'll need to describe how members of your family are related to one another.

Understanding Family Relationships

In essence, this will be *your* family tree, so anything you produce should start with you, with your name placed right at the centre of the blank piece of paper. Everyone else is therefore described in terms of their relationship to you. On this embryonic family tree, your parents' names will be written above you; your brothers and sisters – known as your siblings – will be either side of you, also underneath your parents; and the names of any children you have will be written below you, with their children – your grandchildren – below them. Above each of your parents will be their parents – your four grandparents – and alongside each of your parents will be their siblings, your uncles and aunts.

Every group of people on the same horizontal line represents a separate generation. Most people are familiar with these terms, but these are all close family members, and you will be working many generations back into the past when it becomes harder to keep track of distant relationships; so listed below are some of the key words used to describe relatives from the extended family, and ancestors further back in time, which are perhaps less familiar.

Blood Relations

The direct line in your family tree is made up of all the people who have been biologically crucial to your creation. Therefore they would include your parents and your grandparents, but not any of their siblings and other descendants – these people are your extended family. Each time you move one generation further back, you need to add 'great' as a prefix. Therefore the parents of your grandparents are known as your great-grandparents, and the parents of your great-grandparents are your great-great-grandparents, and so on. Every time you search for another generation in your direct line you will be looking for twice the number of people as the generation that came after that. This is because you have two parents, who each have two parents, so that you have four grandparents. These four grandparents have two parents each, which means you have eight great-grandparents, and then sixteen great-great-grandparents. As you work further back than this, you might find it easier to shorten this description to '2 x great-grandparents'.

Extended Family by Blood

These are the people that are related to you by blood, but are not biologically crucial in your existence today. Where possible, you should include them in your family tree – particularly after your first phase of research – but you might want to focus on your direct ancestors and come back to them at a later date.

Nieces and Nephews

Your nieces and nephews are the children of your siblings. Niece is used to describe a female offspring and nephew to describe a male offspring. Any subsequent children of your nieces and nephews are known as your great-nieces and great-nephews, and another 'great' is added to the prefix each time another generation is born.

Uncles and Aunts

Your uncles and aunts are the siblings of your parents. The siblings of any previous generations in your direct line are described by adding 'great' as a prefix, and each generation you go back another 'great' is added. Therefore the siblings of your grandparents would be your great-uncles and great-aunts, and the siblings of your great-grandparents would be your great-great-uncles and great-great-aunts. Some people use the word 'grand' instead of 'great', and might describe these relations as 'great-grand-uncles and aunts'.

This is your family tree, so everyone else is described in terms of their relationship to you.

Cousins

Your cousins are the children of your aunts and uncles. These are known as your first cousins. Any subsequent descendants of your first cousins are indicated by how many generations they are 'removed' from you. If your first cousin has a child, this child is your 'first cousin once removed'. If your first cousin then has a grandchild they would be your 'first cousin twice removed'. First, second and third cousins can only be used to describe cousins of the same generation. Therefore, if you had a child, they would be second cousins with your first cousin's child. This means that while you refer to that relation as your first cousin once removed, your child would call them their second cousin.

To describe the cousins of any previous generations to yourself, you should refer to them as the cousin of whichever person in your direct line they are of the same generation as. This means that the parent of your second cousin should be called your 'parent's first cousin', and the grandparent of your third cousin would be your 'grandparent's first cousin'. The relationships between cousins is very confusing and you may find that in documents such as wills and census returns people use the word cousin to describe a distant relative without qualifying exactly how they are related. It will be your job to untangle this confusion!

Extended Family by Marriage

Families are complicated entities, and as well as direct blood relatives and their extended family you will quickly discover that there are other relationships that are more complicated to define – usually the result of divorce, subsequent remarriage and an associated second family. Here are some of the more useful terms that you may need to incorporate into your family tree.

In-laws

When one member of your family marries, they are related to their partner's family as a result of the wedding. This relationship is said to be 'in law' due to the legally binding nature of the union. Thus the mother of the bride is the 'mother-in-law' of the groom, just as the father of the groom is the 'father-in-law' of the bride. If either the bride or the groom has siblings, then they become the sister- or brother-in-law of the other party.

Step-relations

The word 'step' is used to describe the relationship to a member of your family that occurs through a subsequent marriage by one (or indeed both) of your biological parents. Your stepfather would be your biological mother's husband from a subsequent marriage, whilst your stepmother would be your biological father's wife if he married again. If either of your step-parents had children from a relationship prior to marrying your biological parent, these children would be your step-brothers and stepsisters; whilst they would be described as your biological parent's stepsons and stepdaughters.

Half-relations

The word 'half' is used to describe a relationship between children who share only one biological parent. For example, if your biological father or mother had a child with your step-parent, this child would be your half-brother or half-sister.

Drawing Your Family Tree

Using these terms, you should be able to build your initial family tree quite quickly; and it should display all the information you've found when you were interviewing relatives and looking through collections of family heirlooms. There are bound to be some things that you don't know yet – full names and biographical dates of more distant ancestors, for example – and you should add question marks against anything you're not sure about. This is one of the main purposes of this first family tree – to show at a glance what you need to verify, check or research more fully. Don't worry if it looks messy at this stage or a bit sparse – you'll be tidying it up and adding new branches when you start your research away from the family.

One thing to decide upon is the presentation style of your family tree, bearing in mind there are several recognized ways of doing this and the final choice will be down to you, based on what you find easiest to work with and how much data you wish to include. Some trees will only show the direct line, whereas others are very large and sprawling, and include all the siblings in each generation and distant cousins.

In the past, genealogy was the preserve of the aristocracy – or those that aspired to higher social rank – who wished to prove their connections to illustrious forebears. They commissioned diagrams, or 'pedigrees', that were as much works of art as family trees, with coats of arms, heraldic beasts and key names circled. A pedigree that shows immediate ancestors as far back as the sixteen 2 x great-grandparents is known as a *seize quartiers*, whilst one that covers all thirty-two 3 x great-grandparents is known as a *trent-deux quartiers*. These linear trees start with the most recent generation at the bottom of the tree, and continue horizontally upwards, with each entry representing people further away in time from the person at the foot of the tree. However, only key relatives, rather than all 16 or 32 direct ancestors, are often shown as they often focused on connectivity to the great and the good, or key marriages, rather than completeness.

Today, family trees that only show a direct line are often drawn not from bottom to top, but from left to right, with the most recent person on the left and their ancestors spreading out from them to the right of the page. There are pedigree templates for trees in this format available from the Society of Genealogists and local Family History Societies that simply require you to write the names and relevant dates in the spaces provided on the form.

While family trees that show just your direct line are a quick and easy way to map your immediate heritage once you have worked back far enough, it makes sense to include your entire extended family to start with as they can provide clues to help you move back further. UK genealogists favour drop-line family trees for this purpose, and these are the most common format you are likely to come across in books. Essentially, they are a diagram that shows how everyone is related to one another; and most people start by placing their own data at the heart of their family tree – sensible really because, after all, they are the ones undertaking the journey and will therefore be describing the people they find in relation to themselves.

You should write your full name, which should be the name you were registered with at birth, rather than a nickname or surname you took later in life. Therefore married women should always be written onto the tree under their maiden name, not their married name. Underneath your name, write a 'b.' to signify 'birth date' and then write your date of birth after that. Draw a horizontal line above your name and a small vertical branch coming down from the line to connect your name to it – a bit like a large 'T' shape with extended horizontal arms. Any siblings you may have should have their names attached to the horizontal line in the same way, which effectively creates an entire branch for your generation. You should start with the eldest sibling first, whose name should be written to the left of the branch, and work along to the right so that if you were the third child, for example, you would appear third on the branch, and the youngest sibling's name is positioned at the far right-hand end of the branch. Write every sibling's date of birth in the same way you did for yourself. If any of them have died you should write a 'd.' underneath their birth date, followed by their date of death.

Above your generation's branch you need to write your parents' full names. Traditionally, the man's name should be written on the left and the woman's on the right. Leave enough space between them to put either 'm.' or '=' to indicate their marriage, and write their date of marriage beneath this. Below the date of marriage you should draw a vertical line that connects their marriage to your generation's horizontal branch, thus showing that you are all related by blood. You can use the same method to add your own marriage date and spouse's name, and the marriages of your siblings if you wish. You may want to include subsequent generations after yours, such as your children, nieces and nephews, grandchildren and so on, in which case you will need to leave enough space below your generation to fit them in. Where space permits, each generation of children should be at roughly the same level on the tree – your nephews and nieces roughly alongside your children. However, if you are only creating a tree of your ancestors and not your descendants, then your name should be positioned towards the bottom of the page to allow more space for you to work back in time, up the page.

▼ You can start drawing your family tree by hand, using lines to connect each generation.

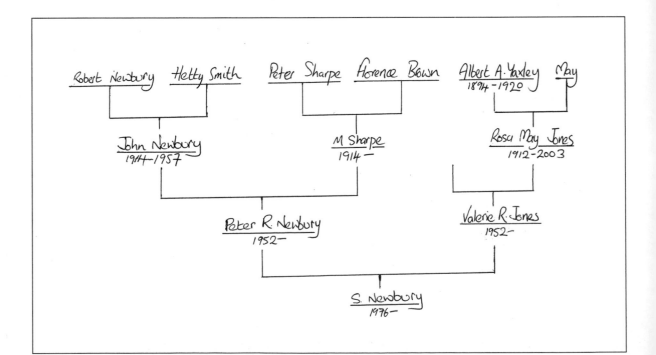

You may come across trees that depict relationships between parents and a child using a dotted vertical line rather than a solid one. This can be used for various circumstances. In the past when illegitimacy was deemed to be a problem, particularly for wealthy families for reasons of inheritance, a dotted line might indicate that a child was born out of wedlock or as the result of an affair. It can also be used to highlight a non-blood relationship between parent and child in cases of adoption.

You can now repeat the process you used for your own generation to put your parents' siblings either side of their names, each set of their parents' names above their branch, and keep repeating the process as far back as you can. The further back you work and the more siblings there are, the more difficult it can become to have them in age-descending order. You may find it more practical to put all your aunts and uncles, great-aunts and uncles and so on in age order but leave the name of the direct ancestor at one end of each branch so as to keep the diagram clear. If you do not know a woman's maiden or unmarried name, leave her surname blank so that you can fill in the space when you discover it. The same rule should apply to any other details you are unsure of, such as dates of birth, marriage or death. These will give you points to work towards, so that every generation has a complete set of details whereby each person's full name and their dates of birth, marriage and death are all known.

Some genealogists include occupations on their trees simply by writing these underneath each person's vital details. Having occupations displayed on your tree can help you to keep your work focused, so that if you are looking for a Jack Brown on the 1901 census you can use your tree as a reminder of his date of birth and marital status, and also of what job he should be described as holding. This can be of assistance if there are lots of people who have the same name in your tree but who can be distinguished by occupation. For example there may be a John Smith who was a woodcutter and a John Smith who was an engine driver. Alternatively, if a particular name was carried down through many generations you may find it useful to add a roman numeral after their name, indicating which generation they belong to. In this way the first William Perry, whose name was passed down to his son, then his grandson and great-grandson, would be known as William Perry I, his son would be William Perry II, his grandson would be William Perry III, and so on.

A family tree is not always drawn in a diagram, but can also be written using indented paragraphs. This requires the use of many of the abbreviations listed in the box on page 25 to explain relationships in

> *It is vital to keep updating your tree after every discovery so that you can see at a glance what your next research step should be.*

place of branches that would otherwise be drawn. Known as the 'narrative indented pedigree', this is not always the easiest method of reading a family tree as it can sometimes be confusing to follow, but it is the most straightforward way of typing up your tree if you are using a word-processing package to record your family tree, which does not allow you to draw branches very easily. It is also very handy to understand this method of describing a tree because some pedigree publications use this style, like *Burke's Peerage and Baronetage* and *Debrett's Peerage and Baronetage* (see Chapter 4). The indented pedigree starts with the earliest known ancestor and their marriage, and then lists the children from this marriage in age-descending order (although sometimes female children are listed after the male children instead). To list the children's offspring an indented paragraph is added under each child's name where their descendants' details are written. Therefore a narrative indented pedigree might look like this:

> James Sherwood *m.* Alice Clarke. Had issue:
>   John Sherwood *b.* 1648 and *m.* Jane Cecily. *dsp.*
>   George Sherwood *m.* Carole Vine and had issue:
>   Simon Sherwood *b.* 1672
>   Joseph Sherwood *b.* 1675 and *m.* Mary Shanks 1699.
>   He *d.* 1722 leaving issue:
>   Katherine Sherwood *b.* 1702
>   Grace Sherwood *b.* 1705
>   Emily Sherwood
>   Sarah Sherwood *b.* 1645.
>   Faye Sherwood *unm.*

This pedigree explains that James Sherwood married Alice Clarke and had four children, John, George, Sarah and Faye. John married Jane Cecily but he died without children. George married Carole Vine and had three children named Simon, Joseph and Emily. These were therefore James Sherwood's grandchildren. His grandchild Joseph Sherwood married Mary Shanks in 1699 and died in 1722 leaving two daughters, Katherine and Grace, who would have been James Sherwood's great-grandchildren.

Irrespective of what style of family tree you eventually decide to use, it will hopefully grow too big for your original piece of paper, so you will probably need to break the tree into sections to make it more manageable. While it is nice to have your entire family tree on one piece of paper, you should be constantly referring to it to help organize your

Abbreviations in Family Trees

Here are some examples of words and abbreviations used specifically in family trees:

b. born
m. or mar. married
= married
2. second marriage
d. died

ob. or **obit.** died
d.s.p. or **o.s.p.** died childless
d.v.p. or **o.v.p.** died before father
l. left descendants

bapt. or **bp.** baptized

chr. christened
bur. buried
lic. licence (marriage licence)
MI monumental inscription

c. circa or about
? uncertain or unknown
o.t.p. of this parish

w. wife
s. son
s. and h. son and heir
dau. daughter
g.f. grandfather
g.m. grandmother
g.g.f. great-grandfather

g.g.m. great-grandmother

inf. infant
spin. spinster (unmarried woman)
bach. bachelor (unmarried man)
unm. unmarried
div. divorced
wid. widow (a woman whose husband has died)
wdr. widower (a man whose wife has died)

mat. maternal or female side of the family
pat. paternal or male side of the family
Distaff female side of the family
Spear male side of the family

research, and for this reason it usually makes more sense to break it down into smaller branches, perhaps with your paternal side on one tree and maternal side on another. Some people find that smaller trees of individual generations are useful for taking to archives. These can then be updated regularly and annotated while you are in the archives, and the new information transferred to your master family tree at a convenient time.

Online Family Trees

Online family trees and family tree software packages are extremely helpful to collate your tree in its entirety so that you can share it with other family members, and to organize the end product of your research. Using these resources saves you the effort of constantly re-writing a large family tree if you run out of space or make mistakes, because you can easily log onto your electronic tree and edit the details as needed. Most genealogy software now saves your family tree and genealogical data as a GEDCOM file, which stands for Genealogical Data Communications. This has been created to make sharing your tree easier, regardless of the software you use.

SUMMARY

The important components of a comprehensive family tree are:

• *Names, including Christian or forename, surname, maiden name and any nicknames*

• *Dates of birth, marriage and death*

• *Place of birth, marriage, death and abode*

• *Occupation*

You can buy computer software packages to upload onto your PC, such as Family Tree Maker, Roots Magic, Family Tree Builder and Family Historian – all are popular and very flexible in how you can organize your data. They will come with instructions on how to print out your tree once you have uploaded it using the software, and most software now shows you how to build your own family history website using one of their website templates. Alternatively, there are many free family tree tools available from genealogy websites, which just require you to register your details on their website to create an account, after which you can share your tree with other enthusiasts online. Most of these also give you the option of keeping most of the details on your tree private or only accessible by users who ask for permission to view your tree first.

The four sample websites described below will be looked at in more detail in Chapter 4 in the context of the sets of data and documents they offer, but their family tree building tools and software are examined here.

www.genesreunited.com

Genes Reunited is a sister site of Friends Reunited and works as a database of family trees, enabling people to find others who are looking for the same ancestors. Other users will not see your full family tree unless you grant them permission after they have emailed a request to view it. You can search the Genes Reunited database of names, years and places of birth to see if any match the people in your tree. To make contact with other researchers you need to upgrade to a full membership for a small fee. When uploading your family tree, each person you add has a fact sheet to complete, listing their names, dates and places of birth, marriage and death, occupation, any notes, and a photo. You can view your family tree in a drop-line format showing all your relatives, just your ancestors or descendants, or your immediate family. It is possible to search the database's collection of historical records and merge these into your tree, and the software will automatically use these to create a lifeline for each person showing the key events of their life. Genes Reunited also has a special print function for printing out your family tree diagram.

www.ancestry.co.uk

Ancestry is an online genealogy company that provides, along with many millions of records, access to family tree building software and

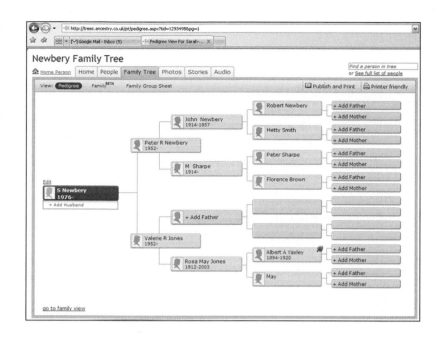

◄ Once you have drawn your family tree, there are many online family- tree-builder packages available to organize your data.

the facility to upload it onto their website. They have created a free online template that can be accessed by clicking the 'My Ancestry' link along the top of the homepage. The Ancestry family tree facility creates a homepage for each person on your tree, where you can enter their dates of birth and death, their spouse's details and children's names, upload photos, write a biographical story, and add events to a timeline. A summary of the information you enter is displayed on a family tree showing the direct line, working from left to right. Ancestry has a search facility that checks the details you enter against its collection of historical records and other users' family trees to see if there are any matches. This can help you get into contact with other people who may be researching part of your tree (which is usually because you are related somewhere along the line). You can change the privacy settings for your tree so that it can only be viewed by those people you email it to, otherwise the default setting puts your tree in the public domain so that other Ancestry users can find the information it contains.

www.myheritage.com

MyHeritage is a genealogy company that provides free family tree software that you can download from their site. This software allows you to create a family tree on your computer and add photos and documents to it using a simple interface. If you wish, you can then publish your family tree online to share it with family members. The MyHeritage.com homepage also lets you create a free family tree online, without down-loading software. This is done in your own family website which you can use for sharing photos, events and news with family members. This approach, often named 'web 2.0', is suitable for users who prefer a web-based experience over using a software program. Special genealogy technologies found only on MyHeritage (both the website and the soft-ware) include face recognition technology that helps you tag people in photos and recognize unidentified people; and tree-linking technology called Smart Matching that can connect your family tree to more than 1.5 million other trees on MyHeritage. Smart Matching helps you enlarge your family tree and find new relatives and ancestors. Also on MyHeritage.com is an extensive genealogy search engine that searches more than 1,500 online genealogy databases.

www.familysearch.org

Family Search is run by the Church of Jesus Christ of Latter Day Saints, otherwise known as the LDS Church. The Church has been gathering and preserving genealogical records from around the world for over 100 years. You can either upload a GEDCOM file of your family tree from your computer onto their website so that it can be viewed by other researchers when they search the website's collection of ancestral files, or you can download free Personal Ancestral File (PAF) software from the website and save your tree on your computer as a PAF file. PAF software allows you to enter information about each person's birth, baptism, marriage, death, notes about the sources you have found, and photographs that can be used to form a scrapbook. You can view your family tree in Family View, which shows you a person's immediate family, in Pedigree View, which displays the direct line in a diagram working from left to right, or in Individual View, which lists each person in your tree and their birth details.

Setting Your Goals

Having drawn up your family tree, you are ready to begin the next phase of your work - setting your research goals, and then working out which archive or resource you are going to use to achieve those goals. To do this, you must first take a long, hard look at all the information you've collected - not just the family tree, but the anecdotes, stories, objects and artefacts - and decide what to tackle first.

HOW TO...

...make your family tree

1. *If drawing a tree by hand, only use ink when you are sure of a fact, having verified it with official documentation such as a birth, marriage or death certificate*

2. *Write information you are unsure of in pencil, and leave a question mark against dubious dates or names so that you know further research is required*

3. *Use your family tree to shape your research plan. Focus on the areas where you want to work further back in time, or are not sure of information you've been told*

4. *Put a date on your family tree each time you revise it, and where possible create a new version each time you've discovered something new – either a new file name if you are working online or with a software package, or draw up an amended tree by hand. That way, you can always go back to an earlier version if you've made an error*

5. *Keep a clean master copy of your family tree, but make copies of sections of it to take into the archives with you. That way you can focus on one branch of the family at a time, which will help avoid confusion*

6. *Once you are more familiar with the practice of compiling your family tree, use a software package to help keep all your notes together. You can add photos, video and audio clips, images of documents and biographical notes to bring the tree to life, depending on the package you have chosen*

7. *Think about adding your family tree to one of the various online communities that link your data to that of other users. You may find that someone has already done the work for you – though beware of simply accepting non-verified data at face value*

Step One: Verify Your Data

Although there are no rights and wrongs, it is strongly advisable to begin by verifying the biographical data you've collected through your initial investigation within the family. This will mean ordering duplicate birth, marriage and death certificates where there are gaps in your immediate family tree – if you are not certain of Great-granny Doris's date of birth, or when your grandparents were married. The good news is that much of this verification process can take place from the comfort of your own home, armed only with a PC and a credit card. As you will find in Section Two, many of the key archive resources (lists of birth, marriages and death, census returns, etc.) are available online; Chapter 4 explains how best to use the Internet during these early stages of your research work.

It is vitally important that you know about more traditional ways of finding biographical information, particularly since a large amount will be stored in archives around the country and may not be available online. You will need to learn how to spot important clues from civil registration (birth, marriage and death certificates) documents (for example, that Great-granddad was a soldier when he got married in 1914); to work out whether there are any relevant documents associated with those clues (there are some army service papers for the First World War, 1914–18); establish where the documents are kept (although some documents can be found online, at www.ancestry.co.uk, the bulk are stored at The National Archives); and then visit the institutions in person, which can be a daunting experience if it's your first time. Consequently Chapter 3 goes through each step of this process and describes the different types of archive that are available. (This is essential reading if you are to make the most of the remainder of this book, particularly Section Three where the topics – military connections, immigration and emigration, social history, occupations and family secrets – cover material that is rarely available online.)

Step Two: Working Further Back

> It is usually sensible to focus on one branch of your family at a time, particularly if they have a more unusual surname which will make them easier to trace.

Having verified the initial data by using civil registration documents as far back as possible – they go back to 1837 in England and Wales, 1855 in Scotland and 1864 in Ireland – most people decide to follow one line of their family further back in time, looking for new ancestors based on the information they've found from these certificates. Once again, the key steps you'll need to take when tracing an ancestor who was born prior to (say) 1837 are covered in more detail in Section Two, where you'll start to work with some of the key sources aside from certificates, such as census returns, wills and parish registers. Essentially, this process means that you'll be adding new branches to your family tree. Although there's a temptation to jump in and tackle all lines at once, it is usually sensible to focus on one branch at a time, particularly if they have a more unusual surname which will make them easier to trace – you'll have more success tracking down Jeremiah Sandwick than John Smith, for example. Once you've got into the swing of things, you can then speed up your research and look at more than one line at a time.

Step Three: Focusing On One Story

Some of the more enterprising among you may decide to focus on one particular family story, which will involve more specialized research in an archive or institution. Depending on the story that you choose to investigate, the period of history in question or geographical location, you will almost certainly have to tackle more complicated archives or record offices. For example, you might want to set out on the elusive tale of your great-grandfather's period of service during the Boer War, or the intriguing story that – somewhere – there's a link to royalty waiting to be uncovered. This will almost certainly involve far more complex lines of research, documents that are less familiar or easy to get hold of, and more sources of frustration if you haven't covered the research basics (certificates, census data, etc.) beforehand. Advice about working in archives is provided in the next chapter, and in particular how to set about locating the relevant archive for the topic of your choice. The more common family history topics, once the basic processes of verification and tree extension are done, are elaborated upon in Section Three.

Set a Budget

There is a cost involved in undertaking a genealogy project, and before deciding which step to take you should consider how much you can afford to spend obtaining information. Travelling to archives, ordering certificates, buying copies of wills, paying for photocopies and signing up to subscriptions for commercial genealogy websites are all a necessary part of the process, but they do all cost money. Nevertheless, if you have planned thoroughly and are careful not to make mistakes (though some are inevitable!) you can avoid unnecessary expense. For example, you should exhaust the resources of any local archive, study centre or near-by family history centre, where you'll find plenty of material that's also contained in a national institution that may be further away. Many local libraries also have free subscriptions to genealogy websites, *The Oxford Dictionary of National Biography* and *The Times* online, which you can use from home if you obtain a library card and PIN number from your library.

Also, when ordering certificates it pays to be patient. By ordering one certificate at a time and waiting for that to arrive to see if the information is correct before ordering the next certificate, you won't waste money pursuing red herrings and false leads.

HOW TO...

... start to plan your research using your family tree as a guide

1. *Note all the ancestors for whom you need to verify key biographical data, and work out which documents you need for each*

2. *Identify one line of the family that you want to work on first*

3. *Extend that line back a couple of generations, updating your family tree as you go*

4. *As you gain confidence, repeat the process for other branches of the family*

5. *Turn to Chapter 3 to learn about which archives you'll need to visit, and which sources to use first*

CHAPTER 3

Working in Archives

This chapter explains how you can make your first foray into the world of record offices, archives, museums, libraries and other research institutions in the hunt for information. You will learn what sort of information exists out there, and how to use it to extract more names and dates, and to flesh out historical information about your ancestors; what an archive is, and how you locate the most relevant one for your initial research; how to work in an archive; and how to organize your research notes.

One of the few drawbacks of making a show such as *Who Do You Think You Are?* is that there simply isn't enough screen time to show all the work that takes place to put together the stories that you see. The actual research takes place behind the scenes over several months – exactly the same work you'll be doing yourself, although you will be able to take as much time as you like.

Once you have read through this chapter, and the research tips and hints in Chapter 4, you should be fully prepared to tackle the next stage of your research with confidence – in which case you can then head to the chapters in Section Two to learn more about how you can trace your family tree further back in time. An introduction to some of the major national archives and institutions can be found in Section Five.

> ' *If your ancestors did not live locally to where you now live, you will need to visit a local studies centre near the place they were from.* '

Gathering Evidence

As outlined in the previous chapter, the route most beginners take is to verify their initial findings, and then take one branch of the family further back in time, generation by generation. To do this, you'll need to use sources outside the family (although you may well have come across some of this material already in the form of certificates, wills and other paperwork tucked away in boxes, drawers and folders). Once these extensions to the family tree have been made, you will be able to put flesh on the bones, so to speak, by using more advanced research techniques to find evidence that puts the lives of your ancestors into an historical context.

Locating this evidence to build a family tree, learn more about these relatives and support the stories that are passed down through generations are the core tasks of a genealogist, so it's time to focus on what material you are going to use to achieve these goals, and where to find it. Roughly speaking, there are two main types of record you'll encounter during your work – *primary* sources and *secondary* sources.

Primary sources come in many shapes and forms, such as contemporary documents that survive from the period, or even oral accounts that are told to you by people who were present at an event. Of most use are officially created sources, such as birth, marriage and death certificates, as their creation and content have been governed and directed by legally binding requirements. These can be more reliable as evidence than personal documents like diaries, which are open to artistic licence and subjective opinion. Official sources are only as reliable as the people filling them in, however, and it is not uncommon for ancestors to 'forget' important details, or deliberately provide misleading information. The lesson here is never to take anything at face value.

Secondary sources are accounts written retrospectively by people who were not present, but may have had access to primary material, and as such can be subject to errors. Examples are history books written about a major event, such as the Boer War or life in a workhouse. While secondary sources will play a part in your research, you should always endeavour to locate primary evidence to back up your suspicions and findings. Stories passed down through the generations also fall into the secondary source category, unless the story-teller was actually present at the event.

Your initial investigations within your family will have already generated both primary evidence, in the form of documents, photos and letters found around the house, and secondary material from relatives in the form of anecdotes told to them by their ancestors. The next task is to find additional primary and secondary material to extend your family tree. Once this is done, you can then proceed to a wider search for information that will place your relatives in their historical context. It is time to turn to record offices, libraries and museums.

SUMMARY

Primary sources consist of:

• *Contemporary documents, such as diaries, letters, photographs, wills and other legal and financial documents*

• *Birth, marriage and death certificates*

• *Oral accounts by people who were there*

Secondary sources consist of:

• *Accounts written by third parties after the event*

• *History books*

• *Stories passed down within families over the years*

SUMMARY

The archival pyramid:

National and specialist collections

Municipal or county archives (area administrative records)

Local studies centres (general material)

Where to Look for Evidence: Archives, Record Offices, Libraries and Museums

What is an Archive?

The majority of primary material will be housed in record offices, libraries and museums, scattered across Britain – or, if your ancestors came from overseas, all around the world. Many people loosely refer to these institutions as 'archives'. Although this isn't the place for academic debate, in technical terms an archive is actually a collection of documents, manuscripts or other primary evidence, although the term is more often used to describe the building or institution in which the collection is housed. It is in this context that the word 'archive' will be used in this book.

For those of you who have never been to an archive before, it can be a daunting experience, but one well worth undertaking. Each archive is unique, will hold a different variety of records, and will have its own way of collecting, storing, cataloguing and indexing its records. Bearing in mind that information about your ancestors could turn up anywhere, the first step of your research strategy should be to work out which archives are going to be of most use to you first. The following notes should help you do this, but don't forget that you will probably need to visit more than one archive over the course of your research, and will often have to return to the same archive many times.

Local Studies Centres

There is a rough hierarchy to archives, ranging from general material held at local studies centres, via the administrative records of a municipal area or county, to national and specialist collections. It is advisable to start at the bottom of this archival pyramid first, and begin by looking for information at a local studies centre. These are often located in a local library, and hold records relating to the immediate area, which may cover a few towns and villages, or all the places situated within a borough. These records can include newspaper collections, rate books, electoral registers, trade directories, photographic material and private family papers deposited by local gentry, as well as maps and plans of the area. You will also find secondary sources here, such as histories of the local area, and if you are really lucky you may also find national collections – indexes to birth, marriage and death certificates, or census returns – on microfilm or microfiche.

The amount of material held varies greatly from one local studies centre to another . Some hold vast amounts of primary material while others are less well stocked. Therefore it is worthwhile contacting your local studies centre beforehand to enquire exactly what type of records they hold. If your ancestors did not live locally to where you now live, you will need to visit a local studies centre near the place they were from. Geography is very important to pinpoint the archives you need to visit.

Family History Centres

If you do not have a local studies centre in your area, then you may want to see if the Church of Jesus Christ of Latter Day Saints (LDS) have set up a Family History Centre in the vicinity. The LDS Church is an American organization founded by the Mormons in Utah. It has been collecting genealogical records from around the world for the last century, depositing them at its Family History Library in Salt Lake City. There are many Family History Centres around the British Isles and the rest of the world where duplicate copies of many of their central records are held, ranging from parish registers to ancestral files deposited by other researchers. You can find your local Family History Centre from the www.familysearch.org website by entering a country of interest in the Find a Family History Centre Near

Your Home search box and then scrolling through the alphabetical list of places for that country.

County Archives

In the hierarchical structure of archives, county record offices (CROs) are the next port of call. As the title suggests, a county record office is a central repository for administrative documents relating to the county, and each county has at least one. (Some have more than one, like Devon, which has three; the Devon Record Office in Exeter, the North Devon Record Office in Barnstaple and the Plymouth and West Devon Record Office in Plymouth.)

In general, most CROs hold census returns, rate books, electoral registers, trade directories, photos and prints, local government documents, maps, parish registers, civil registration indexes, private company and family papers and local history books for every place within the county (rather than just a few towns and villages covered by a local studies centre). In some cases a CRO will store duplicate copies of material held by local study centres, but in other areas the two types of repository will hold completely different sets of records on any given place within that county. It is always worth visiting the local studies centre and the CRO for the area in which you are researching, because there is bound to be at least something extra you will find in the CRO.

Of particular importance are the records deposited by locally important families, who historically would have owned much of the land within the county and therefore played an important part in your ancestors' lives. Their estate records, rent books, employment accounts, correspondence and records as local justices of the peace will contain thousands of names, many of which may be relevant to you and your search. However, it's worth remembering that especially wealthy families owned land in more than one county – so if you can't find what you are looking for in one CRO, it might be worth checking to see if important family papers for principal landowners are deposited elsewhere, possibly in another county where they had their main residence.

If your ancestors lived on the border of a county you should investigate whether the county borders have changed at any time. For example, Bredon's Hardwick, now in Gloucestershire, was for many centuries described as being in Worcestershire. As a result, some records for people who have lived in Bredon's Hardwick are located in Worcestershire Record Office, while other records are held in the Gloucestershire Archives. Equally, if your ancestors lived on the border of one or more counties they may have moved around and spent time living on both sides of the boundary at various times, in which case

there is probably material about them to be found in the record offices for both those counties.

Municipal Archives

It is also worth considering the collections of major cities, which are often stored in their own municipal record offices or archives. Many places have more than one institution for you to visit. For example, London is served by the London Metropolitan Archives, the Corporation of London Record Office and the Guildhall Library, each of which holds important historical and genealogical information.

National Archives

Each country in the British Isles has its own national archive where documents concerning central government are deposited. These are:

● The National Archives (TNA) based in Kew in West London, covering England, Wales and the UK
● The National Archives of Scotland (NAS) at Edinburgh
● The National Archives of Ireland, based in Dublin
● The Public Record Office of Northern Ireland (PRONI) in Belfast

There are also major collections relating to Wales at the National Library of Wales (NLW). Descriptions of each institution are provided in Section Five.

The holdings of each of these archives are not strictly determined

by geography. If your ancestors lived outside England then you may still need to visit The National Archives at Kew as well as the particular country's national archive, as the centralization of administrative records to London has affected all of the countries at some point in time. A change in the location of government does not always mean historic archives have moved to that new location. Each country also has a central General Register Office from which family historians order duplicate copies of birth, marriage and death certificates (see Section Two). The advice provided in Section Three relating to specific topics of family history will explain when you will need to visit each of the national archives, and what records you should use when you arrive.

Specialist Genealogical Libraries

There are other centralized archives that hold some documents that are not accessible at the above-mentioned national and regional archives, as well as duplicate copies of those that are. The Society of Genealogists based in central London has copies of many parish registers from county record offices around the UK, as well as indexes to records held in other archives, documents relating to people around the British Empire, and much more.

Libraries

It is also worth visiting your local library as well as an archive. Not only will many libraries play host to your nearest local studies centre, many have now opened specialist family history services, given the popularity of the subject these days. Furthermore, many libraries hold important manuscript collections that are worth visiting in their own right, as well as important reference works that will play a crucial part in shaping your knowledge of how and where your ancestors lived. This is

especially true at national level. The British Library in St Pancras, London, contains a copy of most books that have ever been published, but – as you will discover in Section Three – it also has a collection of genealogy records for people who lived in the British Empire, including records of baptism, marriage and burial in India. There are similar national libraries for Scotland (in Edinburgh), Wales (Aberystwyth), the Republic of Ireland (Dublin) and Northern Ireland (Belfast). Many academic libraries also hold important collections of primary evidence.

Museums

Finally, do not forget to visit museums, both specialist – such as the National Railway Museum or National Coal Mining Museum (both featured in *Who Do You Think You Are?* when Sue Johnston and Lesley Garrett went looking for their ancestors) – and local, such as the Rochdale Museum where much research into Bill Oddie's ancestors took place. Museums will be full not only of documents, but objects, artefacts, clothes, engines, machinery, books, sporting memorabilia – anything from the past that shows what life was like for your ancestors. This is where you will finally begin to understand the era in which your relatives lived, to encounter history up close, and find out about some of the events they lived through.

Museums can also help you to identify some of the bits and pieces you've found during your own research within the family. Items of clothing or household objects can be taken to local museums or national ones such as the Victoria and Albert Museum (for clothing and textiles especially), where curators can help you date them. Military memorabilia such as medals can often be interpreted at places such as the Imperial War Museum.

How to Find an Archive

There are various online databases to help you find libraries, archives, record offices, museums and repositories around the UK, and even around the world. Most of these allow you to enter the name of a place or use an interactive map to display a list of all the nearest archives to a particular area. The ARCHON Directory is perhaps one of the most

useful databases to start with, and it's available from The National Archives website at www.nationalarchives.gov.uk/archon. It contains addresses, telephone numbers, websites and street maps for local and major repositories all over the UK, the Republic of Ireland, the Channel Islands and the Isle of Man, and you can search by region or by entering the name of the place in which you're interested. Visit the Scottish Archive Network (SCAN) at www.scan.org.uk/directory/index.htm to find an additional alphabetical directory of Scottish archives.

If, as outlined earlier, you want to find out where the principal records of a particular landowning family are kept, the best way is to search the National

Register of Archives, now part of The National Archives. You can view their paper indexes in person at their main search room at Kew, but a quicker route is to key the name of the landowning family into their online database at www.nationalarchives.gov.uk/nra and click on 'Family Name'. You will then be given a list of all archives holding relevant material. So if you were looking for a connection to the Marquess of Sligo, as actor John Hurt was, you could type in 'Sligo' and quickly learn that relevant records were stored in the National Library of Ireland.

▶ A selection of the riches to be found within an archive.

Working in Archives

To recap, your next step after building your family tree is to verify the information you've got. Then you can extend the tree further back in time and, by following a particular branch of the family or story, investigate the historical context. You can start the verification process and extend your family tree online by obtaining certificates, census returns, wills and parish registers, as described in Section Two; but you may find it easier to simply visit the local studies centre or county record office in the area your family comes from to look at paper or microfiche indexes for these certificates and records, many of which are not available on the Internet. In any case, you will certainly need to visit an archive sooner or later to add historical context, so here are some important points to know before you do venture inside.

Step One: Preparing for Your Visit

▼ The National Archives of Scotland in Edinburgh.

If you are unfamiliar with working in an archive, here's a checklist of things to do before you visit. Never just turn up unannounced – it's a sure way to have a frustrating day.

Make Contact

The best thing you can do is to make contact with the archive you plan to visit. Call them, email them or write to them. The archivists there can tell you all about the place, demystify the process of registering as a user (or 'reader'), explain how to search for records, both onsite and online, and – provided you ask simple, detailed and focused questions – may even be able to give specific advice to help you find what you're looking for. If this is the case, you can always ask to talk to the person that helped you when you do eventually visit in person, if they are around and are not tied up with other duties. Don't forget, you can find the archive nearest to you through ARCHON, mentioned above. If in doubt, contact the local studies centre for further advice.

Book a Seat

Family history is big business these days, and unprecedented numbers are flocking into archives as never before. Many institutions are fairly small, with limitations on the amount of space available for researchers, particularly as many of the most popular records are only available on microfilm or microfiche. It is therefore important to check whether you need to book a seat before you visit, otherwise you may be disappointed if you simply turn up on the day.

Registration and Identification

Most archives require you to register as a user before you can view original material or use their search rooms. Usually, you are requested to produce at least one form of official identification, although these requirements will vary from archive to archive. The National Archives asks for one form of official ID, such as a bank card, driving licence, passport, or national ID card for overseas visitors, and then issues a three-year reader's ticket which incorporates your photo, taken on the day you apply. Many county archives also need to see proof of address, and some request passport photos for their records. However, a large number of county record offices have grouped together to form CARN – the County Archive Research Network – and registration at one affiliated archive gives you access to all participating members.

Location and Travel

It is not always evident where an archive is likely to be located. Many form part of council or municipal buildings; some are newly built, just out of town; others may have no parking facilities, or don't have good

HOW TO...

...prepare for an archive visit

1. *Make contact in advance*

2. *Book a seat*

3. *Check ID requirements for registration*

4. *Check location and travel details*

5. *Find out the opening hours*

6. *Make sure you can access the records you want*

7. *Check costs and facilities*

links to public transport. Luckily, the majority of archives now maintain websites, and provide maps or necessary travel details.

Opening Hours

There is no standard pattern to archival opening hours, so don't assume that it will be open when you want to visit. Although many open 9-5, five days a week, some now close for at least one day midweek and offer either Saturday opening, or one late evening, or both. There is usually at least one period each year when an archive closes down for 'stocktaking', when checks are carried out to ensure none of the precious material they hold has gone missing.

Access

It is also dangerous to assume you can simply turn up and expect to see the material you need. Since there are pressures on storage space, many of the less popular documents in large archives are often kept offsite, which means you may not always be able to view material on the day you plan to visit unless you've made prior arrangements. In addition to storage restrictions, there could be other complications. Privately deposited documents – family papers, legal archives or religious collections – often come with their own restrictions. For example, you may need to write to the depositor to secure permission to view material. It is therefore vital to check all these details before you visit to avoid disappointment.

Costs and Facilities

There may be costs involved in visiting an archive. Some charge you an entry fee; others will ask you to leave your goods and belongings in a locker that requires change; and any photocopying you wish to take away with you will have to be paid for. Indeed, you may also want to buy food and drink for lunch and some archives provide snack machines, with larger institutions offering hot drinks, sandwiches or even restaurants.

Step Two: Searching for Documents

Having established which archive you need to visit, and made contact to cover the points listed above, you are ready to search their collections in the hope of finding the key documents you need to supplement your family tree. You should already have set your research goals in advance, but it might be worth writing these out, so you can hand them to the staff at the archive if you need some help. For example, you may have heard that Great-uncle Jeremy fought in the First World War, but don't know where he served. You may therefore decide to restrict your search to establishing his movements during the war. Try to keep this 'wish list' focused and manageable; it is important to be realistic about how much you can get done in a day, and allow time for unexpected discoveries that may lead you into new investigations. Remember, if you don't have time to complete all you originally wanted to do, you can always resume on a later trip.

Catalogues

Each archive 'catalogues' its possessions – that is, it gives a unique reference to every item that it collects. Alongside this unique reference there is usually a description of the item that has been catalogued. These catalogues and document descriptions are the main way that researchers identify documents they need to look at, though you should be aware that archives are complicated places, and there is not one uniform system of cataloguing documents that applies to each institution – each archive will have its own catalogue system, developed over time. Many repositories still have paper indexes to their catalogues, which need to be trawled through in person to find document references even if some of their collections have been uploaded into digital catalogues, although these days many archive catalogues are available to search online.

▼ Access to Archives (A2A) provides a massive amalgamated database of catalogues in which you can search for relevant documents about your family.

Amalgamated Catalogues Online

One important project aims to bring all these disparate catalogues and document descriptions together in one place on the Internet. Known as Access to Archives (A2A), it is an online database containing descriptions of over 10 million documents held in around 400 local archives across England and Wales. It aims to increase awareness of these fantastic resources and facilitate easy access to them. You can search the database by keyword, area, date range and repository name by going to www.a2a.org.uk. Full document descriptions are provided along with references and a note as to where each document is held. Many local and county record offices have submitted their catalogues to the A2A database, but it cannot be stressed enough that if you can't find anything related to your

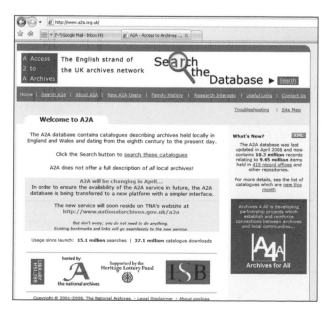

research using A2A you still need to visit the record office itself and consult the original indexes.

The National Register of Archives (NRA) is another treasure trove of catalogue descriptions, but is arranged in a different way to the A2A catalogue. Its database of record descriptions held for around 29,000 businesses, 75,000 organizations, 9,000 families, and 46,000 individuals can be searched using four types of search engine. The NRA database is accessed via The National Archives' website at www.nationalarchives.gov.uk/nra, where you can search under a corporate name, personal name, family name or place name. This database is fantastic for finding out about the location of company archives and the records they hold, especially if your ancestor worked for a major corporate firm.

The Scottish Archive Network is currently in the process of compiling an online, unified index of documents held in many archives across Scotland, which can be searched by keyword from www.dswebhosting. info/SCAN. It so far has descriptions of over 20,000 documents from 52 archives, including the NAS and NLS.

There is even a network allowing you to search the holdings of university archives and libraries, called Archives Hub. It is being added to over time, and can be found at www.archiveshub.ac.uk.

Important Institutional Catalogues Online

It is always worth finding out if the archive you intend to visit has its own online catalogue accessed via its website. If you have located an archive using the ARCHON database, this should have provided you with their website address. Most of the national archives have separate online catalogues containing descriptions of documents not found on A2A or the NRA.

- The National Archives' Catalogue has descriptions of an impressive 10 million documents that you can search by keyword, date range or government departmental code from www.nationalarchives. gov.uk/catalogue. The documents held at TNA are categorized according to the government department they originated from, and a departmental code forms the first part of any TNA document refer- ence. For example, WO is the prefix of all document references for records from the War Office. Searching for document descriptions by government departmental code can help to narrow down the number of results you get if you know what type of record you are looking for.

- The National Archives of Scotland has a database known as OPAC (Online Public Access Catalogue) that can be found at www.nas.gov.uk/catalogues/default.asp. This searches their collections by keyword, place authority, name authority and date, and the NAS website has a list of documents that have not yet been uploaded to OPAC and require a search of the original indexes. You could also consult the National Register of Archives for Scotland (NRAS), which is only available in paper form at the National Archives of Scotland and the National Library of Scotland. This is a survey of papers held by private archives in Scotland.

- In a similar vein, the National Library of Wales has a full catalogue which can be found on its website www.llgc.org.uk under the 'Library Catalogues' page. Here books, periodicals, newspapers, maps, graphics, electronic publications and digitized records can be searched by keyword.

- The National Archives of Ireland's search engine is organized in a slightly different manner, allowing you to search in one go the 19 databases comprised mainly of government departmental records. This is accessed from www.nationalarchives.ie/search but does not cover the entirety of the repository's collections.

- The Public Records Office of Northern Ireland does not have a complete online catalogue either. There is a limited online index available from www.proni.gov.uk by following the links on 'The Records in PRONI' page, where a Geographical Index locates parishes, Poor Law Unions and counties on a map, a Prominent Persons Index finds references for documents relating to individuals, a Presbyterian Church Index and a Church of Ireland Index lists those church records that have been microfilmed, and a Subject Index describes the types of records held at the PRONI.

- The Society of Genealogists, which charges admission for non-members, has its own catalogue known as SOGCAT available from its website www.sog.org.uk/sogcat/access. Here you can search an alphabetical index by parish name, surname or subject to see if they hold copies of the parish registers you are looking for, or records of a particular pedigree you are hoping to find.

Step Three: Working Responsibly in an Archive

Once you have scoured the online catalogues and indexes for all the archives you plan to visit, and have made lists of all the documents that sound useful to your studies, why not see if it is possible to order those

HOW TO...

... work in an archive

1. *Only use pencil and a spiral-bound note pad to make notes*

2. *Don't eat or drink in the reading room*

3. *Handle documents as little as possible*

4. *Respect your fellow researchers: turn mobile phones off and work quietly*

5. *Laptops and digital cameras are usually allowed, but check with the archivist first*

documents in advance of your visit? Most of the websites mentioned have clear instructions on how to do this either online or over the phone if they offer an advanced ordering facility. Your next step is to brave your first visit to the archives ...

Because the material they hold is unique and irreplaceable, there are rules and regulations that you will have to follow during your visit. Actually, these rules are there to help you make the most of your research trip, as well as protect the documents for other users.

Document preservation and conservation is an important part of archival work, and to ensure that documents are not damaged you will find that archives impose strict rules on what you can bring into the reading rooms with you, plus guidelines on document handling techniques. In general, the golden rule of archives is that you must work with pencils only – biros and pens are forbidden due to the potential harm they can cause to original material. Similarly, erasers and pencil sharpeners should not be used or placed near documents, as they can cause damage. There is usually a no eating or drinking rule in place for similar reasons, and this extends to cough sweets and chewing gum.

If you are unsure about how you should be handling an item, or you feel it is delicate, please ask an archivist to assist you. Most archives have a store of foam wedges, supports and weights to help set the document out in a way that carries a minimum risk of harm. Try to limit your own contact with the item; for example, if reading a line of text, do not run your finger along the document, as grease from your skin can cause damage. Instead, place a piece of white paper under the line of text to help you keep your place. If you are having difficulty reading faded text, ultraviolet lamps can often help pick out lost words. Similarly, maps and plans are often covered under clear protective sheets, and you should always ask before you attempt to trace a document.

The amount of material you can bring into the reading rooms will also be limited. Apart from banning pens, erasers and pencil sharpeners, it is likely that you will be asked to leave the majority of your research notes in a locker outside the reading room area, and bring in only spiral-bound note pads or sheets of paper stapled together; and what you can bring in will be probably searched on the way in and the way out. This is to prevent document theft; sadly, many items 'go missing' each year.

Finally, you should, wherever possible, respect your fellow researchers and work in silence. If you do need to confer with a friend or colleague, try to talk quietly and leave the reading room to do so. Mobile phones should be turned off or left in silent mode – there's nothing more annoying than having your concentration disturbed by someone's phone ringing! Most archives allow you to bring your laptop into the reading room, but you should also set them up so that they are silent when turned on. Digital cameras are also largely welcomed these days, though you need to obtain clearance first from the archivist if you want to take photos, as there could be copyright implications and not all cameras are 'document friendly'.

Step Four: Avoiding Potential Pitfalls

Here are a few tips to help you avoid common mistakes, and make the most of the material at your disposal.

Physical Labour

Visiting archives can sometimes be more physically demanding than you might think. The increase in the amount of material that has been digitalized and made available online is gradually changing the process of archival research, but you may still need to spend a proportion of your time on your feet or lifting heavy books and large documents. If you are not very robust you might want to consider taking a friend or relative along to help you out. Many documents are also stored as duplicate copies on microfilm or microfiche, which some people find difficult to read for long periods of time if they have poor eyesight. Many archives are aware of the difficulties faced by elderly visitors or those with special needs to cater for, and have invested in specially designed computers for people with poor eyesight.

It might also be worth considering coming to an archive in old clothes. Many old documents are quite dirty, particularly ancient leather-bound tomes whose spines have decayed to an old, red powder that can make quite a mess on clean, white clothes! You should always wash your hands before and after visiting an archive, as you never know what old microbes you might have picked up from the documents during your visit.

Latin

If you are fortunate enough to be able to trace the history of your family back before 1733, you may well encounter difficulties interpreting relevant material, as the language of official documents was Latin. So material such as manorial court rolls – a highly important source for a family historian – will need translating, as will any official record of deeds or land transfers that were enrolled in the central courts. The exception to this is the Interregnum period (1649–60), when the

> *It's not unusual to find variant spellings of the same word, particularly personal and place names, in a single piece of text.*

Ask yourself why the document was created, and what information it was originally intended to provide.

Parliamentary regime decreed that all official documents should be written in English, and you will also find that some official documents were written in the English language before 1733.

Reading Old Handwriting

Another potential problem will be that scribes tended to employ abbreviations when recording entries, so you will not necessarily be working from easily identifiable Latin words. Handwriting changed over the ages, and even if a document has been written in English it may be difficult to decipher. Official sources can be easier, as scribal technique – the way someone wrote a document – tended to change more slowly as writers adopted the handwriting of their predecessors. However, private hands varied widely, even within a relatively short period, often employing idiosyncratic shorthand techniques. Spellings also differed widely between authors, and it is not unusual to find variant spellings of the same word, particularly personal and place names, in a single piece of text. All of these problems can make interpreting documents difficult. However, there are ways to make documents seem less intimidating.

Most archives stock Latin dictionaries to help you translate key phrases, whilst there are similar publications to help you understand palaeography, which is the technical term used to describe the handwriting and abbreviations employed in the documents. Furthermore, there are specialist volumes written for family and local historians that provide translations and explanations of the formulae for the most commonly used documents that you will encounter. If you are still unsure, try selecting a similar document from the Interregnum period, which will be in English. Most documents follow standard patterns, with only the details of individuals and places altering. This will enable

▶ Websites such as www.scottish handwriting.com can help you to decipher some of the older documents that are tricky to read at first glance.

you to decide where you should be looking in the document for key phrases, and assist with translation. In addition, some local history societies provide transcriptions and translations of important document series, with the added advantage that they are usually indexed. These too can be used to aid interpretation of difficult original material.

Many archives and institutions have created resources online to help you to teach yourself Latin and palaeography. One of the best, since it's linked to their own documents, is provided for free by The National Archives on their website at www.nationalarchives.gov.uk/palaeography, and there is a site where ancient Scottish handwriting can be demystified at www.scottishhandwriting.com.

Dates

Not everyone is familiar with the way documents are dated. Many dates are given in the form of a Regnal year – the year is described in relation to the date the monarch ascended the throne and the number of years for which they had reigned, rather than the familiar reference to the number of years since the birth of Christ. For example, 20 Henry VIII covers the period 22 April 1528 to 21 April 1529 – the twentieth year of his reign, which began on 22 April 1509. Similarly, a large proportion of legal documentation also incorporates a legal term date – Michaelmas, Hilary, Easter or Trinity – which signifies a particular part of the year in which business was conducted.

Furthermore, you may come across dates such as 28 February 1700/01, which refer to the old-style dating technique employed by the Church following the Julian calendar, which started the New Year on 25 March, rather than on 1 January as we do today. The practice was dropped in 1752, the same year that the Gregorian calendar was adopted. The best guide to the many and varied ways of writing dates is Cheney's *Handbook of Dates*, which provides tables giving Regnal years, Easter days and saints' days, which were also used as ways of giving a date.

Context

When looking at the material you have selected, it is very tempting to jump straight in to identify references to your family hidden within the pages. Understandable though this is, given all the procedures you've had to go through to get to this stage, it would be a mistake to launch straight in without first checking what you are looking at, and why. Before you can usefully extract information from a document, you will need to understand why that document was created in the first place, how it would have been used, and what message it contained at the time

HOW TO...

...be prepared for pitfalls

1. *Be aware that searching archives can be physically demanding and dirty work*

2. *Older official records will be in Latin, and often abbreviated Latin at that*

3. *Handwritten records can be hard to read and spellings can be erratic, especially of names*

4. *Documents may be dated using Regnal years, legal terms and saints' days*

5. *Before 1752, years started in March, not January*

6. *Always bear in mind a document's original purpose and context*

it was written. If you do not do this, then you may be taking the information it contains out of its historical context and therefore run the risk of misinterpreting it. After all, documents were not initially created for the purpose of helping family detectives locate their ancestors in the twenty-first century. The records might not easily lend themselves to modern research techniques – for example, indexes may not survive, or you may need to identify the property where people lived rather than the person themselves. For example, electoral lists are a great way of tracing people's movements, particularly in the twentieth century; however, they are rarely indexed by surname and so you need to work out their place of residence, for example from a certificate of birth, marriage or death.

Ask yourself why the document was created, and what information it was originally intended to provide. This will allow you to read it in its own context, and thereby understand why it is arranged the way it is. It may therefore be necessary to corroborate the source with one or more others before you can extract useful information from it. Most archives provide information leaflets about documents and why they were created, so set aside some time to read these useful articles so that you fully understand why you need to look at the documents. That way, you will come away with new names to add to your family tree, and a greater understanding of what they did to end up in an historical document.

A good example is the search for a relevant death duty register (described in Chapter 8). Initially, they were created to provide information about the estate of a deceased person so that tax could be levied; but family historians now use them to track down the place where the will was registered, or to obtain further information about some of the beneficiaries in the will. At first glance, the notation used in the death duty register can be confusing or hard to read. Closer inspection, coupled with information contained in the accompanying research guide provided by The National Archives (where the records are stored) makes it easier to decipher the content of the document and allows you to extract the necessary data from the various sections of the register. This can then be used to find the will, and work out where some of the beneficiaries named in the will were living.

General Organization

When you start working in an archive for the first time, you'll need to be properly prepared. As well as following the above steps to locate an archive, locate documents within the archive, and ensure you interpret them correctly, you will also need to devote some time to the way you record and write up your findings. Here are some tips to help you.

Note Taking

Good note taking is an essential part of your research. If you spend a whole day in an archive, you could be wasting your time if you do not bother to record the exact searches you did, which indexes you looked at, the references of the documents you examined, what information these documents contained, and the names of any books you took copies from. You will find that the next time you go to the archives you will more than likely end up redoing searches you have already conducted simply because you cannot be sure whether you have done them or not.

You should establish a way of recording the parish register, civil registration and probate searches you have completed so that you know exactly which parishes, years and quarters you have looked at in case you need to extend these searches at a later date. Decide on a note-taking system that works for you. Most people use abbreviations for the terms that are repeated often throughout their work. You will probably find the abbreviations used in many family trees (see Chapter 2) are handy to learn. However, consistency in the way you write your notes is important so as not to confuse yourself. For example, if you start using 'b.' to indicate 'born', you should then decide on another abbreviation for 'baptized' and 'buried' – don't use 'b.' for all three as you will soon get confused!

Some people prefer to take a laptop with them to the archives so they can type their notes straight into electronic form. But there will be occasions when a laptop will not be allowed in certain areas of an archive, so be prepared for this. You should always have a set of notes, whether written or electronic, which you can take into the archives with you to work from. Keep hold of your research plans and ensure you record how much of it you achieved so that you know how much you need to do on your next visit. Date your notes so that you can keep a chronological track of your progress and can work from the most recent set of notes, and record the name of the archive you visited to avoid confusion, just in case two archives use a similar referencing system for their documents.

When you are taking notes it is important to record the source of absolutely everything, whether it is a person, an archive, a website or a book. When writing document references be sure to include the exact page and folio numbers where you found the correct entry so that you

> *Good note taking is an essential part of your research. Record the exact searches you did, which indexes you looked at, the document catalogue references and the information those documents contained.*

> *It is important to distinguish between evidence and analysis in your notes, otherwise mistakes will start to creep into your work.*

can find it again easily if you need to, even the line on which it was written. If you consult a document that turns out to be of no use, make a note of this so that you do not go back to it again.

Secondary sources also contain valuable information for family historians, so when you take a photocopy from a book or write out a paragraph from it, record its full title, the author's name, the publisher and year of publication, which should be found on the inside cover, as well as the relevant page numbers. Recording the year of publication for books will be surprisingly useful to your research. You may find a fascinating paragraph in a local history book describing the house your ancestors used to live in, but if you don't bother to look at when the book was published you won't be able to put that description into its own historical context. Many history books were published in the nineteenth and early twentieth centuries, a lot of which are still in our libraries today, and their descriptions of events and places will differ from those of more recent authors.

It is important that your notes distinguish between what is evidence and what is analysis. When you return to your notes at a later time, perhaps to type them up or to remind yourself what you found on your last visit to the archives, you need to be able to rely on them. Therefore, if you transcribe a passage from an original document, put that paragraph in speech marks so you know that was what was written word-for-word. If there are any phrases you are unsure of put them in square brackets, because assuming the meaning of a few words could alter the entire context of a piece of text. Anything you have scribbled down as presumption, analysis or ideas should be labelled clearly as being so, because these opinions may change as you find more documents. You want to avoid confusing fact with ideas, otherwise mistakes will start to creep into your work.

Filing

A genealogy project produces an enormous amount of paperwork, from your research notes and photocopied documents to the photographs you find and the family papers your relatives give you. Sooner rather than later you will undoubtedly find that you don't know what to do with it all! It is very tempting to just pile it all into a big box and hide it under the bed, but that would be a huge shame after all your hard work, not to mention that it makes finding documents and notes you need to work from a bit of a nightmare.

There are various filing systems you can adapt to suit your own purposes. It is worth investing in a decent expandable file with plenty of dividers and labels. And you may prefer to showcase some key documents in a portfolio file to keep them pristine. There are many ways of organizing your paperwork. You might want to classify your notes alphabetically by surname, or perhaps keep all the notes from one archive visit together, but it does help to keep copies of original documents together with the relevant notes. It can be useful to separate your notes for each side of the family, and then by surname and branch, especially if the same surname appears in two different branches.

If you are looking after original documents on behalf of your family, it is important to keep them away from heat, damp and direct light to prevent them from deteriorating. Where possible original documents, particularly photographs, should be stored in strong, acid-free boxes, but if you do decide to keep them in a plastic folder with the rest of your notes, you should place each document inside two thin sheets of acid-free paper to prevent the plastic from damaging them. There are companies that advertise in family history magazines or that can be found online who specialize in products that preserve fragile documents for family historians. If you are unsure, have a chat with your nearest archivist for further advice.

If you have a computer, you may want to consider typing up your notes so that you can create a new folder for each surname. It does help to keep the paper copies of your notes, though, in case you need to take them with you on a future visit to an archive. If you do opt for keeping an electronic record of your research, it's handy to have one central document that compiles all of your notes, with references to where each piece of information comes from.

There is a multitude of genealogy software packages on the market with the aim of making this easy to do. They also enable you to organize the data you enter into different styles of family trees, charts, reports and indexes, which is almost impossible to do if you are just using a word-processing package. If you choose the right type of genealogy software, you will find that you can not only use it to store all your research, including a fact file of each individual linked up with images of photos and documents, audio recordings of interviews with family members, family holiday videos, and notes about the sources you have found, but you can simultaneously do some of your research online by connecting to genealogy websites that are compatible with your chosen software and migrate the online records you find into your family file. Don't be scared to invest in a package to play around with, and learn how to get the most out of it as you go along.

CHAPTER 4

Research Tips and Hints

The tips, hints and advice in this chapter will help you to achieve far more when you actually start looking for documents, and to make the most of your time in archives and research institutions. We also introduce the amazing array of resources now available online, give some advice about the pitfalls of Internet research, and what to do when you get stuck with your research and need a little shove in the right direction.

However, before you start worrying about where to look for help, here are a few useful tips to help you avoid making mistakes in the first place!

Avoiding Mistakes Early On

There is nothing more exasperating than spending several hours in an archive and a small fortune on certificates only to realize you've been following the wrong branch of people, simply because a small mistake was made early on. This can easily happen if you don't order every registration certificate for a person and ensure each name, date, place and occupation on your family tree is substantiated with as much documentation as possible.

Never Assume ...

Cross-referencing sources is essential, but if you cannot find conclusive evidence then do not just assume that links to earlier generations

> *Try to locate more than one source to corroborate information you have already found.*

are correct, even if they look likely. Just because the name and date appear to be right doesn't mean you've found the right person.

If you have run into problems, leave that branch for the time being and keep pushing back on neighbouring branches that might give you more clues. For example, it is possible to link witnesses' names on a marriage certificate to family members who appear on earlier census returns to strengthen the case that you have found the right person, and therefore help you to fit the jigsaw together.

Where possible try to locate more than one source to corroborate information you have already found. If you discover that there is no concrete evidence whatsoever to verify a link, make a list of all the circumstantial evidence that led you to your initial assumption, and continue forward, making a note that you have not found firm supporting documentation.

Question the Evidence

You should always question the reliability, or at the very least the historical context, of every document you encounter. Primary evidence can contain errors, but if you have enough different sources available so that you can compare the vital details for each ancestor, then you should be able to work out which sources are accurate by a process of elimination. For example, census returns can sometimes give the wrong ages and can contain misspellings if names have been inaccurately transcribed from the original forms. Death certificates are also known to contain mistakes, especially if a young and distant relative, or doctor who was unsure of the facts, registered the death.

Our ancestors were prone to stretching the truth when asked about their age, or were themselves unsure of their own year of birth in times when paperwork and the process of filling out forms was far less common than it is today, which can explain discrepancies between a birth certificate and an age given on the same person's marriage certificate. Any evidence of our ancestor's existence is important, but you should be cautious when using this evidence.

Concentrate!

Simple mistakes are easy to avoid just by staying focused and alert. Keep checking the exact spelling of the names you are searching for so that you don't waste half an hour looking for the birth of James John Clark, when it should have been John James Clarke. And it might sound

CASE EXAMPLE

Question the evidence

*When researching **Ian Hislop's** family tree, two people called Murdo Matheson were found living at the same time, in the same place, and who joined the same regiment in the late eighteenth century. Painstaking research was required to work out which Murdo Matheson was related to Ian, only solved by comparing the clasps on a medal awarded to Ian's relative, which had been passed down through the family, with the movements of each battalion and therefore eliminating the 'wrong' one.*

obvious, but you'd be surprised how easy it is to confuse the marriage indexes for the birth indexes, so always double-check you are looking at the right set of documents.

Has Your Tree Been Researched Before?

Don't be afraid to look for short cuts – they can save you enormous amounts of time and money! Here are a few places you can investigate to see if someone has done some work on your family tree already.

Society of Genealogists

Part of your initial research should include checking whether any part of your family tree has already been published or recorded. One of the best places to start looking is the Society of Genealogists (SoG), one of the country's premium research institutions. The SoG is based in Clerkenwell, London, and for an annual fee you can access its impressive research library as well as copies of key datasets, attend lectures, seminars and workshops, and examine their extensive collection of manuscripts. Further information about the SoG can be found from their website, www.sog.org.uk.

Among the SoG's records are pre-researched family trees, pedigrees and associated research notes left by family historians in the past who wanted to share their work with others. You check the SoG's online catalogue at www.sog.org.uk/sogcat/access to see if your surname is listed. And don't forget, it is also worth asking at your local studies centre or county record office to see whether they have copies of pedigrees deposited by other local researchers.

It's worth asking at your local studies centre or county record office to see if they have copies of pedigrees deposited by other local researchers.

Published Pedigrees

If you suspect – or even know – that you have blue-blooded relatives, then one of the volumes that publish pedigrees of the aristocracy will be of use. Surprisingly, this is more likely than it sounds, given that many aristocratic families can trace their lineage back hundreds of years, and as each new generation is born, the distribution of wealth and status thins out among the younger branches. You may not realize at the start of your journey that one branch has noble roots, in which case it will probably not become apparent until you have been investigating your genealogy for a while.

Alternatively, if there is a family story that a certain ancestor was descended from a specific duke or lord, then it is worth tracking down the pedigree of that family to see if it could link in with your own research at some point in time. *Burke's Peerage and Baronetage* and *Burke's Landed Gentry* have recorded the genealogies of titled and landed families throughout the United Kingdom and Ireland for over 175 years. Their content includes information on the extended family and deceased distant relatives of each noble name. *Debrett's Peerage and Baronetage* has been published since 1802, and is sometimes considered to be more dependable because it does not rely so heavily on information obtained from the family in the way *Burke's* does. On the other hand, *Debrett's* contains far less information about extended branches, preferring to concentrate on direct ancestors, descendants and living relatives.

Both publications are regularly updated and most libraries and archives have copies of these volumes, which include editions detailing the ancestry of extinct titles as well. Alternatively, you can purchase a subscription to browse the database of entries in *Burke's* from the website www.burkes-peerage.net. If you do manage to trace a line of your tree back to a titled family, then you can also look them up in *The Complete Peerage* by George Edward Cokayne, copies of which are held on open shelves at the National Archives in Kew and at other notable research centres. *The Complete Peerage* cites all of its sources and gives a bit of background about some of the more distinguished characters, such as their involvement in certain battles; however, it only follows the direct line of heirs.

In addition to the popular pedigree publications mentioned above, there are plenty of other editors who have printed pedigrees throughout the last few centuries, although these volumes are generally less well known because they are no longer in print. Check the shelves of your local library and archives for these. You may also find that a local historian has published genealogical records for families who lived in your local area.

There are some indexes to published pedigrees arranged by surname so that you can locate the relevant books that may contain information about your family tree. These are:

- *The Genealogist's Guide* by G.W. Marshall of 1903; indexes a large number of pedigrees published between 1879 and 1903

- *A Genealogical Guide* by J.B. Whitmore, published in 1953; continues this cataloguing for pedigrees published between 1900 and 1950
- *The Genealogist's Guide* by G.B. Barrow, published in 1977; for pedigrees published between 1950 and 1975
- *A Catalogue of British Family Histories* by T.R. Thomson; the most recent index for pedigrees, published between 1975 and 1980

It is highly worthwhile consulting these indexes for all of the surnames in your family tree, and continually referring to them as you find new names. Copies are held at most major libraries and archives.

College of Arms

The College of Arms, located in central London, has records of the visitations conducted by royal heralds in the sixteenth and seventeenth centuries. They were sent out by the Crown to review the claims of families whose status gave them rights to bear a coat of arms each time a new generation was born. (This need to check the family trees of the nobility became increasingly important after the English Civil War, 1642–46, at which time illegitimate claims were made to some titles.) Heraldry, or the system of displaying personal symbols on shields such as coats of arms, is of great use to family historians because of its hereditary nature. Heralds kept pedigrees of the families that were entitled to bear coats of arms so that they had a record of the line of descent the coat of arms could be passed down, which was usually to the male heir. Since the cessation of heraldic visitations, the College of Arms has been responsible for issuing coats of arms and holds updated copies of pedigrees for many distinguished families from around the British Isles. You can visit their website at www.college-of-arms.gov.uk to find out more about their history and the services they provide.

The College of Arms' records can only be searched by members of staff, who are still known as heralds today. However, Frederick Arthur Crisp and Joseph J. Howard published a series of pedigrees based on the heralds' visitations that

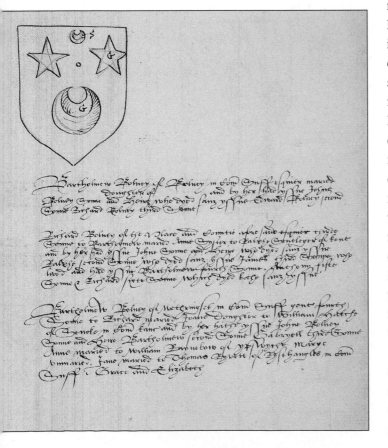

▼ During the 16th and 17th centuries, pedigrees of thousands of families throughout the English and Welsh counties were recorded. Here four generations of the Bolney family of Suffolk are shown, together with a sketch of their Arms.

include twentieth-century descendants, entitled *Visitation of England and Wales* and *Visitation of Ireland*. In addition, in 1952 Sir Anthony Richard Wagner released *The Records and Collections of the College of Arms*, which may be worth consulting if you believe a branch of your family may have been entitled to bear a coat of arms at one time.

Online Pedigrees

For those of us who are not so lucky as to have had our family history already published, it is worth scouring the many genealogy websites that enable researchers to share their family trees online. As has been explained in Chapter 2, there are many websites where you can upload your tree as you go along, including www.genesreunited.com, www.ancestry.co.uk and www.myheritage.com, so why not use these resources to find out if there is somebody else out there looking for some of the same ancestors as you? Many of these sites allow you to search their database of records for free simply by entering the name of a particular ancestor you would like to find. Usually you will be provided with a limited amount of detail about all the people in the database that match your criteria, and if you subscribe to the website you can email other users who seem to be researching the same people to ask their permission to view their research in full. Some websites provide free access to other people's online pedigrees, such as www.familysearch.org and www.genealogy.com.

Guild of One-Name Studies

The Guild of One-Name Studies is an organization that supports researchers keen to investigate the origin of a particular surname. Its members are interested in everybody who has the surname they are studying, though they might restrict their study to a certain geographical area, which means they are not looking at one particular pedigree. Nevertheless, their records sometimes include lineages of many families. Visit www.one-name.org to find out if a one-name study has been established for any of the surnames on your tree. For example, the Izzard surname is listed, so if you suspect you have a connection to Eddie Izzard, you can visit the site, click on the link and learn more about the origins of the surname and its derivatives, as well as how often it appears in historical documents through time.

The website provides a useful profile for some of the registered One-Name Studies, including their aims and the data that has been

> *Treat all second-hand evidence as merely a guideline to follow, rather than gospel truth.*

SUMMARY

Check the following to see if some of your proposed research has already been done:

- *The records of the Society of Genealogists*

- Burke's *and* Debrett's Peerage and Baronetage, *and* The Records and Collections of the College of Arms

- *Indexes of published pedigrees*

- *Genealogy websites*

- *The Guild of One-Name Studies, for a particular surname (especially if it is unusual)*

gathered so far, along with contact details for the Guild member who posted the information. You can contact that member if you have a specific question you would like to ask about their findings. The Guild supports projects designed for experienced researchers, so if a surname you are interested in is not registered with the Guild, it is wise to research your own family tree first and then build up a portfolio of information about that particular name before registering the surname and starting a study. There are guidelines about how to begin a one-name study on the website.

Remember ... Check the Evidence

It is imperative that you find out the sources of any research under-taken by other people before you even consider incorporating this data into your own family tree – it is vital that you can double-check their accuracy. Even if somebody else's research has been published, their work is still liable to human error and it is not unusual to find a pedi-gree published in two books that has different dates of birth or death cited in each version. Therefore, check as many editions as you can for each published pedigree so that you can compare the information they contain, and then follow up the document references in the footnotes and examine the original sources.

Just as you did when you were collecting information from relatives, you should treat all second-hand evidence as merely a guideline to follow, rather than gospel truth. Use it as a short cut to the records that will allow you to verify the information, rather than taking the data at face value as proof of the past. Family trees published on the web are even less reliable simply because anybody can add to genealogy websites without needing to authenticate their entries. If you do contact another researcher who has seemingly investigated one of your family branches, do not be afraid to ask them how they came to their conclusions and what sources they have used. You can then follow up these sources yourself to see if you find them convincing as evidence.

Despite these words of warning, family tree sharing facilities are fantastic genealogical tools, particularly the online versions that enable researchers to share ideas and learn from each other's work. Once you have completed your research you should consider making it into a book, perhaps using one of the family tree software packages suggested in Chapter 2 – many of which include a publishing suite – so that you can deposit your work at the Society of Genealogists or your local record office for other researchers to benefit from.

Researching Your Genealogy Online

Genealogy as a pastime is at its most popular in the twenty-first century, partly thanks to the wealth of resources that have been made available online to millions of people at the click of a mouse. Whereas family history was once the domain of the upper classes who had access to (or were keen to prove) their pedigrees, and die-hard genealogists prepared to spend hours scouring reels of microfilm in search of each new name on their family tree, now anybody with a vague interest in their roots, regardless of status or origin, can start investigating. Because there is such an enormous interest in the subject, there will always be somebody online who can be of assistance if you are struggling with your research.

There are literally millions of genealogy websites out there – just try typing 'family history' into Google and see how many hits you receive! This can make it difficult to know where to begin, so here we'll de-mystify the process and highlight the most useful sites you'll need to visit. Basically, the core resources you will be working from can be broken down into the following categories:

- Commercial or institutional websites supplying access to datasets, images of records, or indexes to documents ('dataset' websites)
- Websites of genealogical organizations that provide advice, or links to other resources ('portal' websites)
- Websites where you can link with other users and join social networks ('network' websites)

Dataset Websites

Commercial Sites

There are many commercial websites that have worked in conjunction with archives to provide reliable historical material online. For example, The National Archives at Kew have teamed up with commercial companies such as Ancestry and Find My Past to enable online access to census returns, military records, ships passenger lists and more, whilst the National Archives of Scotland have established a partnership with the Scottish General Register Office and the Court of the Lord Lyon to create an official website – www.scotlands people.gov.uk – where Scottish parish registers, civil registration,

> *Datasets are digital collections of particular records that can be searched using a name-based index or search engine.*

probate and census records are available to those unable to travel to the archives.

As a result, there are many commercial subscription-based websites that offer access to datasets, namely digital collections of particular records that can be searched using a name-based index or search engine, and you will often find more than one website offering access to the same records – civil registration indexes and census returns being prime examples (the relevant resources are listed in Chapters 5 and 6 respectively). The differences between the various websites are usually the type of search engine they offer and the cost of accessing the records. For instance, you may find two websites that will give you online access to census records, but one is better for finding ancestors by name and one is more useful for locating addresses on the census if you cannot find an ancestor by name. Equally, one website may offer a better deal for pay-per-view subscriptions and one a better deal on yearly subscriptions, so shop around and work out which ones better suit your needs.

Archives

In addition to the commercial organizations, major archives around the British Isles also give online access to digital copies of some of the documents in their holdings – though of course hard copies can still be ordered if you prefer. Therefore if you cannot get to these archives very easily you may be able to access some material from home. The types of records available from each archive's website varies widely; for example, The National Archives' website has a Documents Online area where you can pay to view copies of wills from medieval times to the mid-nineteenth century, as well as many military and naval documents, and you can order images of specific document references that are emailed to you using the online Digital Express facility. In contrast, the Public Record Office of Northern Ireland does not yet enable any access to its holdings over the Internet. It is worth checking the websites of both the national and local archives that cover the regions your family were from to see if they have digital collections of any of their holdings, though this should not be considered a substitute to visiting the archives. Despite the growth in online availability, the bulk of the documents you'll need will almost certainly have to be viewed on-site.

Newspaper Collections

It's not just archives that are putting their holdings online. Online newspaper archives are increasingly useful for family historians, helping you to locate articles about your ancestors that may otherwise have never been found. It is always worth checking the local newspaper that covered the area your family lived in if you have a date of birth, marriage or death for a prominent ancestor that may have warranted a mention in the personal announcements or obituaries section. In the majority of cases you will still be required to visit the local library or county record office or go to the British Newspaper Library in north London to scroll through the original chronological records.

The most important records are the digitized collections of national newspapers, many of which can be searched by keyword, article type and date range. *The Times* is one of the most prominent of these; its collections are available online from 1785 to the present day. To access the entire collection of *The Times* an InfoTrac database connection is needed, to which most major archives and local library networks provide free access. Your local library may be able to give you a PIN number and instructions about how to use the database for free from home. The *Scotsman* also has an online database of its backdated editions from 1817 to 1950, accessed at http://archive.scotsman.com, and the historic archive of the London, Edinburgh and Belfast *Gazettes* can be searched and viewed for free at www.gazettes-online.co.uk.

Local newspapers tend to be more fruitful when searching for newspaper articles for the majority of our ancestors, but you may be surprised to locate an entry in a national newspaper in the form of criminal reports, court summaries, advertisements, changes of name and address notices, in addition to the many birth, marriage, death and funeral announcements and obituaries they contain. These ever-growing online newspaper archives are so useful to modern genealogists purely because they enable us to find articles about our ancestors by name rather than by searching through months' or even years' worth of original newspaper reports.

Portal Websites

There are numerous professional bodies in the field of genealogy that are at hand to aid you in your quest to find as much material as you can about your heritage, and most of the larger organizations attempt to do this as best they can via the web so that they reach the widest group of people. The Federation of Family History Societies (FFHS) has a Research Tips section on its website www.ffhs.org.uk, and the Society of

HOW TO...

... make the most of the Internet

If you are not an experienced computer and Internet user, it is probably worth investing in a good guide, like Peter Christian's The Genealogist's Internet *or* How To Trace Your Family History on the Internet *published by* Reader's Digest. *These books provide detailed explanations of how to research each aspect of family history using online resources and will guide you through the most popular sites. They also suggest a good range of websites for your online research.*

Genealogists' website www.sog.org.uk has genealogy leaflets that can be downloaded giving general advice about how best to go about tracing your roots. Many of the major archives also have subject-related research guides on their websites that guide you through locating and interpreting documents when at these institutions. One of the best places to find out more about the key resources you'll be using is at www.familyrecords.gov.uk, which describes civil registration indexes, census returns, wills and probate documents and other material.

GENUKI

GENUKI is a web-based charitable organization that offers information on all aspects of genealogy in the UK and Ireland. Its database of information and links is organized geographically and by theme on the www.genuki.org.uk website. While this site provides links to many other websites where you can seek help with any problems, the website itself aims to serve its users as a 'virtual reference library' and is not designed to answer specific research questions. It is supported by the FFHS and its member societies, who collaborate with GENUKI to provide them with much of the information contained on the site. It is therefore extremely useful for locating rare online indexes and transcriptions for records held in local archives that cannot be found on some of the large commercial dataset sites. As soon as you can pin a branch of your ancestors to one particular place, it is definitely worth visiting the GENUKI website because its sophisticated geographic index will show you all the key online tools and organizations that specialize in that area. Its topographical index is just as useful if there is a certain genealogical subject you are struggling with.

Cyndi's List

Cyndi's List is a web-based directory to help you find websites relevant to all areas of family history. Established by Cyndi Howells in 1996, Cyndi's List is an excellent finding aid for genealogists, giving you links to thousands of useful websites. Go to www.cyndislist.com and search the Topical Index to find a website for whichever area of your research you need help with.

Network Websites

The benefits of data-sharing websites, where you can upload your family tree and search other researchers' entries to find common ancestors, have been explored in Chapter 2, but there are other types of websites which help you to share and learn from other people's research and expertise by posting messages about a specific ancestor, branch or topic of interest and awaiting responses from other users. Forums and chat rooms are particularly good for this purpose, and you will find that many sites with access to datasets also provide a forum for their members to communicate by.

Forums

Forums are great for accessing as many people with a like-minded interest as possible. If there is a family myth that you have been unable to unravel, it is possible that posting a message in a subject-related forum will attract replies from people who may be able to help, either because they have faced a similar scenario in their research, or perhaps because they recognize the names as some of their own ancestors. This way, if a family story has been passed down more than one branch distant cousins may be able to embellish with more detail about what they have heard and you can compare the two versions of events. Place your query in a forum category that best fits the subject of your enquiry, and be as specific as possible, mentioning the names you are interested in finding out more about and any relevant places and dates so that other users will be able to tell straight away if they can help you.

The British Genealogy website has links to forums covering all topics and counties at www.british-genealogy.com/forums, but also you should remember to look out for a forum section on the other websites mentioned in this book. The more forums you post your query in, the more likely you are to get a response.

Communities and Mailing Lists

The Internet is home to a range of online communities that are worth joining in order to communicate with other people who are equally as enthusiastic about their research as you. Being part of a community that has an interest in a genealogical subject you are really passionate about gives you access to a goldmine of knowledge. Whenever you are stuck you can turn to your online community and ask them for their help or opinions, whilst keeping abreast of all the latest news about your area of interest. If there is a special convention coming up, or a really important dataset is due to be released online, you should be among the first to know about it by keeping in touch with your fellow enthusiasts.

These communities come in various guises. Joining a mailing list is one way to be part of an online community. Mailing lists allow subscribed members to debate a chosen topic and exchange information; each time

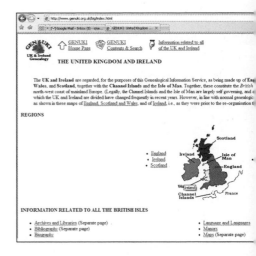

▲ If you are not sure where to start looking for information, try a portal website such as GENUKI.

Being part of an online community gives you access to a goldmine of knowledge.

SUMMARY

*Online sources of
information include:*

• *Dataset websites, such as those
of national and local archives,
newspapers and commercial
organizations, providing digital
collections of records*

• *Portal websites, for advice,
information and links to other
websites and collections*

• *Forums, communities and
mailing lists, for subscribers to
share information and research*

somebody writes a comment an email is sent out to alert the other members of the list so that they can read it and reply. One of the most popular genealogy mailing lists is at www.rootschat.com, but a large selection of lists covering an extensive range of genealogical subjects can be found listed at Rootsweb – www.rootsweb.com – whilst GENUKI also has a directory of mailing lists at www.genuki.org.uk/indexes/MailingLists.html.

If you would rather avoid the constant emails generated by mailing lists, but still like the idea of joining an online community, take a look at the Nations Memory Bank (NMB) website where you can become a member of one of the Family, Military, House, Fashion, and Local, National Trust or Food communities at www.nationsmemorybank.co.uk. NMB is a digital archive of all of our memories, not just family history, where photos can be uploaded and memories of different events relating to the images are placed on a memory map and discussed by other users in the forums. For example, you can post a picture of your family and ask other users of the site to help you name the people in it, or provide stories about what they were like. This is a great website for learning from other people's experiences, and you can search for key words to find memories about a topic or place of interest relevant to your research. NMB is also a brilliant space for storing your own research. If you have recorded an interview with an elderly relative, why not transcribe that interview and store it as a memory on the site so that other users can read and learn from their recollections? (However, remember to seek the permission of the interviewee before putting their life story in the public domain.)

Internet Etiquette and Problems Associated with Online Genealogy

Experienced genealogists have voiced their concerns over the past few years about the increasing reliance on Internet resources as opposed to traditional methods of research. Whilst this largely stemmed at first from a reluctance to adapt to new technology, they have raised some very valid points.

Millions of original documents, catalogues and indexes have been digitized so that they can be searched online quickly and from the comfort of your own home, but this has led to a misapprehension that it is now possible to research an entire family tree on the Internet. This is certainly *not* the case, and although most of the key 'first step' resources are now online, only a very small percentage of the entire range of original documents useful to genealogists is available from the Internet. You will be required to make many trips to various archives, but this is all part of the fun.

Before you can even begin taking advantage of the many records that are online, you will probably need to research your family tree offline for at least a few generations before the Internet records are of any real use. The majority of records online pertain to the nineteenth century, mainly because many more recent sources (such as the vast majority of twentieth-century censuses) are 'closed' – that is, they are not available to be made public – because they contain sensitive personal information about individuals who could be still alive. More datasets for the twentieth century are now becoming available, however, such as phonebooks and directories, though these alone will not be enough to trace back a branch – only place them in one location at a given time. Online records for earlier eras are also scant, mainly because there are no centralized indexes for resources like parish registers and wills proved in county courts, making the task of putting them all online a very large-scale and time-consuming one.

Common Problems Working Online

Aside from the documents you may not be able to locate on the Internet, the ones that are available online come with their own set of problems. For a start, do not expect to locate online sources for your ancestors by simply typing their name into a major search engine like Google, Ask Jeeves, or Yahoo! While this will sometimes yield results if you are very lucky or if an ancestor was particularly famous, most will need to be found using a site-specific search engine, which requires that you first locate the relevant website. This is best done by starting with a source such as GENUKI or Cyndi's List, described earlier.

Specialist search engines allow you to search the website's online indexes either by keyword, surname, date or place. They will often have been created by people who have manually transcribed words from the original documents into computer software. While this is part of the beauty of Internet research, allowing you to find what you are looking for instantly without the need to consult card indexes or scroll through every page of a document to find a particular name, databases create problems of their own. There are many errors in the transcriptions that can make locating the entry you want difficult. These errors are usually because the

> *The majority of records online pertain to the nineteenth century.*

person doing the transcribing was not able to read the original old handwriting very well. For example, if you were looking for Adam Benny on the census, his name may have been wrongly transcribed into the computer's index as something similar, such as 'Alan Remy', in which case it would be difficult to locate this entry using a name search. There are tips on how to overcome some of these common errors throughout the book under the relevant subject headings, but if you find it very difficult locating records that should be online, then you will have to consult the original indexes or records at the archives.

Reliability: Checking Sources Online

As has been mentioned many times before, if you do find other researchers willing to share their findings with you online, whether through a family tree sharing site or a forum, make sure to always ask them how and where they found their sources so that you can double-check them yourself and ascertain their accuracy. There are thousands of people working in the online community who will hopefully be able to help you when you are stuck, but there is always the possibility that they have made errors too. This advice goes for websites set up by enthusiasts as well – plenty of people have now mastered the art of compiling their family history onto a personalized website, but there are no official checks to ensure all information published online is accurate, so it is important to carry out your own checks on their data. You can usually establish whether or not a website belongs to an accredited organization or a private individual from the URL address (i.e. the www. website address). If the address ends in 'gov.uk' this means it has been set up by a government organization, and those ending in 'ac.uk' belong to academic institutes, therefore their content should be reliable. Look out for the website administrator's contact details so that you can get in touch with them should you need to qualify the validity of their data.

Citations and Copyright

When incorporating online sources into your research it is vital that you cite those sources with the same attention to detail as you would for original documents from the archives. Include the full web address and details about the dataset or the owner of the site's material. If you are considering publishing your family tree on a website or online using a family tree sharing facility, you must be cautious of the various copyright laws that protect information supplied to you by other

> *If you publish your family tree on a website or online, be aware of the various copyright laws that protect information supplied to you by other researchers or publications.*

researchers or publications. Any information you have found from databases online or on CD-ROM, or that has been supplied to you by other researchers' websites, forum or mailing list postings, is protected by copyright, and therefore you should not replicate this data without first obtaining the owner's permission.

The laws on Crown Copyright have recently been modified so that you can transcribe extracts from unpublished original documents found in archives as long as the full reference is quoted. The laws pertaining to copyright are complex and official advice can be found at www.ipo.gov. uk. You should also be very wary of posting information in your family tree about anybody who is still alive without first asking their permission. The laws on data protection are unlikely to be a problem to you, but you have a duty to protect the privacy of those who have been kind enough to help you with your research, and those relatives who may be unaware of your investigations into their past. Most family tree sharing sites automatically hide details about living relatives from other users until you have given your permission for specific users to view your tree, and Cyndi's List has links on its GEDCOM page under 'Privacy in GEDCOM files' to programs that will remove living relatives from a GEDCOM file before uploading it to an online database.

Getting Help Offline
Problem Solving

Section Four provides a unique resource for all family historians – a structured route through some of the more popular but often technically difficult topics that you are likely to encounter during your investigations into your past. However, if the suggestions included in the Section Four guides can't answer your questions, here are a few more tips and tricks to help get you back on track again – bearing in mind, of course, that there may not actually be a solution!

Each type of document will present its own unique set of obstacles that may hinder you from finding the person you are looking for. A birth in England or Wales in 1846 may not be found in the civil registration indexes, for example, because although civil registration began in 1837, the rules governing its enforcement were not tightened up until 1875, and many people simply didn't bother. These specific problems will be addressed when each subject is explored in detail in Sections Two and Three, but there are more general issues that affect most types of records.

SUMMARY

Problems associated with online genealogy:

• *Not all documents are available online; you'll still have to visit archives*

• *Transcription errors in online indexes will make your search more difficult at times*

• *The Internet has no quality controls on entries; double-check all your sources*

• *Copyright law will still apply to all online sources you quote*

> *If you cannot find an ancestor in alphabetical indexes under the name you were expecting, think of all the variations of spelling that name could sound like and conduct a search under those options too.*

Changes in the way names and places have been spelt over time are a common hindrance to family historians. When registering the name of a birth or baptism, or even when filling out a census return, the priest, registrar or enumerator would write a name how they heard it said, and very often the informant was illiterate and so would not be able to correct them if it was spelt wrong. Therefore, if you cannot find an ancestor in alphabetical indexes under the name you were expecting, think of all the variations of spelling that name could sound like and conduct a search under those options too. Common variations occur when there is a silent letter, such as 'e' at the end of a name.

Similar rules apply to the spelling of place names that have been known to change frequently over time, but you should also be wary of places around the country (and even the world) being called the same name. There are indexes, like F. Smith's *A Genealogical Gazetteer of England*, that can help you to find in which counties a place name is found, and thus help you continue your research in the correct area.

There are plenty of useful reference books to help you with a particular line of historical research. Your local library will stock a range of publications on the subject, particularly specialist volumes that include indexes. There are also many genealogical journals and magazines released weekly and monthly, like *Ancestors Magazine*, *The Genealogists Magazine*, *Family History Monthly* and *Family Tree Magazine* and, of course, the *Who Do You Think You Are?* magazine. They are all packed with fascinating articles and top tips, discussing the latest finding aids, computer software and issues that affect the modern genealogist. A lot of these magazines also have a genealogy agony aunt who will answer readers' research questions.

Family History Societies

If, at any point during your research, you feel daunted by the next step, or have hit a brick wall on a certain line of enquiry, there are a multitude of individuals and societies out there that can be of assistance. Whatever your problem, it is likely that others have ground to a halt for similar reasons before you but have eventually found a route forward. Very few research problems are unique in genealogy, and as you find your way around one obstacle you will be able to use that experience as a lesson for the next time you get stuck. Even if you are confident researching your family tree alone, it is still advisable to join one of the many family and local history societies that we are lucky enough to have access to in the UK. You will always learn something from the experience other members have to offer, and be able to utilize the indexes, transcriptions and local projects that they have worked upon and which may not be available anywhere else.

There is a small charge to join a society, but the talks they offer to members, the networking they provide between researchers, the regular journals that are issued and their access to indexes unique to their subject matter make it all worth while. You should join a history society local to the area your ancestors were from so that you can benefit from the expertise of others who have researched that area and are compiling indexes for records relevant to that location. The Federation of Family History Societies (FFHS) is the umbrella organization that unites and represents all of the smaller societies around England, Wales and Ireland. You cannot join the FFHS itself, but you can consult their website at www.ffhs.org.uk to find a family history society local to you. The Scottish Association of Family History Societies (SAFHS) provides a similar network for family history societies in Scotland, and its membership list can be searched from www.safhs.org.uk. The GENUKI website also has a page for locating societies geographically, with links to each society's website.

The range of history societies open to genealogists from around the world is staggering – take any topic you can think of and you are likely to find that a society has already been established to unite and aid researchers in that field of study. In addition to the hundreds of regional societies, there is a range of organizations that interest themselves in particular industries, professions and occupations. The Railway Ancestors Family History Group may be of benefit if you find an ancestor who worked in that profession; the Society for Army Historical Research could be worth joining if you have a long line of military ancestors in your tree. Societies dedicated to researching certain ethnic or religious groups also exist, such as the Romany and Traveller Family History Society, the Jewish Genealogical Society of Great Britain, and the Catholic Family History Society.

There is likely to be more than one society that can assist you with your investigation, and the more you join, the wider your network will be when you do need advice.

▼ Always look to join a family history society. Further information can be found on the Federation of Family History Societies website – www.ffhs.org.uk.

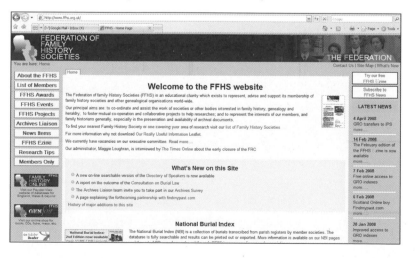

Archives run induction days, lectures or tutorials at both local and national level, often with a particular theme as the focus.

Lectures, Courses and Workshops

If you find you have a burgeoning passion for family history and want to immerse yourself from the very beginning in all the research skills you'll need, it might be worth investing in an Adult Education programme or Workers' Educational Association course. These range from full-time courses to evening classes and are advertised by local libraries, colleges and some universities. The Institute of Heraldic and Genealogical Studies (IHGS) promotes its own genealogy courses on its website at www.ihgs.ac.uk. Obtaining a diploma or certificate in family history will not only help you to fully understand the more complex aspects of genealogy, but will be a good investment for the future if you decide to go on to teach the subject to others. Indeed, many universities now offer distance learning courses in the subject, as well as qualifications at degree level.

Archives also run induction days, lectures or tutorials at both local and national level, often with a particular theme as the focus. Some archives publish details on their websites of the tutorials that are planned for the forthcoming year, as do other organizations.

The Society of Genealogists holds lectures on the documents that can be found for various types of occupation and circumstances, from researching ancestors in the brewery industry to finding out if your great-great-grandfather was sent to debtor's prison. Lectures held at the SoG usually explain how to locate original documents from a few different archives if the relevant sources are not all held in one place. Booking your place at a lecture not only gives you the opportunity to ask the speaker specific questions they may not have covered in their talk, but you also get to meet other researchers who are interested in the same topic, so you can discuss ideas and problems with others after the talk.

Some professional genealogists and historians organize day-long workshops at archives using case studies to illustrate how to go about ordering and interpreting original documents at the archive. You should keep an eye out for advertisements for these workshops in family history magazines and local newspapers, and your family history society should be able to update you with the main events in the calendar. Although they usually cost quite a bit more than an hour's seminar, these workshops are worth the money if you are having serious difficulties, because they offer a far more comprehensive lesson on researching the subject.

Many organizations, such as the Federation of Family History Societies, hold annual conferences which include keynote speeches, workshops and seminars where you can also learn a great deal about the subjects you are researching. There are also regular annual family and local history day events run by regional family history societies and groups, as well as the annual National History show in which family history features prominently, including the Society of Genealogists' family history fair, and of course the *Who Do You Think You Are? Live* stage.

Professional Researchers

The last port of call if you really cannot overcome a research problem should be to turn to the professionals. It might also be worth paying somebody to research specific documents in an archive that is far away from where you live if it works out to be more costly to travel there yourself. There are many professional genealogists with years of experience working in archives who provide their services at a charge. Because very few archives offer their own research service, most will have a list of private researchers who have a thorough knowledge of their records. The National Archives has a list of independent researchers on its website where you can browse each person's area of expertise and find contact details.

When approaching a professional researcher for a quote, you should always try to be as clear as possible about what you know already, what is fact and what is hearsay, and what you are hoping they will find out for you. It is better that you give too much information rather than too little to ensure they do not cover work you have already done yourself. However, you do not need to give them your entire family tree if you are only asking them to help you with one branch, especially as some researchers will charge you for the time it costs them to read through all your notes.

A lot of professional researchers charge by the hour, though some will have a daily fee. If you do want to go down this route, shop around to get some quotes before going with the first researcher you find, and make sure a set number of hours or a price is agreed before you commission them to go ahead with the work. While most researchers will advise an estimated quote for a job, it can be very difficult to judge exactly how long it will take to get to the bottom of a mystery or to conclude a job if you haven't simply requested a set list of document searches, so your researcher may suggest they spend an initial few hours looking into the case so that they have more of an idea what documents survive. This way you pay a smaller amount, and will be updated about further avenues that could be explored.

SUMMARY

Sources of offline help:

• *Check out reference books at your local library*

• *Agony aunts in genealogical magazines and journals can help with research queries*

• *Family and topic-specific history societies provide local indexes, networking opportunities and expertise*

• *Courses, tutorials and workshops can give in-depth help and tips on particular subjects and archive collections*

• *Professional researchers*

Basic Sources

This section examines in detail the most important sources used to construct a family tree: civil registration or birth, marriage and death certificates, census returns, wills and probate material, and parish registers. These are used in combination. As a rough guide, civil registration certificates will be your first port of call. From the beginning of the twentieth century you can also use census returns. These will be available as far back as the early nineteenth century, before which time parish registers should be consulted. Prior to that you will only be able to rely on parish registers and probate documents.

CHAPTER 5

Civil Registration

Some of the most important sources for any family historian are the records generated by civil registration – birth, marriage and death certificates. They are, essentially, the 'building blocks' for any family tree and can be used to verify initial information gathered from your relatives, or extend your family tree further back in time. This chapter explains what these sources are, where you can find them, how you can order them and various ways you can extract relevant information to help with your research.

Birth, marriage and death certificates are crucially important for anyone wishing to research

The journey from cradle to grave has been officially recorded by the state since the nineteenth century, when the civil registration of births, marriages and deaths was first introduced. The government passed legislation making it mandatory to register the birth of every child, the marriage of each couple and the death of every person from 1 July 1837 in England and Wales, with the subsequent issue of paperwork – birth, marriage and death certificates. Similar legislation enforcing the same was enacted in Scotland from 1 January 1855 and in Ireland from 1864 onwards (although Protestant marriages in Ireland had been registered since 1845). These monumental changes to everyday life came about through the government's desire to monitor population trends more effectively, following a Parliamentary Report in 1836. Previously, the established Church of England had collected some of this information through its parish registers – a subject tackled in Chapter 7. However, in the late eighteenth and early nineteenth centuries, penalties against non-conformist religious bodies were relaxed, which led to a growth in these movements, and the number of people whose journey through life was not recorded by the Church of England increased dramatically. Therefore by the beginning of the nineteenth century the information held by the established

Church could no longer be deemed accurate, and so a parliamentary committee was set up to investigate the problem.

The introduction of a centralized system whereby birth, marriage and death certificates were generated is crucially important for anyone wishing to research their family tree, as it is possible to obtain copies of every certificate issued going back to the earliest records in 1837. Each type of certificate will give different clues, depending on which one is viewed, and this chapter explains how the system worked; what each certificate contains, and how you can obtain copies for your ancestors; common problems in tracking down certificates; what material is available online; and a summary of civil registration in Scotland, Northern Ireland and the Republic of Ireland.

Civil Registration in England and Wales

In 1837, England and Wales were divided up into 27 registration districts, based upon the contemporary Poor Law Unions. Each district was administered by a Superintendent Registrar and was further subdivided into local districts staffed by local registrars. The original registration districts were reorganized in 1852 and their number increased to 33, with a further revision taking place in 1946. A Registrar General was appointed to be responsible for the entire system and was originally based in London.

The local registrar would record each birth or death and originally it was the responsibility of the official to collect this information. He would be expected to travel through his local district and record each birth within six weeks and each death within five days. As there was no onus placed on the family to report this information there may be some gaps in the early registers. The situation changed in 1874 with the passing of the Births and Deaths Registration Act. The burden of responsibility for reporting the information now lay with the family; fines were payable for late or non-registration from 1875 onwards.

Each event was recorded on a special form, with one copy retained by the registrar and one copy issued to the informant. The information compiled locally would then be sent to the superintendent registrar, who would in turn send a copy of all registrations in his district to the Registrar General in London on a quarterly basis.

The situation was slightly different for marriages. The clergy for

Making the Most of Civil Registration Certificates

Every birth, marriage and death is recorded at a local Registry Office and a certificate is produced to confirm the details of each event, although the information on each type of certificate varies according to the country it was registered in. Each country has a centralized registration index arranged chronologically so you can research all of your ancestors from one place regardless of their geographic spread. It is essential to have evidence of at least each person's birth and marriage on your tree. Even if you are starting with yourself, make sure you can locate your birth certificate and compare it with your parents' marriage certificate to ensure all the names, occupations and dates match up.

This process should be repeated for every person on your tree. For example, if you have a *birth certificate* for Mary King, born in 1912 in Rotherham, South Yorkshire, which told you that her father was Herbert King, a railway fireman, and her mother was called Thirza King, formerly Payling, then you would expect her parents' *marriage certificate* to be dated prior to 1912 and contain similar details. This marriage certificate would then tell you Thirza's father's name and that of Herbert's and their occupations, giving you new information to work with. You would also expect Mary King's own marriage certificate to

confirm her year of birth, father's name and occupation.

Death certificates are of less genealogical use than birth and marriage certificates because they tend to only really give information about the deceased individual. That is not to say it isn't worthwhile ordering death certificates. They can tell us the deceased person's age, which enables you to establish when they were born if you have nothing else to work from. Death certificates can be more helpful for ancestors who died shortly after the introduction of civil registration, because it will be more difficult to find information about them from other records. Apart from details about the cause of death, notes given on death certificates can lead you to other sources by giving details about a coroner's inquest that might have taken place. If you know when and where an ancestor died (which will be recorded on the certificate) it also makes the hunt for a will and burial record easier.

Look out for the names of witnesses and informants on civil registration certificates. These people are often close family members and if you know their names, even if you are not yet sure exactly how they are related, you may be able to identify your ancestors in other documents, such as household census returns.

Research hints

There are general rules you can follow when searching for the births, baptisms, marriages, deaths and burials of ancestors you have never known even if you only have a rough idea of when they were alive:

1. If you start from the last known birth on your tree for which you have a birth certificate, say your grandmother's birth certificate dated 1917 for example, this should give you her parents' names. You can then search for a marriage under their names back from 1917. You may have to work back as much as twenty years to 1897 if your grandmother was the youngest of a large generation of children, but once you have found your great-grandparents' marriage in the indexes you can order the certificate to find out their ages, which will enable you to then search for their birth certificates over a range of a few years.

2. Some marriage certificates do not give exact ages and will state 'full age' instead, meaning a person was over 21 years old, or will say 'minor' if they were less than 21 years old. Where this is the case you can search for that person's birth date starting from around 16 years prior to the date on their marriage certificate and working back perhaps as many as 20 more years, if they married late in life. Starting a birth search 16 years prior to a marriage date also works well when searching parish registers, which rarely give ages.

3. If you are keen to find out when an ancestor died, the only way to do this from death and burial indexes is to establish the last known time they were alive and work forward from then. Perhaps your grandmother was a witness on her daughter's marriage certificate in 1965, in which case you can conduct a search for your grandmother's death from 1965 onwards. If you are looking for the death of a person who was born over 100 years ago, you would usually only need to search up until they would have been 100 years old. It is important to conduct a search for the longest period of time over which an event was likely to have occurred, particularly if you are looking for somebody with a common name.

4. When searching the birth, marriage and death indexes you will often come across more than one possible match, and the only way to find out which one is correct is to order the certificates for the most likely options and compare them against other information you have gathered for that person. If you are confident that you have conducted a thorough search of the indexes, then you will know that you have not missed anything.

churches that were officially authorized to record marriages were expected to send the quarterly returns straight to the Registrar General in London. Non-conformist churches had to have their buildings licensed to perform such ceremonies, with the local registrar being legally obliged to be present to record the details. However, from 1899 the situation changed thanks to the Marriage Act of 1898, and non-conformist clergy from these churches could also record and submit the information themselves.

As mentioned above, it was the duty of the local superintendent registrar to forward the information to the Registrar General in London. Therefore there are two sets of records: the original records held at the local registrar's office and the copies held by the Registrar General. Once the records arrived at the Registrar General's office in London, clerks would reorganize them. They made alphabetical indexes for the certificates, broken down on a quarterly basis. Currently, the general public has no legal right to view the original certificates held locally but only the copies held by the Registrar General, though you can order duplicate copies of the original records from local register offices. The records of the Registrar General for England and Wales are now in the General Register Office (GRO), which is a department of the Office of National Statistics, and duplicates can also be ordered online at www.gro.gov.uk. Separate arrangements exist for Scotland and Ireland, and are discussed later in this chapter.

What Do the Certificates Contain?

Birth Certificates

Birth certificates are the official record of the individual's place and date of birth. As mentioned, each birth had to be recorded within six weeks of the event, although this would not always happen, particularly if the family were travelling at the time of the birth, and waited to register it until they returned home.

The GRO birth indexes include all of England and Wales. Each entry is entered in alphabetical order, annually and then in the relevant quarter – March, June, September, December. All births registered between 1 January and 31 March are included in the March quarter; between 1 April and 30 June in the June quarter; between 1 July and 30 September in the September quarter; and between 1 October and 31 December in the December quarter. After 1984 the registers are arranged annually and not on a quarterly basis. An appropriate index

▲ Indexes to civil registration in England and Wales can be accessed online from various commercial websites.

reference number is also provided, which is the key piece of information needed to order the certificate. From the September quarter of 1911 the maiden name of the mother was also included in the index entry.

The actual certificates provide the following information:

● *Where and when born:* The precise date and location of the birth; if the exact time is given it signifies that it was a multiple birth (possibly twins or triplets). In this case you may wish to search for the other sibling(s), who should have the same surname and registration reference.

● *Name (if any):* This should be the full name given, including any middle names (the index will only give the initials of any middle names given). Some parents would change the name (this was allowed up to one year following registration). In such a scenario both the original and the altered name should appear.

Sometimes a birth would be registered even though no first name had been chosen. This explains the 'if any' in brackets on the certificate. In the indexes there are also entries at the end of surnames for 'male' or 'female', used when the first name had yet to be decided.

● *Name and surname of the father:* The full name of the father.

● *Name and maiden surname of the mother:* The full name of the mother, including her 'former' (maiden) name; this last piece of information is particularly useful when trying to trace the maternal line further back. You may also find evidence of a prior divorce in this section too.

● *Rank and profession of father:* This provides the occupation of the father. This is a good genealogical clue, determining the social status of your ancestor. You may also be able to use this piece of information to search for employment records for your ancestor. Bear in mind, however, it would not be that uncommon for people to 'inflate' the status of their occupation.

● *Signature, description and residence of informant:* This is the individual who registered the birth. In most cases it would be the father, but not always. Sometimes there is a mark instead of a signature, indicating the informant was illiterate.

● *When registered:* The date the birth was officially registered; don't forget, this could be up to six weeks after the actual birth, so if you

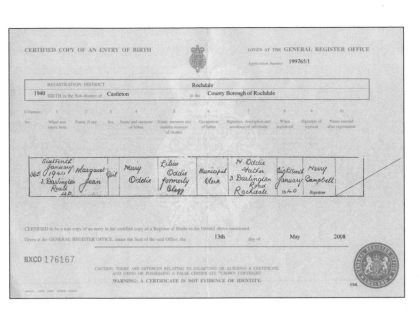

Birth certificates

Bill Oddie's story was one of the most poignant told on Who Do You Think You Are? *as he wished to investigate the background to his mother's ill health and rumours that he had a sister. The story of his missing sister was quickly established by tracking down his parents' marriage certificate of 1938, establishing Bill's mother's maiden name (Clegg) and looking in the national GRO indexes for the birth of any children with the surname Oddie, mother's maiden name Clegg, in the Rochdale area, where the family lived at the time.*

A fairly quick search revealed that a Margaret J. Oddie, mother's maiden name Clegg, was born in the March quarter of 1940. On ordering the certificate (above, left), her parents were listed as Harry Oddie and Lilian Oddie, née Clegg – the same as Bill's. This therefore was his missing sister, and a further search of the death indexes showed that she had died as an infant the same quarter, explaining why Bill never knew about her existence.

think your ancestor was born in late March, June, September or December and can't find an entry in the relevant quarter, it might be worth checking the indexes for the following quarter too.

- *Signature of registrar:* The name and signature of the registrar.

Marriage Certificates

A marriage certificate is the official record of when and where a marriage took place, in addition to the record that would have been compiled in the relevant religious institution (a parish register, for example; these have been kept since the sixteenth century, and continue to be compiled today – see Chapter 7 for more details). They are a particularly rich source for the genealogical researcher as they give lots of clues for various ancestors. As mentioned above, from 1837 onwards, marriages of individuals of the Church of England, along with Jews and Quakers (where buildings were licensed to hold marriages), were recorded by the priest or responsible clerk and sent to the General Registrar's Office in London. For other non-conformists, the local registrar recorded the marriage. This requirement was relaxed in 1898 and an 'authorized person' from other religious denominations could also record this information and send it forward to the appropriate bodies.

It is important to remember the age of consent before conducting a marriage search. In 1929 it was raised to 16 years for England, Wales, Scotland and Northern Ireland. Prior to this it had been 14 for boys and 12 for girls. In the Republic of Ireland the age of consent was only raised to 16 in 1975, having been kept at the ages of 14 for boys and 12 for girls until then. It's worth bearing these ages in mind when searching for marriage certificates, making sure you search back far enough. The age of consent, however, differs from the legal age at which people could marry without parental consent, and in England and Wales this was 21 until 1969, after which it was reduced to 18 – though in Scotland it is as low as 16.

The GRO indexes include an entry for both the bride and the groom. If you know the names of both parties who were married it is advisable to search for the least common surname. The registers are arranged annually and then on a quarterly basis. They are then indexed alphabetically by the surname and then forename of the bride and groom. From the March quarter of 1912 the surname of the spouse is also given. Lastly there will be a numerical reference for the marriage.

The actual certificate will provide the following information:

- Above the columns there will be a section stating exactly where the marriage took place (which church or other place) and in which parish and county. This is very useful as it can indicate whether your ancestors were non-conformists or not.
- *When married:* The exact date the marriage took place.
- *Name and surname:* The full names of both parties getting married.
- *Age:* The given age of the bride and groom; it is important to note that these may not be entirely accurate. It was not uncommon for people to state that they were simply of a 'full' age or even a 'minor' age. Prior to 1969, full age would be someone aged over 21 years and a minor anyone younger than 21. Thus it could be problematic working out the exact ages when trying to find the birth records of these people. Also, it was not uncommon for people to lie about their age, depending on the circumstance. People under 21 may claim to be several years older to avoid the need for parental consent. Alternatively, older women marrying men significantly younger than them may give a younger age to minimize any potential scandal.
- *Condition:* This column states whether the marrying party was a bachelor, spinster or widow/widower. You may be surprised to find out that your relatively young ancestor was already widowed and marrying for a second time. However, mortality rates would have been significantly higher in the early period of civil registration and sometimes people lost their spouses quite soon after marriage, particularly in childbirth. Remarriage was therefore a viable practice, especially with young widowed men with small children who needed a maternal figure to look after them.

- *Rank or profession:* The occupation of the two parties is stated here. In the same way as with the occupation section in birth certificates, be aware of exaggerations and, depending on the given occupation, whether you will be able to locate their employment records. Female occupations were not regularly detailed until the twentieth century.
- *Residence at the time of the marriage:* The address of the bride and groom at the time of their wedding; the usual custom was to be married in the parish of the bride and sometimes the groom would have a temporary address in that same parish, as he would have had to have been living in the parish for a month to be married there.
- *Father's name and surname:* A vital clue for the genealogical researcher, helping one get one generation further back at the same time, although the mother's name would not be given. If the father was deceased it would often (but not always) be recorded as 'deceased' in brackets after the father's name.
- *Rank or profession of father:* The occupation of the father of both bride and groom is provided. Again, be aware of the same inaccuracies in the given occupations.
- *The type of marriage:* Whether the marriage was performed by a marriage licence or banns, announced in the parish church for the three weeks preceding the wedding.
- *Name and signature of two witnesses:* At least two people are required to witness the marriage. These would often, but not always, be family members of the bride and groom.

CASE EXAMPLE

Marriage certificates

*To trace **Bill Oddie**'s family further back in time, it was necessary to verify the personal details of each generation and order the necessary certificates. This meant starting at the beginning by obtaining the marriage certificate (left) of his parents, Harry Oddie and Lilian Clegg, in 1938. The GRO indexes were examined, and an entry quickly found in the December quarter. The certificate was ordered, and contained the following information.*

Marriage at the Methodist Church, William Street, Rochdale
Groom: Harry Oddie, 28, bachelor; occupation municipal clerk; address 17 Russell Street, Rochdale; father Wilkinson Oddie (deceased), loom overlooker
Bride: Lilian Clegg, 24, spinster; occupation shop assistant, general store: address 82 Grove Street, Rochdale; father Joseph Peter Clegg (deceased), engineer
Witnesses: Edgar Oddie, Marion Oddie

With this information other searches could be started. Harry and Lilian's ages at marriage meant that their birth certificates were relatively easy to locate – Harry was born in 1910 and Lilian in 1914 – whilst searches could be started for their fathers' death certificates as well, prior to 1938.

Death Certificates

These certificates record the time and cause of death. They are perhaps less obviously useful for people trying to take generations further back, but can give a useful picture of the social standing and life conditions of your ancestor (usually indicated by cause of death).

The recording and indexing of death certificates is done in the same way as births, with a record being kept locally by the superintendent registrar and one nationally at the GRO. A death would be recorded locally where it occurred, rather than on the actual residence of the deceased. Again, the GRO indexes are organized annually and then subdivided into quarters. Within these quarters the individuals are listed strictly alphabetically, surname first and then forename. From the March quarter of 1866 to the March quarter of 1969 an age of death for the deceased also appears. Hence there is no need to order the actual certificate if you only require this information (and if you are sure this is the correct ancestor, as with common names you may only be able to verify if it is the right person by obtaining the actual certificate). From June 1969 an age at death was replaced by the date of birth of the individual. As with birth and marriage certificates, the indexes were organized annually from 1984, and not further subdivided on a quarterly basis.

The actual certificate will give the following information:

- The exact registration district where the death was recorded.
- *When and where died:* The exact location of where the death occurred, which may not be where the deceased lived as he or she may have been visiting family or died in hospital.
- *Name and surname:* The full name of the deceased.
- *Sex:* Whether the deceased was male or female.
- *Age:* The age of the deceased; this is perhaps the most useful piece of information for those wishing to take their family tree further back in generations. Once you have an age it is possible to start searching for the individual's birth certificate and parentage. Bear in mind, however, that this information would not always be accurate as no proof of age was required. Not everyone remembered their age with absolute certainty, especially those born prior to civil registration or in the early part of the nineteenth century.
- *Occupation:* A good genealogical clue in helping trace appropriate employment records if relevant. Women who were married or widowed would usually have the name and maybe occupation of their husband provided, which can be useful.

- *Cause of death:* The more modern records of death may give quite specialized medical terms that may need to be researched. The early certificates could be somewhat vaguer in the medical terms used. The cause and age of death are good indicators of the living conditions of the deceased, with poorer people generally having shorter life spans then the wealthier classes. Additionally, a sudden death or accident would often require a coroner's inquest before the death certificate could be issued, and the date of the inquest should be stated on the certificate. These inquests may well have been reported in local newspapers and it is worth pursuing this line of enquiry. You may also be able to find the actual coroner's report in your local record office. However, they are subject to data protection for 75 years and not every report would have survived, as there is no legal requirement to retain the information after 15 years have elapsed.
- *Signature, description and residence of the informant:* This can also be a useful piece of genealogical information as sometimes it would be family members who would register the deaths. However, after 1874 the law changed and it was compulsory to have a doctor's certificate before a death certificate could be issued, and hence doctors would sometimes appear as the informants.
- *When registered:* The date the death was registered, the legal requirement being five days after the death of the person. However, if a coroner was involved, there may well be a considerable delay in registration.
- *Signature of registrar:* The signature of the local registrar.

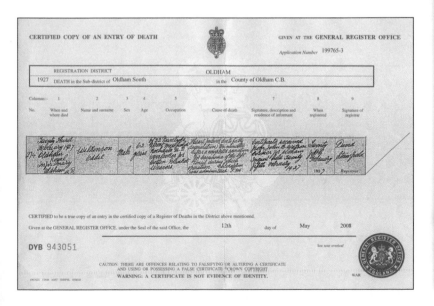

CASE EXAMPLE

Death certificates

Whilst pursuing Bill Oddie's family tree, his parents' marriage certificate revealed that both their fathers had died by the time the couple married in 1938. This information was used to search for the death certificate (below, left) of Wilkinson Oddie, Bill's grandfather. Starting with 1938, a search was made backwards in time, and an entry was found in 1927 for Wilkinson Oddie, aged 62, whose death was registered in the Oldham district. This important biographical information made it easy to look for his birth certificate, which was found in Rochdale in 1864. Given Bill's father Harry was born in 1910, Wilkinson would have been 46 at the time of his son's birth, which seemed quite old. Having found his birth and death certificate, a search was made for Wilkinson's marriage certificate prior to 1910, which was found registered in Rochdale in 1907, to Emily Hawksworth. On the certificate, Wilkinson's age was confirmed at 42 and his marital status was listed as widower. Clearly, further stories remained to be uncovered in Wilkinson's background ...

How to Locate and Order a Certificate

As mentioned above, information has been recorded at a local and a national level so there are two sources you'll need to consider. The national indexes have been compiled and retained by the GRO, whilst there are also indexes to the certificates available locally. It is crucial to remember that the index entries in the local registers are not the same as the ones available at the GRO as each office would use their own indexing system.

Local Registration

The original certificates for each registration district are held at the superintendent registrar's office. Each major city would have one of these offices and there would be numerous superintendent offices per county. However, due to some boundary changes throughout the nineteenth and twentieth centuries some of these offices may have been abolished and their records transferred to another office nearby. Each local register office is likely to have indexed the information by local district, year and then alphabetically (probably by first letter of surname only and not in strict alphabetical order). Hence in order to begin searching you will need to know the superintendent district and then the sub-district. The advantage of searching in the local registers is that it will be a quicker search to conduct, especially if the name you are looking for is relatively common. However, if you are not certain where the registration occurred, it is advisable to turn to the national GRO indexes.

National GRO Indexes

To order a duplicate certificate, you need to identify the relevant entry in the national indexes and note several pieces of information:

- The name of the individual (arranged in strict alphabetical order by surname)
- The name of the local district where the registration occurred
- The two-part numerical reference (the first being a code for the superintendent district and the second number a reference to the page where the certificate will be found)

Until October 2007, the national paper indexes were held at the Family Records Centre in Islington, London, before they were moved to Christchurch, Dorset; but they are no longer available for public inspection. Two projects are underway to create an online digital index service known as MAGPIE, linked to the Digitization of Vital Events (DoVE) project whereby the actual certificates would be made available as well. However, many commercial companies have created their own digital images and searchable databases of the GRO indexes – a topic that will be covered shortly – whilst the national GRO indexes have also been copied onto microfiche, and many local libraries and record offices hold copies. All duplicate certificates located on these national indexes have to be ordered online via the GRO website, www.gro.gov.uk, where you'll also find details of how to complete the necessary forms and pay for the certificates and the expected length of time it will take to deliver.

This is where you are likely to incur the most cost when building your family tree. At the time of going to press, each certificate will cost you £7 to purchase from the GRO, and takes a minimum of four days from receipt of order to dispatch of duplicate certificate. You can order a certificate on 24-hour turnaround, but these cost £23 so patience is probably a virtue! Despite these costs, you will need to order (where possible) a birth, marriage and death certificate for each direct ancestor, as the clues they contain will not only allow you to work back generation by generation but will also give you important information about their social status, place of residence and occupation.

> *Certificates give vital information about social status, place of residence and occupation.*

Common Problems

Although it was a statutory obligation to register all births, marriages and deaths from 1 July 1837, you may well experience difficulties in finding an entry even though it should be included. There are numerous reasons behind this:

Late Registration
Often people would not register punctually. If you do not find an entry in the appropriate quarterly index, keep searching as it may well turn up later. A common mistake is to assume a marriage occurred at least nine months prior to a birth. This is by no means always the case, with people rushing to marry before a birth to avoid the stigma of having an illegitimate child.

Lack of Registration
Unfortunately, not every single event was registered. This was particularly the case in the early period of civil registration as some people treated the legal requirement to register with a degree of suspicion. Additionally, until the 1874 Act it was the responsibility of the local

registrar to note down the event rather than that of each individual, and many people did not bother to report events to the registrar.

Some studies have estimated that as many as 15 per cent of births would not have been registered in the early years until the rules were changed from 1875, rising to as high as 33 per cent in some urban areas. Indeed, parents would attempt to hide the age of their children in order to send them to work as young as possible (child labour was being regulated by statute through various acts in the nineteenth century). Ignorance also played a part, as it was often not realized that registration was still required even if the child had been baptized, many people believing the church ceremony should be adequate. Hence, if the birth is not found, you should check the relevant parish records.

There are fewer gaps in the registration of marriages, although again it may be worthwhile consulting the local parish registers (see Chapter 7 for more information) to try and find a marriage this way, as some marriages in the early days of civil registration may have been recorded by the Church only. Additionally, some people lived as man and wife without actually ever marrying (legally it was the responsibility of those accusing the couple of having an 'invalid' marriage to prove it). This could be the case when people had separated but not formally divorced and remarriage was not an option.

The most complete set of registration certificates should be for deaths, but even some of these were missed in the early years of civil registration. Again it might be worthwhile searching for the burial in the appropriate parish, if known.

It is possible that the birth, marriage or death being searched for did not occur in England or Wales, and you may have to search in the Irish or Scottish records, discussed below. Alternatively, events may well have occurred overseas whilst a member of your family was on board a ship, serving in the armed forces or working in a colony in the British Empire. Information about looking for overseas civil registration is also discussed below.

Incorrect Index Entry

This is possibly the most common reason for not finding an entry, the mis-transcription by the clerk originally entering the information. Unfortunately, this was not so uncommon, especially in the earlier registers when everything was handwritten, making it difficult to read original certificates and therefore entering an index entry in the wrong place was an easy mistake to make. Certain letters are easily confused and this should be borne in mind when thinking of variant spellings:

> *An incorrect entry into the index is the most common reason for not finding an entry.*

- A capital handwritten B, P, D or even K can be easily confused
- It can be difficult to distinguish a 't' from an 'l', an 'm' from an 'n' or an 'e' from an 'i' when handwritten
- As letters were often handwritten with large loops they could be easily misread and confused
- Some surnames have common variant spellings. For example 'Matthews' may be spelt 'Mathews', 'Doherty' as 'Docherty' or 'Johnson' as 'Jonson'. Certain forenames may also have alternative spellings, such as 'Sarah' for 'Sara', 'Conor' for 'Conner' or 'Coner', or 'Jane' for 'Jayne'.

Each step in the registration process could lead to a misspelling. Hence, by the time an entry has been placed in the national indexes the name could have altered a great deal. Thus if you have encountered a problem in the national indexes, try searching the local registers.

Another problem is that in the nineteenth century spellings were not necessarily uniform and some people spelt their names differently at various times. The relatively low level of literacy would also lead to inaccuracies as it would not be possible for people to ensure their names were spelt properly. In such circumstances the individual writing down the information would have to spell the name phonetically, which could lead to problems with uncommon surnames.

The last thing to remember is the use of nicknames, as information may be recorded either as the full correct name or as the more informal nickname. Hence, when looking for the birth of an 'Anthony', 'James' or 'Nicholas', remember to search for the shortened versions of these names – 'Tony', 'Jim' or 'Nick' – if you have no joy.

Online GRO Indexes

The growth of the Internet in the past 10 years has seen a huge growth of genealogical websites. Many commercial ventures have invested a large amount of time and money in digitizing many genealogical documents, including the GRO indexes and some local registers. It is now possible to search for your certificates online and, depending on which website you choose, many of these searches are also free of charge. Below is a list of some of the most useful online sources.

www.freebmd.org.uk

This is a free-to-view website run by volunteers who have been manually transcribing each single index entry in the GRO indexes. At the time of going to print the team has transcribed over 135 million records, with entries being relatively complete from 1837 to about 1915. It is an ongoing project and it is hoped that the whole period of civil registration will eventually be covered.

The main advantage of this site is that you can search for a particular name through a number of quarters all at once, rather than having to search through each quarter one by one. If you do find a relevant entry on the website it is advisable to double-check the entry with the original entry before ordering, in case of any transcription error.

www.ancestry.co.uk

This is the largest commercial genealogical website geared to the UK market currently on the Internet. Although many of its databases are only accessible upon payment, it is possible to search the GRO indexes online free of charge after registering your details on the website. Ancestry has scanned images of each page of the GRO indexes for every quarter, which means you need to search for an entry by going through each quarter at a time, as there is no single-name database or digital image of each individual entry.

www.findmypast.com

This is another commercial genealogical website. It has placed the GRO indexes on its website in a similar fashion to Ancestry, by digitally scanning each page of every quarter for the entire civil registration period. Again, due to the digitization process you still need to search through each quarter as not every entry has been individually scanned. It is not free of charge but runs a pay-per-view service.

▼ Family Relatives provide a range of searches for civil registration records, arranged by type and date period.

www.familyrelatives.com

Family Relatives have also provided online access to digitized GRO indexes. There are fully transcribed searchable indexes for the periods 1866–1920 and 1984–2005, whilst it is possible to search the periods 1837–65 and 1921–83 by surname and browse the GRO index images. To use this service, you need to register as a user, log in and buy credits.

Along with placing the GRO indexes online, certain local archives and record offices are investing in placing their local registrar indexes online too. There are websites such as www.ukbmd.org.uk, which lists which local indexes have been transcribed or placed online.

Civil Registration of Britons Overseas

Millions of Britons have worked overseas in the armed forces, as civil servants in one of the colonial administrations that comprised the British Empire, or on board a vessel travelling between foreign parts. Although they were not incorporated in the main national or local civil registration indexes, attempts were made to register as many of these people as possible, and the records are analysed here.

Overseas registers have been kept by the GRO and duplicate certificates can be purchased from the GRO website www.gro.gov.uk once you've found the correct registration reference. There are indexes available on microfiche at The National Archives at Kew and other archives, or online at www.findmypast.com. These are broken down by period and type, covering:

- General indexes from 1966
- Colonial and ex-colonial indexes, 1940-81
- Civilian indexes 1849-1965, consular registers of births, marriages and deaths
- Civilian indexes 1837-1965, marine registers of births and deaths
- Civilian indexes 1947-65, air registers of births and deaths
- Civilian indexes, various foreign registers
- Military indexes 1761-1924 regimental births
- Military indexes 1796-1880 chaplains' returns of births, marriages and deaths
- Military indexes 1881-1955 army births, marriages and deaths
- Military indexes 1956-65 army, navy, RAF births and marriages
- War deaths 1899-1948

In addition, there are separate consular records for people who were baptized or married or whose death was recorded at a British embassy or consulate. These records are predominantly held at The National Archives in a variety of record series. For a full list of countries covered, and where the records are stored, you should consult *The British Overseas: a guide to records of their births, baptisms, marriages, deaths and burials available in the United Kingdom* (3rd edition, 1994) published by Guildhall Library. Further information is likely to be held in consular correspondence, which is also held at The National Archives in series FO 83 and FO 97, with an index available in document FO 802/239. Records of non-statutory registers, many of which relate to births overseas and on board ships, can be found in the collected archives of the Registrar General at The National Archives in series RG 32–36.

Civil Registration in Scotland

Scotland has its own civil registration process, and the records are known as Statutory Registers. Civil registration was begun slightly later than in England and Wales, in 1855, but the certificates are more detailed then their counterparts across the border. Indeed, the earliest ones in 1855 are particularly detailed, though the sheer amount of information requested proved very difficult to record and thereafter the list of questions was simplified somewhat.

Additionally, the civil registration records are held in the same place as the parish records, in the General Register Office of Scotland (GROS) in Edinburgh. It is therefore possible to conduct a large amount of your genealogical research in the same place, which can simplify things greatly. The GROS levies charges for anybody using their services. These charges vary depending on whether you wish to visit the office for one day, one week or annually. At time of going to print the daily rate is £17 and the weekly rate £65. It is also advisable to book an appointment before visiting as there are only a limited number of spaces and the office may be fully booked.

Another advantage that the Scottish records have is that the indexes are fully computerized, which means you can search for a specific entry by name across the entire period. The computer database contains summaries of microfiche registers that contain the entire entry, and this latter entry is the one required for ordering copies of the certificates at GROS.

Birth Certificates

Birth indexes include the mother's maiden name from 1929 onwards. The certificates themselves are similar to the English and Welsh certificates in giving the full name, the child's sex, when and where (including time) they were born, the father's full name and occupation along with the mother's name (and maiden name), and similar details are provided relating to the registration details. However, where Scottish birth certificates differ is that they provide the details (time and place) of the marriage of the parents, including any other married names. In 1855 the

certificates also stated the birth details of the parents along with details of other siblings, but this was quickly deemed too much information and was not given from 1856 onwards. The years 1856 to 1860 do not give marriage details of the parents either.

Marriage Certificates

Indexes are arranged separately by bride and groom. However, from 1855 to 1863 and then from 1929 you can find entries for brides in both their maiden names and their married names. The certificates themselves also note where and when the marriage took place, the type of marriage ceremony, full names, ages, marital status and occupation of the bride and groom. The additional details peculiar to Scottish certificates are the occupation and maiden names of the mother of each party. In 1855 only the certificates also state where the bride and groom were born and any previous marriage (along with names of any other children from the previous marriage).

▲ One of the best websites is www.scotlandspeople.org.uk, a one-stop shop for Scottish genealogy covering civil registration, census returns and old parish registers.

Death Certificates

The indexes for these certificates are arranged thus:

- Age at death is provided from 1866 onwards
- The maiden name of the deceased's mother from 1974 onwards
- Deceased married women are indexed by their married and maiden name after 1858

The registers themselves detail the name, age, exact time and location of death, the person's occupation and their marital status. Medical causes of death are also provided. Scottish certificates give the name of the deceased's spouse from 1861 (and in 1855) and the names of the deceased's parents (including the mother's maiden name). The earliest certificates of 1855 state where the deceased was born and how long they lived there, along with the details of any children (their ages and if they were still alive). From 1855 to 1860 burial details of the deceased are also provided.

Other Hints

There is a central website - www.scotlandspeople.gov.uk - that allows you to search by name the statutory registers of births (1855–1906), marriages (1855–1931) and deaths (1855–1956), as well as the old parish registers (which are covered in Chapter 7). You can search and download images of these registers for this period without needing to travel to the GROS in Edinburgh, although the website does charge to access this information.

The Society of Genealogists holds copies of indexes for Scottish statutory registers for the years 1855 to 1920. You may also be able to find indexes at various local family history societies.

Civil Registration in Ireland

Although civil registration for Protestant marriages began from 1 April 1845, the comprehensive civil registration of births, marriages and deaths was introduced to Ireland later, on 1 January 1864, on a similar model to that of England and Wales. These events would be recorded locally at the district registrar's office and then copies passed on to the General Register Office in Dublin. However, in 1922 the country was partitioned and divided into Northern Ireland and the Republic of Ireland. Subsequently, civil registration was also divided between those two territories and a separate office was opened in Belfast, Northern Ireland.

Republic of Ireland General Register Office

The General Register Office of Ireland (GROI) is located in Dublin and houses the national indexes along with microfilm copies of the originals for the entire country from 1 January 1864 until 31 December 1921. It also has the copies of all the early Protestant marriages, which were recorded from 1 April 1845. From 1 January 1922 the office records all events that occurred in the Republic of Ireland. You will be charged to search through the indexes and a further charge will be incurred for ordering any certified copies of the registers.

Birth Indexes

The indexes are arranged alphabetically by the child's surname and then forename. After 1902 the mother's maiden name is included in the indexes. The actual date of birth of the child can be found in the indexes from 1903 to 1927.

> *Comprehensive civil registration was only introduced in Ireland in 1864.*

Marriage Indexes

The indexes are arranged by name of both the bride and groom. The registers themselves contain the same information as to be found on the English and Welsh certificates.

Death Indexes

The age of the deceased is included in the indexes. There is no date of birth of the deceased in the modern registers. All other details tally with the information found on English and Welsh certificates.

General Register Office of Northern Ireland

Civil registration records are housed in the General Register Office of Northern Ireland (GRONI), Belfast, for all events in the six counties of Northern Ireland since 1922. The original registers for births and deaths from 1864 can also be found here.

Birth Indexes

Births are indexed in this office. From 1903 to 1921 the date of birth of the child is also provided. They are arranged in a similar manner to those for England and Wales.

Marriage Indexes

Marriages are only to be found from 1922 onwards. Prior to that, if you are searching for a marriage in the six counties you may be able to find it in the applicable district registrar's office.

Death Indexes

You can find records covering the entire period since 1864, and they are arranged in a similar manner to those for England and Wales.

CHAPTER 6

Census Returns

Along with civil registration certificates, census returns are the other vital genealogical source for tracing people in the nineteenth century. Since they cover an entire household at a time, they enable you to extend and broaden your family tree to include the extended family. This chapter will explain what census records are, what they contain, how to find them and extract their information, and various ways of using this data to start other lines of research.

> Censuses provide snapshots of entire families at a particular moment in time.

Although there had been sporadic population surveys at various times in this country (such as the Domesday Book, commissioned in the late eleventh century by William the Conqueror), it was not until the introduction of the census in the early nineteenth century that collecting detailed information about the size and nature of the country's population became a regular event. Censuses are of vital importance to a genealogist because they provide snapshots of entire families at a particular moment in time, linking relatives and different generations together in the same household as well as providing information about where they lived, their social status and their line of work.

The first census was conducted for England, Wales, the Isle of Man, the Channel Islands and Scotland in 1801 and there have been censuses conducted every 10 years since that date (except in 1941 due to the Second World War). Censuses started in Ireland a bit later, from 1821. The decision to collect the information followed much debate and controversy in Parliament, as many people feared the process could infringe individual freedoms and liberties. The surveys were not intended for family history purposes and their usefulness in this sphere was realized only later.

The censuses for 1801 to 1831 were simple headcounts, which were used to produce accurate population figures and trends for the country. Indeed, the census for 1801 was primarily conducted in response to the threat of invasion during the Napoleonic Wars as an attempt by the government to ascertain how many potential soldiers would be available for conscription. Interestingly, however, the census returns for 1821 and 1831 for a very small minority of places contain more detailed information other than simple headcounts, with the names of the head of each household included, as happened in Hackney, London, for example. This was for no other reason than the enthusiasm of the people who went round collecting the information – little-known men such as

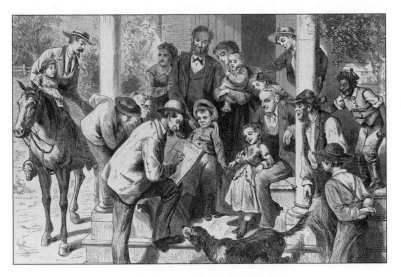

▲ Early censuses were seen as a novelty and were often greeted with great curiosity and, in some cases, suspicion (*Harper's Weekly* magazine, 1870).

Richard Stopher, who lived in the Suffolk village of Saxmundham most of his life and added notes to his census returns based on his local knowledge of the village's occupants.

In 1840 the responsibility for collecting census information became part of the remit of the General Register Office (GRO) and subsequent censuses contain more details. From 1841 the censuses started listing the names of everyone in each household, and after 1851 even more detailed information was provided, including exact place of birth (providing researchers with the vital clue to trace these people further back in time). For reasons of privacy, censuses are not released into the public domain for 100 years. Hence, it is currently possible to view all returns only up until 1901.

The census for 1911 will be released to the public in its entirety on 3 January 2012. However, due to the passing of the Freedom of Information Act (2000), the Information Commissioner ruled in 2006 that people were entitled to view parts of the census information now upon request. Currently The National Archives (TNA) holds all census returns, and will answer specific requests relating to particular addresses (it is not possible to do a name search) using its paid research service. TNA also hopes to offer a comprehensive searchable service for the census from 2009 onwards. However, this will exclude certain personal information (such as mental deficiencies or handicaps) until 2012.

Ireland has already released its censuses for 1901 and 1911. Unfortunately, however, no full censuses exist for Ireland prior to 1901 as they were destroyed in 1922 by a fire in the General Register Office in Dublin during the Irish Civil War. Those wishing to trace their Irish ancestors will have to rely on other sources for the nineteenth century, such as the Griffiths' Valuation.

How Census Information was Collected

Censuses record all residents living in a particular property on one specific night (which varied depending on which census is being viewed – see below). A week or sometimes a couple of days prior to the given date, census enumerators would deliver census forms to each household within their enumeration district. The head of the household was obliged to fill in the required information as accurately as possible and the enumerator would then collect the forms the day after census night. As illiteracy levels were high in the nineteenth century, the

Making the Most of Census Returns

Most census returns show us the names of everybody in a household, usually including how they are related to one another, their ages, occupations, places of birth and where they lived. Combining this material with that of civil registration certificates and parish registers gives you a fuller picture of your family's background, so that you can see how their occupations changed over time, how they migrated around the country, as well as giving you a better idea of how each generation interacted as a family. You might want to use the information gathered from these sources to locate the addresses where your ancestors lived and see if their houses still stand.

Research hints

The data found on census returns can be used to narrow down searches using other records:

1. If you know from your great-grandfather's birth certificate that his parents must have married before 1899, you can immediately reduce the number of years you have to search for their marriage if you find the family on the 1901 census and work out that their eldest child was born around 1892. You can then start searching for their marriage back from 1892 rather than 1899.

2. Deaths can also be traced with the help of census returns. If you find a couple living together on one census but on a census return

taken ten years later one spouse is missing and the other is listed as a widow or widower, you will know to conduct a ten-year death search for that period.

3. Use the details given on the census returns to corroborate information found on certificates. Check the addresses, ages and relationships on the returns to see if they match those given on civil registration certificates of a similar date. Equally, if you find part of your family living in a particular town on the census returns, you should find out what civil registration district that town was covered by so that you can look out for that place when locating those ancestors in the birth, marriage and death indexes.

enumerator would often assist the head of the household in filling out the forms.

The next step would be for the enumerator to use these 'schedules' and transfer the gathered information into his 'enumerator's book'. He would also record which houses lay uninhabited within his district. These completed books would be checked by a supervisor and then sent to London to allow the statisticians to compile the information they wished. It is these enumeration books that form the census records now available for the general public to view. Unfortunately, the original forms completed by each household were destroyed.

As the records are handwritten, the returns often have the enumerator's notes alongside the entries, sometimes obscuring the actual information. An important notation to bear in mind is the practice of separating each household by slashes on the top left corner of the head of the household's name. A single slash on top of the name would indicate a separate household within the same property and a double slash separate households in different properties. These slashes are particularly useful when individual house numbers have not been noted.

The information on the census was organized by distinct registration districts for England, Wales and Scotland. These were initially identical to the registration districts created in 1837 for civil registration purposes, based on existing Poor Law Unions that had been set up in 1834. Each registration district was a subdivision of a county and its size was dependent on population. These registration districts would be divided into smaller sub-districts and the sub-districts would be further divided into individual enumeration districts. The size of the enumeration district was an estimate of how many houses the enumerator could visit in one day. Inevitably, enumeration districts would be geographically larger in rural areas where the population was less dense. Additionally, each enumeration district book would have a cover page giving in detail the area and exact roads included in the district, along with parish, hamlet, village, town or county details.

These enumeration districts were roughly the same for the years 1841 to 1891 in order to make valid comparisons of data collected on specific censuses. However, the large increase in population and the industrialization of urban areas meant it was not always possible to adhere to this. Any such alteration would be recorded in the summaries of the returns, so it is worth looking at these cover pages if you want to find out more about the area in which your family lived – an important part of your work, if you remember the advice about historical context from Section One!

▲ 1950s, Manchester: the census enumerator comes calling, asking for information about who lives in the house.

England and Wales: Census Returns 1841–1901

Information Contained on the 1841 Census

> *Census returns add real colour, as they provide additional information besides biographical data which allows you to investigate the social history surrounding your ancestors' lives.*

The first detailed census was taken on Sunday, 6 June 1841, and recorded every individual that spent the night in a property; therefore family visitors and boarders would be recorded as living in that property, and not at their permanent place of residence. The format of the form was a two-sided columned page, with information running across the top of the page that stated the hamlet, village or borough plus parish details on the right-hand side. Both pages would have the following columns recording information about:

- *Place:* This would usually be the street, with occasionally the house name or number. However, house numbers were rarely recorded.
- *Houses: Uninhabited or building / inhabited:* The enumerator would mark each new house on the street. He was also expected to indicate where a house was uninhabited.
- *Names of each person who abode there the preceding night:* It was common for middle names to be unrecorded. As stated above, each person who had slept in the property on that night had to be accounted for. No relationship to the head of the household was given and it is not always possible to work out family relationships.
- *Age and sex:* Ages of children up to the age of 15 years were recorded accurately. However, adults' ages above 15 were usually rounded down to the nearest five years. Hence, an individual whose given age appears as 40 could, in fact, be aged anything from 40 to 44 years old.
- *Profession, trade, employment or of independent means:* This could be misleading as in the nineteenth century people would often have more than one occupation and not every job was noted. The abbreviation 'M.S.' or 'F. S.' was for male or female servants.
- *Born: whether born in the same county? Whether born in Scotland, Ireland or Foreign Parts:* This is the closest information relating to place of birth provided. It would simply state whether an individual was born in the same county as the one they lived in, or in Scotland, Ireland or 'foreign parts'. These would be abbreviated as 'S.', 'I.' or 'F.' accordingly. The abbreviation 'NK' may also be used for 'not known'. Although in rural areas people tended to be living in the parish of birth, this would by no means be universal (especially in

urban areas), and hence finding a birth or baptism record would be difficult from the information provided here.

Two other problems are worth bearing in mind when searching this census. First, unlike later censuses, the original enumerator books were filled in using pencil not pen. Thus many pages have now become faded and can be difficult to read (especially the microfilm copies that are often held in county record offices). Secondly, there are some counties where the returns do not survive in their entirety. A complete list of missing and incomplete returns can be found online at www.ancestry.co.uk (see below).

Archive References for the 1841 Census

Every census return now has a modern archive reference, based on the government department that had responsibility for organizing the census at the time it was carried out. The original returns are now held at The National Archives at Kew, and no matter where you are viewing the returns – at TNA, a county archive or online – the archive references form an important part of either finding the correct return or creating your own referencing system when you download information from the Internet into your own files. Wherever archive references appear in this book, they will be accompanied by an explanation of what they mean, and how you should use them in your notes or files. Further information about locating census returns follows shortly.

The 1841 census had a different form of organization and referencing than later censuses and was not based simply on registration districts. It was administered by the Home Office, and has been given TNA series classification HO 107. Individual parishes in each county were grouped together into hundreds, and the census returns were subsequently sorted by county on an alphabetical basis, then by hundred, and lastly by parish. These hundreds were given unique piece numbers, which you can see on the scanned reference slip that appears alongside each census image, either online or on the relevant microfilm.

Each enumeration district was grouped together to form books. Each book would contain approximately five or six enumeration districts and would also have a unique number, given after the piece number on the reference slip. The books themselves would be broken down further, by folio number and individual page number. Folio numbers were stamped on every other page before the returns were microfilmed. Page numbers were printed on the original returns along with the columns.

▼ The 1841 census was the first to record details of everyone who lived in each house.

Thus an example of an 1841 census reference would be HO 107/910/2 whereby HO 107 would signify the 1841 census, 910 would be the piece number (in this case Condover hundred in Shropshire) and 2 the book number. The next relevant number would be the folio number and lastly the page number. However, the latter two would not be on the reference slip itself.

Information Contained on the 1851–1901 Censuses

These six census returns all record roughly the same pieces of information and can be grouped together. The dates the censuses were taken moved from June to either March or April, depending on the census:

- 1851 census: Sunday, 30 March 1851
- 1861 census: Sunday, 7 April 1861
- 1871 census: Sunday, 2 April 1871
- 1881 census: Sunday, 3 April 1881
- 1891 census: Sunday, 5 April 1891
- 1901 census: Sunday, 31 March 1901

Far more information was provided, giving precise birth details along with relationships to heads of households. Although not intended for genealogical research, the information is vital for anyone trying to trace their ancestors during the nineteenth and early twentieth centuries. The top of the page has parish, hamlet and township details along with the relevant borough. The columns are roughly the same for all censuses between 1851 and 1901, and are explained below:

- *Number of house, indenture or schedule:* This is not to be confused with the house number, but is the number of the property being assessed in the enumeration district.
- *House inhabited or uninhabited / building:* This question was omitted in 1851 but included afterwards.
- *Name of street, place, or road, and name or number of house:* As stated, the number or house name is provided along with the street. Unlike the 1841 census, house numbers and names were meant to be provided.
  - ➤ *From 1861: Road, street etc., and no. or name of house:* More complete details of the address of the property were included from 1861, though many houses simply didn't have a number or name; details are likely to be more complete for urban areas.

- *Name and surname of each person who abode in the house on the night:* By 1851, it was more usual for the middle name to be included or, at least, the middle initial, making it easier to identify the correct individual. As mentioned previously, every person who had spent the night in the dwelling place was recorded, regardless of whether it was their usual place of residence.

- *Relationship to head of household:* This is an additional column compared to the 1841 census, which is very useful for genealogical research. It detailed how each person in the household was related to the head of the household and so helps place people accurately on the family tree. It is not uncommon to find a niece or aunt or grandfather living in the household, thereby giving extra clues about your ancestors. It is also possible to identify how many servants were in the household as they were also noted separately, which gives an indication of social status.

- *Marital condition:* This column denotes whether the individual was single, married or widowed. Sometimes unmarried people were simply listed as U, with married people denoted M or Mar.

- *Age:* The ages were no longer rounded down and therefore should be more accurate. Bear in mind, however, that some individuals would not remember their ages with complete accuracy and so there can be errors, with a margin of a year or two either way.

  ➤ *From 1881 to 1901: Age at last birthday:* This was intended to make the age data more accurate.

- *Sex*: Denotes the gender of the individual, usually given as M or F.

- *Rank, profession or occupation:* What the occupation of the individual was. Children at school would be noted down as 'scholars'.

  ➤ *From 1891 to 1901: Employer, employed or neither:* This was intended to establish statistical information on the nature of Britain's working population. You will often see the number of employees that worked for an employer noted here.

  ➤ *From 1901: Whether the individual was working at home:* New information to ascertain how many people still worked at home, and the numbers who regularly went to a place of work.

- *Where born:* People were required to note down exactly where they were born, usually stating the parish of birth. This information

▲ From 1851, far more census information was collected, including full address, name, age, marital status and relationship to head of household.

enables current researchers to find the birth or baptism details of those born prior to the onset of civil registration in 1837 and, therefore, to trace back the family tree further still.

- *Whether blind or deaf and dumb:* Such physical disabilities were to be noted.
  - ➤ *From 1871 to 1901: Imbecile, idiot or lunatic:* Additional disabilities were to be included.
- *From 1891: Language:* Anyone in Wales or Monmouthshire was required to state whether they spoke English only, Welsh only, or English and Welsh (listed as 'both').
  - ➤ *From 1901: Language:* The language spoken section was extended to the Isle of Man census.

As well as information on people living in households across the country, people in various residential institutions – schools, prisons, workhouses, hospitals and asylums – were also noted, though to preserve the anonymity of some of these categories, initials only were used instead of full names, making it tricky to identify a relative who you feel might be away from home. Data on the crews of ships docked in British ports are also included in the returns, as are soldiers in barracks and sailors in naval bases, establishments and ships in port.

Archive References for the 1851–1901 Censuses

1.

1851 census

The same prefix code as the 1841 census is used, HO 107 (a National Archives reference). Registration districts were now used and were further divided into smaller sub-districts. The returns were organized by registration district. Each sub-district was given a piece number to follow on from HO 107. The first piece numbers were for the London area and then they were organized on a rough south-to-north basis. After all of England had been allocated piece numbers, subsequent ones were allocated for Wales and then the Isle of Man and, lastly, the Channel Islands. Each county, depending on its size, could include numerous piece numbers.

There was also a folio and page numbering system similar to the one mentioned for 1841. Once a new enumeration district started within a sub-district the page numbers would start from number 1 again. The reference slip is now on the bottom of a page and an example of a reference would be HO 107/2036. HO 107 is the standard reference and 2036 would be the piece number for the registration district of Stourbridge in Worcestershire. To find the exact page you need the folio and page number, although this is not be found on the reference slip itself, but on the top of the census return page. The folio page was stamped on every other page and the page number was printed on every page.

2.
1861 census

From 1861 onwards the TNA prefix is different. Instead of 'HO 107', each census return is prefixed with 'RG' (Registrar General) and, depending on the year of the census, an appropriate number. Hence, for 1861, each census return has the initial prefix RG 9, the number 9 signifying the year 1861.

Other than that, the numbering system is similar to that of 1851. Each registration district was given a unique piece number and these numbers were organized on a similar geographical basis as those of 1851 (with the returns for London coming first). The reference slip is found at the side of the page; a typical one would be RG 9/602 where RG 9 would signify the 1861 census and 602 would represent registration district 85 for Brighton. Again for a complete reference you would need the appropriate folio and page number (described above).

3.
1871 census

The referencing system is the same as that used in 1861. The only difference is the first prefix is now RG 10, signifying it is the 1871 census.

4.
1881 census

Again, the referencing system is the same as used in the previous three censuses. The TNA prefix is now RG 11, as it is the 1881 census.

The 1881 census for England, Wales, Scotland, the Channel Islands and the Isle of Man was fully transcribed by the Church of Latter Day Saints in the 1980s. The Church has made access to this census in particular free of charge on its website, www.familysearch.org.

5.
1891 census

The referencing system is the same as the previous censuses, RG 12 being the appropriate prefix code for this series.

The request for information about the employment status of individuals, where appropriate, was first made in this census. Additionally a column has been added detailing the number of rooms that were occupied in the dwelling house if less than five.

6.
1901 census

This is the last publicly available census until the release of the majority of the information in the 1911 census in 2009. The appropriate prefix for this collection is RG 13.

Accessing Census Collections for England and Wales

As already mentioned, the original householders' schedule forms were destroyed for 1841 to 1901. The original enumerator books, which form the census returns, are held at TNA for England, Wales, the Channel

Islands and the Isle of Man. The census returns for Scotland and Ireland (discussed below) are held at their appropriate record offices.

The census returns for 1841 to 1901 have all been microfilmed and it is these microfilmed versions that are available to view. However, a far easier way to access and view census returns is via the Internet, as the records have been digitized by different commercial websites that offer access to them for a fee. Below are details of the many different ways you can access the censuses.

The National Archives

All census returns from 1841 to 1901 are available, free of charge, at TNA in its reading rooms at Kew, South West London, on microfilm or microfiche. Reference guides, leaflets and indexes are available to help you locate the relevant TNA reference. Online access to the returns for 1841-91 via TNA's commercial partner Ancestry is also available in the reading rooms for free, though you have to pay for any copies you make. You can also search the 1901 census database for free, though access to the actual digital images still costs money. The 1881 census has been fully transcribed, and an index is available in the reading rooms. Parts of the 1851 census have been indexed by family history societies, and these indexes are also available.

Local Record Offices and Archives

As the censuses were microfilmed, many local libraries, record offices, family history societies and archives were able to purchase copies that cover the local vicinity. Most of these institutions will only have information for the relevant county or place, but they will also have useful local indexes that might not be available nationally, particularly if they were prepared by a family history society. These will include many local projects to catalogue and index the 1851 census, as well as the complete 1881 census index. Some indexes for the 1841 census are also available, and some companies have produced CD ROMs for local census returns for 1861 and 1871, and for 1891 as well. Additionally, staff will have specialized knowledge of the census for their area and can inform you of any missing areas. They will usually be available on microfiche or film and may suit people who are not IT literate.

Online

There are many commercial genealogical websites on the market, most of which have indexes and digital copies of the census online, although they are seldom free of charge. Here is a list of the most complete collections:

Local record offices will have useful local indexes that might not be available nationally.

- **www.ancestry.co.uk** This is one of the largest genealogical websites, with numerous databases, including a comprehensive census collection for England, Wales, the Channel Islands and the Isle of Man from 1841 to 1901. It is a payable service, either by a monthly/annual subscription or a pay-per-view system. Each census has an index and hence it is possible to do simple name searches (including various other details if required) when conducting a search. It is possible to search the index for free, although viewing the entire entry can only be done at a cost. Additionally, the index for the 1881 census can be searched in its entirety without cost as it has been previously transcribed (see above). It is through Ancestry that TNA provides access to the census records onsite. Another useful aspect of the census collection of Ancestry is that it details missing or incomplete registration districts for the 1841, 1851 and 1861 censuses. Hence, if you think you know where your ancestor should have been living, you can run a check against the list if you are having difficulties finding the individuals.

▲ Ancestry have a full set of census returns linked to a searchable name database.

- **www.1901censusonline.com** This was the first website to offer a census online in collaboration with TNA. It was a joint venture to release the 1901 census for England and Wales in January 2002 (after the 100 years closure period). However, the website now offers searches for all other censuses apart from 1881. The index is free to search although payment is required to view the original record. You can search by name and the website also offers other useful search functions. For example you can search by address, vessel (Royal Naval ships amongst other things) or institution (such as a hospital or prison). To view the original images you will have to purchase pay-per-view vouchers from the website.

- **www.findmypast.com** Formerly concerned with providing access to birth, marriage and death indexes, Find My Past has a growing collection of censuses. At the time of going to print it was possible to search the 1841, 1861, 1871 and 1891 censuses for England and Wales free of charge, although viewing the transcriptions or the originals costs a number of units which have to be purchased in advance.

- **www.origins.net** This is another large commercial genealogical website, with a number of databases, including census collections for England and Wales. Its census collection is not complete, however. At the time of print it covered 1841 and 1861 in their entirety, but its database for the 1871 census was incomplete, only covering certain counties (listed individually on the website). Again it is a payable service and it is only possible to do a very simple search without first subscribing.

- **www.freecen.rootsweb.com** This is the sister site to www.freebmd.org.uk that provides transcriptions of the national GRO birth, marriage and death indexes for free, and this site for census returns works on the same principle. It is run by a team of volunteers who are transcribing various parts of the census free in an attempt to make as much information available on the website without cost to the researcher. It is an ongoing project working on particular counties of England, Scotland and Wales for all censuses from 1841 to 1891. No census has been completely transcribed but the website does provide a graph showing which counties are covered for each census, along with the percentage of coverage for each county. The project is constantly recruiting volunteers to assist with the process.

Troubleshooting: What to Do If You Can't Find Your Ancestor

It can be difficult trying to find your ancestor on the appropriate census, even if you're pretty certain that they should be there. Here are a few reasons why you may experience problems and solutions to help you overcome these.

Data Wrongly Indexed or Transcribed

This is by far the most common reason for difficulties in locating an ancestor. It could be that, in cases of illiteracy, the enumerator at the time had to interpret and record the information given to him by the head of the household. There was no other way of verifying this data and therefore inaccuracies would not have been picked up.

Additionally, most of the websites listed above have created their own name indexes and searchable databases for the census returns, and

CASE STUDY

Bill Oddie pt 1

Bill Oddie's family tree was examined for the first series of *Who Do You Think You Are?* The story primarily focused on his search for information about his mother, and why she disappeared from his life when he was a child. However, another part of the programme examined conditions facing other members of his family as they grew up in the industrial North West. Key to this storyline was Bill's grandfather, Wilkinson Oddie, who was born in 1864 and worked in cotton mills for most of his life. His marriage certificate of 1907 was tracked down, when his age was given as 42 and he was described as a widower. This information permitted a search of the 1901 census, based on the fact that he would have been 36 at the time and possibly living with his first wife, in the hope of finding out more about his background.

The search concentrated on census returns in or around Rochdale, where he was living in 1907 at the time of his second marriage to Emily Hawksworth. Initially, no reference to Wilkinson Oddie could be found; however, there was a Wilkinson Oddy of the right age in the right place, and further investigation of earlier census returns in 1891, along with a check of relevant civil registration documents for his birth and first marriage, showed that this was indeed the correct person. Clearly, the census enumerator had written down a phonetic version of his name, transcribing Oddie as Oddy. This highlights one of the most common pitfalls when working with census records – you can't rely solely on a surname, but have to incorporate all sorts of other data such as age, place of birth and occupation.

The return for 1901 showed that he was listed as a widower, living with his children Betsy, aged 12, John 9 and Mary 7 in their house in Castle Court, Rochdale. Wilkinson was a cotton loom weaver – as was Betsy.

This information allowed a search for his first marriage prior to 1889, when Betsy was born, and for the death certificate of his wife after 1894, when his youngest child Mary was born.

The relevant certificates were quickly located thanks to this data, and showed that in 1888 Wilkinson Oddie married Cecilia Heneghan , a 21-year-old cotton weaver. She died in 1897 aged 31 and the cause of death shows it was a result of childbirth, a common danger at that time.

Once you've found an ancestor on one census return, it should be possible to locate them on earlier ones. Therefore, given the amount of information already gleaned about Wilkinson Oddie, it was fairly easy to locate him, aged 16, living at home with his parents in 1881. They were John Oddie (listed as Oddy in the census records) and his wife Mary. Wilkinson lived at home with six siblings, the oldest being 21 and the youngest only 1 month old.

By repeating this pattern, it was possible to track down John Oddie in earlier census returns as well, such as 1851. He was found living at home in Over Darwen, Blackburn, aged 15 and already working in a cotton mill, with his parents Wilkinson Oddie senior, aged 41, and his wife Mary. Wilkinson Oddie senior was born in Mitton, Yorkshire, in about 1811, and armed with this information a further search of parish registers was possible to locate earlier branches of the family.

errors may have crept in whilst compiling these resources. There are several reasons why this might have happened, ranging from the poor condition of the original returns to the difficulty that the modern transcriber faces in reading nineteenth-century handwriting. Bear in mind all possible variant spellings of forenames and surnames and how easily some letters can be confused for each other. For example, if your ancestor's surname was 'Parker', it could easily have been indexed as 'Barker', 'Darker' or even 'Porker'. Lateral thinking is often needed in overcoming such mis-indexing.

Always remember that there have been many stages that the data has gone through to reach your computer screen and errors could have occurred at each step. If you have a valid street address for your ancestor, perhaps from a certificate, it may be worthwhile searching microfilms manually under this address if you are having difficulties locating your ancestor online.

Inaccurate Information Provided by Ancestors

It was not uncommon for ancestors to provide inaccurate details when filling in the schedule forms. This could be for a variety of reasons:

- A different name was given. Although your ancestor's official name might have been Jennifer Sarah Marks, she could have been commonly known as Sally and be recorded under that name. People could always use their middle names as their first names or vice versa. Check both if you are having problems.
- Ages were inaccurate. Sometimes teenagers would register as older than they were so that they could work at an earlier age than was legal. Other times, people would give younger ages for reasons of vanity (especially when there were large age gaps in marriages). Alternatively, people might simply not remember their exact year of birth.
- Covering up family secrets. The most common secret a family might wish to conceal would be illegitimacy, and information might have been tweaked to hide this. For example a child may appear to be a year or two younger on the census than was the case, to mask a birth outside wedlock. Another possibility is an untruthful relationship, whereby an illegitimate grandchild of the head of the household may be recorded as their child to avoid scandal.

Bear all this in mind when searching for your ancestor. Most of the search facilities on the various websites allow people to filter the results with as much or as little detail as possible so you can allow for such inaccuracies when searching.

Your Ancestor Was Not Recorded on the Census

Although the theory was that every person living in the country at the time should be recorded, sometimes people slipped through the net. The early censuses attracted much suspicion as to why the state needed to record such personal information in the first place. Consequently people were often very reluctant to fill out the relevant forms. For example, a census return for Westminster in 1841 includes a margin note written by the enumerator stating that the head of the household refused to provide any information.

There may always be a simple explanation, in the sense that they were simply not present at home on the night the enumerator came to collect the forms, and failed to fill out a form where they were staying. Of course, the opposite is true as well, and some people appear more than once in the same census – once at their formal place of residence, and another at their lodgings whilst travelling around the country.

Additionally, babies and young children may be omitted as some parents felt that information should not be provided until the child was baptized. Alternatively, parents may not have detailed every child to avoid accusations of overcrowding. There is also the factor of human error; where enumerators had to record the information themselves (in cases of illiteracy) they may have simply missed the odd person out, particularly at the end of a long day tramping through the streets of the parish meeting hostile and suspicious householders!

Missing Census Returns

Unfortunately, census returns for certain districts have not survived to the present day for various reasons. If every other search has failed, check to ensure that the return for your sub-district does survive. Ancestry's website has a list of missing census returns for the early censuses (1841 to 1861). Otherwise, check with your local record office. Of course, you need to be sure your ancestor was living in that area in the first place. Sometimes, people would be visiting relatives on census day and would not be recorded in their home town anyway.

Don't forget that there might be another reason your ancestor isn't listed at home – they might be residing in a school, workhouse, prison

Although every person should be listed on the census, people did slip through the net.

or ship, or perhaps had enlisted in the army or navy, in which case they could be located far away from their place of birth.

Census Returns for Scotland

Censuses were taken in Scotland on a similar basis to England and Wales from 1801 onwards. Scotland was divided into registration districts and sub-districts and enumerators were responsible for circulating and collecting the schedule forms. The enumerators would then enter the information into the census return books, which form the Scottish census returns that you can see today.

The original returns are now held at the GROS in Edinburgh. These records are also subject to 100-year closure rules. Nevertheless, it is possible to view the returns from 1841 to 1901 at GROS, which has computerized indexes for these returns for the entire period along with some other indexes, such as street indexes for certain areas and some privately produced indexes. As microfilm and fiche copies were made of the original returns, various local archives, record offices and libraries will hold copies for their area as well, so a trip to Edinburgh might not be necessary.

The returns are also available online on the following websites:

- **www.scotlandspeople.gov.uk** This is the main website for those conducting genealogical research in Scotland, being the official government source of genealogical data. Most of the key sources have been placed here, including all publicly available census returns for Scotland. The information is not free of charge, although you can search the databases after registering for free. However, you will have to pay to view the results and see digital images.
- **www.ancestry.co.uk** Ancestry has also placed Scottish data on its website. At the time of writing, searchable indexes for all census returns from 1841 to 1901 were available. These indexes relate to transcriptions of the original entries on the forms. It is not possible to view a copy of the original entry, only the transcription.

Census Returns for Ireland

The process of producing census returns for Ireland began slightly later, in 1821, and they were compiled every ten years subsequently.

' Most of Ireland's early census records were destroyed by a fire in 1922. '

Unfortunately, most of the records were destroyed during the Civil War in 1922. Very little now survives for the nineteenth century. However, the returns for 1901 and 1911 are available and survive in their entirety.

Nineteenth-century Returns

The small proportion of census returns that do survive for this period can be found at the National Archives of Ireland, Dublin. They are organized county by county and some name indexes for certain counties have been prepared. Local county record offices may have copies of surviving records for their local area and it is worthwhile contacting these institutions first.

1901 and 1911 Census Returns

As mentioned, both these returns are already available to the public and there are plans to make them available online from late 2008 onwards. Details can be found on the National Archives of Ireland website.

The records were collected on a similar basis to those of England, Scotland and Wales; forms were duly completed by the head of the household and given to the enumerator to compile the returns. There is no complete name index for the entire country and the records are organized by Poor Law Union, district, parish and town. Hence, it is necessary to have an approximate idea of where your ancestor was living before you can search.

The returns contain all the details that are to be found in their English, Welsh and Scottish counterparts, as well as additional information. The 1901 census provides details about the condition of the house in which your ancestor was resident, as well as their religious denomination. The 1911 census also includes information on how long women had been married and the number of children born of this marriage that were still alive.

The census records can be found for the whole of Ireland until 1921 at the National Archives in Dublin, which also holds the census returns for the Republic of Ireland post 1921, though they are not yet open for public inspection. The Public Record Office of Northern Ireland, Belfast, has the census returns for the six counties prior to partition and also after 1921, though these later records are similarly closed to the public. Additionally, local centres should have copies of returns for their particular area.

CHAPTER 7

Parish Records

The third of the four main sources that genealogists use are ecclesiastical records generated at local level by parish churches and various other religious organizations. Registers of baptisms, marriages and burials contain some biographical information that you can use to extend your family tree further back in time – theoretically to the sixteenth century, when parish registers were first introduced. This chapter explains what the records are, how you can use them, and where they can be found. It also lists some of the non-conformist records generated by religious groups outside the authority of the Church of England.

> *Parish records are among the longest continuous sets of records available.*

It is possible to make significant progress in building your family tree using the two sources discussed in the previous chapters – civil registration certificates and census records. However, if you want to work further back in time, pre-nineteenth century, you will have to turn to records generated at a local level, not by the State but by the Church. Together, these sources are loosely described as 'parish registers' and they record key events in a person's life, such as baptism, marriage and burial. Since parish registers were introduced in the mid-sixteenth century, and continue to the present day, they are one of the longest continuous sets of record available – though a large degree of luck is required to find an ancestor in the earliest surviving registers.

Historically, Christian Britain was divided into dioceses, each administered by a bishop and consisting of smaller territorial sub-divisions known as parishes. A parish is a geographical unit under the administration of a local priest or pastor, and they have existed in England, Wales, Scotland and Ireland since the end of the sixth century. By the nineteenth century, England and Wales had approximately 11,000 parishes, varying considerably in size and population. Although there have almost always been religious minorities such as Jews in England, the vast majority of the population belonged to the Established Church of England from the sixteenth century onwards, when the country broke away from the authority of the Papacy in Rome. Those Christian minorities that did not thereafter subscribe to the Established Church, for example Quakers and Roman Catholics, came to be known as 'non-conformists'. They remained small in number until the nineteenth century, in part due to the persecution they faced. Surviving records for these non-conformist groups will be discussed separately below.

Parish Records of the Church of England

In 1538, Henry VIII's chancellor Thomas Cromwell introduced legislation that required every priest to record all baptisms, marriages and burials within his parish, and it is these surviving records that enable genealogists to trace their ancestors beyond the start of the great record series of the nineteenth century – civil registration certificates and census returns. Few records survive this far back – on average, most English parishes have records that start around 1611 though there are some examples in Wales from 1541 – because many of the early records were not kept with any degree of care, being written on loose sheets of paper which have not survived the passage of time. A further royal proclamation was issued in 1558 instructing that these parochial events be written on parchment rather than loose paper, which increased the chances of survival; therefore 1558 is generally recognized as the start date of parish registers.

Another Act of Parliament, passed in 1597, is also important as it led to the birth of what are known as 'Bishops' Transcripts'. As well as compiling their own parish registers, local clergy were instructed to make annual copies of each register and send them to the bishop of the diocese in which they served. Therefore these are very useful duplicate copies of the original parish registers and can be used as an alternative if the original does not survive (or is partly or wholly illegible). However, Bishops' Transcripts sometimes contained less detail than parish registers, or recorded slightly different information, so it is worthwhile examining both sources where possible.

There are other factors to consider when viewing the earliest registers.

HOW TO…

…make the most of parish registers

1. *Prior to the commencement of civil registration, parish registers that recorded baptisms, marriages and burials on a local level are our only way of confirming the births, marriages and deaths of our ancestors. These can be more difficult to trace, because there is not one centralized index and you usually need to know which parish your family was living in to be able to locate their entries. Parish registers also contain less detailed information than civil registration certificates, making it more challenging to compare details from record to record. Nevertheless, the fact these registers survive for some parishes as far back as the sixteenth century means there is a wealth of information about your ancestors waiting to be discovered, with a little patience and determination!*

2. *If you find civil registration documents for your ancestors that do not give the information you were expecting to find, it may be worth looking for the parish register of whichever religious ceremony would have marked the event to see if the two records corroborate one another.*

Other than the possibility that they may no longer be legible (ink may have faded or pages rotted), early registers were usually written in Latin. However, this shouldn't cause too many problems. The nature of the information is fairly formulaic and is usually contained in a single sentence. Moreover, most archives have Latin dictionaries that enable you to translate words such as calendar months into English, as well as Latin versions of English names.

Another point to bear in mind is that the modern Gregorian calendar was adopted in the mid-eighteenth century instead of the old Julian calendar. Until 1752 the New Year did not begin on 1 January but 25 March. Hence, for example, all events occurring after 31 December 1675 to 24 March of the following year would belong to the year 1675, even though today we would consider them as belonging to 1676. Genealogists refer to this in their dating by using the formula 'February 1675/6'.

Further, prior to 1813 there was no uniform method of registering events. Instead, individual clerks recorded information in their own unique ways and consequently the amount of information contained in parish registers varies considerably. Some parishes kept separate registers for baptisms, marriages and burials, whereas others prepared annual registers recording all these events together. The Rose Act was passed in 1813 and from this date all baptisms, marriages and burials were written in pre-printed books issued by the 'King's Printer'. These books ensured that the same details were recorded by each parish throughout the country. Often there would be more detail than had been recorded previously.

Lastly, do remember that parish registers were still kept after the introduction of civil registration in 1837 and, in theory, should exist to the present day. This is very useful when you are having problems locating events in the national GRO indexes as an equivalent record may be found in the local parish register instead, especially during the early days when civil registration was less popular or thought to be an unnecessary inconvenience.

How to Interpret Parish Registers
Baptisms

Registers recorded the baptism of the child in the local church, rather than its birth. The earliest registers in the sixteenth and seventeenth centuries can be very basic in detail, often giving the name of the child and only the name of the father (with no mention of the mother). As time went on more details started to be recorded, including the mother's Christian name, either the child's date of birth or its age (in days or months) and sometimes additional information such as the occupation of the father and where the parents lived. The Rose Act introduced more uniformity and from 1813 the following information was routinely recorded:

- When the child was baptized
- Child's Christian name
- Parents' names (Christian and surname)
- Abode of family
- Quality, trade or profession
- By whom the ceremony was performed

Marriages

The earliest parish registers, prior to 1754, would usually record the full names of the bride and groom along with the date of the marriage. Some registers may give extra information such as the name of the bride's father, the groom's occupation and whether the marriage was by 'banns' or by licence. It was customary for marriages to be in the bride's home parish.

Marriage by Banns
Before 1754, most people married in their local parish church, and the intention to marry was announced in the church by the vicar or priest, usually for three consecutive Sundays prior to the wedding. These were known as 'banns', and are often recorded in the parish register prior to the marriage ceremony itself.

Marriage Licences
The other legal option to being married by banns would be to marry under licence, particularly if you wished to marry outside the parish where the bride or groom lived. As this was the more expensive option, it carried a certain prestige. Parish registers should indicate if a marriage was conducted in this manner and the documents relating to marriage licences are useful extra sources for genealogists. The licences themselves rarely survive but 'allegations' and 'bonds' do. Both are genealogically useful:

- *Allegations* were an intention to marry, and record the names of the bride and groom, parish of residence, age, groom's occupation and, if either party was under 21, the names of parents.

▼ *Wedding Feast in the Village* (oil on canvas) by Nicolas Lancret (1690–1743).

• *Bonds* were required until 1823. They were sworn by two bondsmen and guaranteed that the information provided was correct under forfeit of a certain amount of money. The bondsmen were usually the groom and perhaps the bride's father or other family member.

Licences were issued by the local archbishop or bishop and the surviving documentation should be amongst the ecclesiastical records of these bishops. These are often at the local county record offices. Always remember that licences were issued for people intending to marry; not every couple would then get married.

Irregular or Clandestine Marriages

Irregular or clandestine marriages occurred when people did not wed according to the Church's requirement of being married in your own parish church by either banns or licence. Instead couples chose to marry under different circumstances. Until 1753, one could marry simply by exchanging vows in public or private, without the need of a priest or witnesses. Such marriages were legally recognized even though there was no documentation to prove the marriage. Alternatively, many people chose to marry in churches away from their own parishes, since they could also avoid the expense of acquiring a licence or banns, and avoid the costs of a wedding party – though many irregular marriages were criticized for their rowdy and 'unchristian' behaviour. These marriages were popular for many reasons: couples could be married with great speed, no parental consent was required and bigamists could also marry again this way. Certain institutions became famous for conducting clandestine marriages. Perhaps the most famous is the Fleet prison, where many thousands married prior to Hardwicke's Marriage Act.

Hardwicke's Marriage Act

Hardwicke's Act was passed in 1753 and affected the registering of marriages. From 1754, it became compulsory for marriages to be held in either the bride's or groom's home parish, and parental consent was compulsory for marriages of those aged under 21 years (though the legal age for marriage were still 14 for boys and 12 for girls once parental consent had been granted). Hence, from this point forward marriages for most individuals should be found recorded in the appropriate parish registers. Only Jews and Quakers were exempt from this new law and could have separate ceremonies, but other non-conformist bodies had to comply.

▼ An extract from a marriage register, showing the wedding of Esther Ennis and Thomas Albert Kidd – Jodie Kidd's ancestors.

Burials

The requirement to bury the dead has been with us since the dawn of civilization, with prehistoric burial mounds, Roman cemeteries, Anglo-Saxon burial sites and Viking ship burials. However, Christian burial has generated its own records, some physical such as tombstones and crypts, and others on paper such as parish registers.

Parish Burials

The most important thing about burial records, similar to death certificates, is that they can provide an age at death and can therefore greatly assist a search for the baptism of the individual. They can also provide clues when searching for wills. However, some of the early records give very basic information, usually only the name, with no age details. The later records start recording age and may also give further information, such as residence or occupation. Records of married or widowed women may also include the name of their husband, and young children would often have the name of the father (and sometimes mother) entered alongside their name.

Cemeteries

As church graveyards in towns and cities became increasingly full during the early 1800s, due to the accumulation of bodies over the years and the increase in the urban population fuelled by the Industrial Revolution, an alternative option to parochial burial was introduced, namely interment in extra-parochial cemeteries. The earliest cemeteries were privately built, with plots sold to the rich who often built elaborate monuments to themselves and their families, creating necropolises of the dead. Municipal cemeteries were soon to follow throughout the nineteenth century as urban expansion continued, often taking up many acres of space on the outskirts of towns and cities. Consequently, maps and plans showing the plots where people were buried were created by the authorities in charge of each cemetery. Many of these are now deposited at local record offices, though some material may still reside with the cemetery today.

Monumental Inscriptions

Monumental inscriptions are the engravings and writings found on the gravestones and tombs of the buried. As such they are of unique importance, sometimes providing vital clues about your ancestors. As they are not official documents, the information varies but can give exact birth or age, next-of-kin details (names of wives and children), residence and occupation of the deceased. Many family history societies have made efforts to transcribe memorial inscriptions found in churches and graveyards and deposit them in local archives. However, this has not been done for every burial place, and it may be worthwhile visiting the appropriate location to physically search through the graves or memorials. The very earliest inscriptions would be found in the actual church, and wealthy parishioners would have such inscriptions from the medieval period onwards, in the form of tombs, brasses and other monuments. Graves in the churchyard started to be marked from the early seventeenth century onwards by the better off on stone, as opposed to wooden markers, although they may not be legible today.

Location of Parish Records for England and Wales

The majority of parish registers deposited into the public domain in England are held locally in county archives, local studies centres or metropolitan record offices. They are mostly available as copies on microfilm or fiche, as the originals are usually too fragile or delicate to be handled by the public, given the huge demand to see them. Welsh parish records can be found predominantly at the National Library of Wales, with the remainder located at local repositories. The Society of Genealogists has collections of microfilm or fiche copies of parish records for various parishes throughout the entire United Kingdom and continues to expand its collection. Local family history societies are also likely to hold copies or transcripts of parish registers, whilst family history centres run by the Church of Jesus Christ of Latter Day Saints (LDS), otherwise known as the Mormons, can order up copies of parish registers from around the UK on request.

Parish Records for Non-conformists

The above has been a guide to the parish records for the Anglican Church. However, there have always been minorities in the country that have not been part of the Established Church. If you are unable to find your ancestor in the Anglican parish records, it may be because they were non-conformists.

One important and large non-Anglican group were the Roman

CASE STUDY

Bill Oddie pt 2

Having begun an investigation into the family background of Bill Oddie, a combination of civil registration certificates and census returns has pushed the knowledge of his ancestors back to the early nineteenth century. His 2 x great-grandfather, Wilkinson Oddie senior, is listed on the 1851 census aged 41, with a place of birth listed as Mitton, Yorkshire. This was sufficient information to search for details of his origin.

Some geographical research showed that Mitton was one of three closely linked villages in the lordship of Clitheroe, and searches of the parish registers revealed his marriage to Mary Slater in Clitheroe in 1833. A search of records of the neighbouring village, Grindleton, revealed Wilkinson's birth on 21 January 1810 to John Oddie, a weaver. Eight further children were baptized in the parish to John and his wife, Ellen, up to 1828, including twins Elizabeth and Richard.

Having established Wilkinson Oddie senior's parentage, a further search was made in the marriage registers for John and Ellen's marriage prior to Wilkinson's baptism in 1810. However, the limitations of working with parochial material are made clear with the absence of any surviving marriage registers before 1844 for Grindleton, and no record was found in the larger parish of Mitton, of which Grindleton was a chapelry. It is possible that the marriage took place in Ellen's home parish, but without further information about her background, searches were limited to the parishes that surrounded Mitton – always a good tactic in a period when people rarely strayed too far from where they were born. However, in this instance no matches were found.

However, it was possible speculatively to search for one generation further back in time, given the repetitive naming patterns that occur in the Oddie family. A wider trawl for Oddie baptisms was made, covering all parishes that neighboured Grindleton, starting 30 years prior to the verified baptism for Wilkinson Oddie in 1810. A baptism of a John Oddie to parents Wilkinson Oddie and his wife Betty was recorded in Grindleton on 11 December 1785, on the same day that John's elder brother Joseph was also baptized aged nearly three. A further two children were baptized to Wilkinson and Betty after this date. Without further data, it is impossible to state with any certainty that this is indeed the correct lineage, but the circumstantial evidence is compelling, and the appearance of other branches of the Oddie family in the Grindleton area suggests this was the family heartland.

Parish records can also assist with more modern lines of research. Information from civil registration certificates had already shown that Wilkinson Oddie junior, born in 1864 – Bill's grandfather – was married twice, and that his first wife Cecilia died in childbirth. A search of the parish registers for St Clements Church, Spotland, Rochdale, revealed that Cecilia died on 14 December 1897 and was buried three days later. The daughter that she had given birth to, Alice, did not survive either, and was buried on 13 December, the day before her mother died.

Catholics, who became a persecuted minority after 1559 when Elizabeth I tried to ban the rituals of Catholicism. Of course, until Henry VIII's break with Rome the Roman Catholic Church was the 'established' Church.

Other non-conformist groups also started to appear from the sixteenth century onwards. The term derives its origins as a label applied to those who did not 'conform' to the Act of Uniformity, passed by Charles II in 1662 requiring the rites and ceremonies prescribed in the Book of Common Prayer to be used in Church of England services. Anyone who refused to use them either left the Anglican Church, or was ejected. All these groups of non-conformists were subject to discriminatory legislation until the nineteenth century. Depending on which non-conformist group your ancestor belonged to, their records may, or may not, exist in record offices.

Protestant Dissenter Records

The majority of records generated by Protestant non-conformists can now be found at The National Archives. This is because, with the onset of civil registration, an attempt was made by Parliamentary commissioners to collect together all non-conformist registers in 1837, with a second collection made in 1857. These were deposited in the newly created General Register Office. Nominally, these registers cover the period 1567 to 1970 (although very few cover the period after 1837) and include Presbyterians, Baptists, Congregationalists and Methodists, amongst others. They are now found in The National Archives series RG 4 to RG 8.

Baptism Records

Although infants were baptized in non-conformist chapels, only Anglican baptism certificates were recognized legally and therefore many infants were also baptized in Anglican parishes. In 1742 the law changed, allowing non-conformists to register their baptisms in the General Register of Births of Children of Protestant Dissenters at Dr William's Library in London, which contains records from *c.* 1716 to 1837. These certificates are now available on microfilm in the National Archives under series RG 5. The series contains records of approximately 50,000 births. Similarly, the Wesleyan Methodist Metropolitan Registry of births was founded in 1818, with records dating from *c.* 1773 to 1838. These records, covering 10,000 births, are also at The National Archives in series RG 4 and RG 5.

Marriage Records

Hardwicke's Marriage Act of 1753 ensured that marriages could no longer be conducted in any place other than Anglican parish churches. Only Jews and Quakers were exempt from the new Act. Prior to 1754, it had been possible to marry in non-parochial establishments, and some registers exist before this date. These records now are mostly found in The National Archives, and occasionally at county record offices.

Burial Records

Non-conformist churches were forbidden their own recognized premises and burial grounds until the Toleration Act of 1691. Before then they would have been buried in local church graveyards (where allowed). After 1691 separate burial grounds were established and their registers can be found at The National Archives, series RG 4 and RG 8.

Other Religious Communities

Aside from the Protestant non-conformists, other religions and groups kept records that can be found in a variety of archives.

The Society of Friends (Quakers)

The Society of Friends, or Quakers, was established in 1650. They kept detailed records of births, marriages and deaths and were not subject to the restrictions placed on other non-conformist groups. Their registers can be found either at The National Archives (series RG 6) or at their own centre, the Religious Society of Friends' library in Euston, London.

Roman Catholics

Roman Catholicism is the oldest Christian denomination in the United Kingdom. However, political and religious unrest in the sixteenth century made the practice of Catholicism illegal by 1559, when Queen Elizabeth I passed the Acts of Supremacy and Uniformity. The situation lasted until 1778 when the Catholic Relief Act was passed. As Catholicism was so strictly prohibited, parish registers are incomplete. There are very few, if any, records from the sixteenth to eighteenth centuries, as events would not be recorded due to fear of prosecution. Alternative sources need to be used to locate Catholic ancestors, for example criminal records for those tried as recusants or tax records for those fined for their beliefs. These are held at local archives or in The National Archives. As parts of northern England remained Catholic, local record offices for those areas may have relevant records.

> *Few Catholic records exist because of the fear of prosecution after 1559.*

There was a surge in Roman Catholicism in the nineteenth century due to the repeal of punitive legislation and large numbers of Irish Catholic migrants arriving in Britain. Bear in mind that your ancestor may well be an Irish immigrant of this time period, rather than an English Catholic. Roman Catholic churches started keeping registers from this period onwards although, due to religious concerns, few churches wished to deposit their registers with the Registrar General in 1837 (unlike the Protestant non-conformist bodies). Nevertheless, The National Archives does have some registers in series RG 4 and others may be found in local record offices. Otherwise, the local church may still have the historical registers. The Catholic Family History Society (www.catholic-history.org.uk/cfhs) and the Catholic Record Society (www.catholic-history.org.uk) may offer useful guidance.

Huguenots

The rise of various Protestant movements in Europe during the sixteenth and seventeenth centuries caused a great deal of religious and political strife, with many groups having to flee their homelands for fear of religious persecution. One of these groups was the Huguenots, French Protestants who fled to Britain and Holland throughout the sixteenth and seventeenth centuries. Two events in France led to a large exodus of French Protestants: the St Bartholomew's Day Massacre in 1572 (when approximately 70,000 Huguenots were murdered on order of the King, Charles IX), and the revocation of the Edict of Nantes in 1685 (the Edict had promised religious tolerance to French Protestants). It is estimated that approximately 50,000 Huguenots arrived in England (settling in London and southern England) and a further 10,000 in Ireland.

If your ancestor was a Huguenot or descended from a Huguenot family, you can search for their records through a variety of sources, mostly found at The National Archives or at the Huguenot Society of Great Britain and Ireland. The National Archives has calendared lists of immigrants amongst its State Papers series (SP). You may also find records for Huguenots applying for denization or residency from the seventeenth century onwards amongst State Papers too. In 1708 an Act was passed making it far easier for French Protestants to become British citizens. Records of those being naturalized in such circumstances can be found in the Court of Exchequer Rolls (series E) at The National Archives.

▼ Britain has welcomed many overseas visitors to its shores, including Huguenots who compiled their own registers of marriage and baptism.

The Huguenot Society has transcribed and published many of these series. They are available at their library and larger reference libraries. Huguenot communities established their own churches upon their arrival in England. Surviving registers of such churches can be found amongst the non-conformist parish registers deposited in The National Archives (in series RG 4). Additionally, the Huguenot Society has transcribed these registers and made them available in their own library and throughout larger libraries in the country. Further information about tracing immigrant Huguenot ancestors can be found in Chapter 22.

Jews

The oldest non-Christian minority in Britain is the Jewish community, who originally arrived shortly after William the Conqueror but were later expelled by Edward I in 1290 in an Edict of Expulsion. Oliver Cromwell relaxed the enforcement of the Edict during the Commonwealth period, and other Jewish communities started to arrive from 1656 onwards. However, as they are not part of the non-conformist Christian community, specific details about tracing Jewish ancestors can be found in Chapter 22.

Non-conformity in Wales

Wales has always had a large non-conformist population and by the nineteenth century the majority was non-conformist. These included a variety of groups, including Baptists, Independent and Calvinistic Methodists. The records can be found at either the National Library of Wales or in local archives.

Name Indexes for Parish Registers

Although there is no single comprehensive name index covering all parish registers, various collections do exist. The most useful ones are listed here.

International Genealogical Index

Created by the Church of the Latter Day Saints (Mormons) the International Genealogical Index (IGI) remains the most extensive collection of British parish register transcripts, with over 70 million entries listed. The IGI predominantly covers baptisms for the British

Isles from the sixteenth century to about 1885, though later entries are included. It also contains some marriage records but by no means as many, and very few burials. Coverage is not complete, and some parishes are excluded altogether. The means of access is usually via microfiche at county record offices, but the IGI is now available online via www.familysearch.org, described below.

When searching the IGI, always bear in mind that you are working from someone's transcriptions, and that errors can occur. You should always use the IGI as a short cut to locate a record, and then go back to the original parish register to check for accuracy, as well as scan for related entries. Similarly, if you can't find an entry in the IGI it is worth checking to see if the parish you need is actually covered in a publication such as the *Philimore Atlas and Index of Parish Registers*.

British Vital Records Index

Linked to the IGI (and also prepared by the Mormons) is the British Vital Records Index, a CD ROM which contains millions of transcriptions from British parish registers. Many record offices, libraries and family history societies have bought copies and make them available in their search rooms. As with the IGI, you should always try to verify the transcription from the original, where possible.

Boyd's Marriage Index

Boyd's Marriage Index is an attempt to cross-reference the names of the bride and groom contained in marriage records in each county, from the introduction of parish registers in 1538 to the introduction of civil registration in 1837. It was compiled by Percival Boyd and his team from parish registers, Bishops' Transcripts and licences, and covers all English counties – though none are complete. Over 7 million transcriptions are covered in the index, and the area with the best coverage is East Anglia, which is particularly useful since this region is poorly served by the IGI. Most county archives provide access to printed editions of the index, but you can access more extensive collections at the Society of Genealogists in London. It is also available at www.origins.net (see below).

Pallot's Marriage and Baptism Index

If you are looking for a marriage or baptism in the City of London between *c*. 1780 and 1837 you should consult Pallot's Marriage and Baptism Index, which covers all but two of the 103 parishes covered by the City's historic jurisdiction. There are, for example, over 1.5 million marriages listed, and if you find an entry that matches your search, you are likely to find the name of the spouse, date of marriage, and parish where the wedding took place. Although the index started out focusing on London, there are entries from Middlesex and other parishes. The index can now be searched on www.ancestry.co.uk.

The National Burial Index

When searching for the burial place of your ancestor, the National Burial Index may be helpful. The Federation of Family History Societies is running a project to index all burials in England and Wales. The project was begun in 1994 and it aims to cover burials from Anglican parishes, non-conformist and cemetery burial registers from 1538 to almost the present day. It is a vast project and the FFHS is releasing its index in stages. To date the second edition (published in 2004) is available and contains over 13 million burial records, though it is not yet complete. Most local family history societies and archives should have access to the Index in CD format. It should be noted that it is an index of burials only; memorial inscriptions are not included.

Searching for Parish Registers Online

As many parish records are scattered through various archives and libraries throughout Britain and Ireland, it is worthwhile checking whether the information may be online; there are a number of websites that have transcribed parish registers and made them available online. Below is a list of the largest websites.

▼ FamilySearch has one of the largest collections of transcriptions of British parish registers that can be searched online.

www.familysearch.org

As mentioned above, Family Search is run by the Latter Day Saints and is crucial for anyone conducting research into parish registers. It incorporates several key datasets:

- Ancestral File
- Pedigree Resource File
- The International Genealogical Index
- British 1881 census

There are also other sources that are probably not relevant to most researchers in this country, such as US census returns.

Ancestral files and pedigree resource files are voluntary submissions by other users and members of LDS, and are based on private research and material stored at the LDS library in Salt Lake City. As with any non-verified material placed online, you need to exercise caution before incorporating it into your family tree. Where possible, try to contact the submitter and ask where their sources were drawn from, and then go back to those sources and check them. Further advice about the way material is submitted, and the datasets themselves, can be found on the website. It is free to use.

www.familyhistoryonline.net

This is another very useful website for people needing to search parish registers. It is run by the Federation of Family History Societies, the umbrella organization for individual family history societies throughout the country; the Federation currently has over 210 member societies. Many of these family history societies have transcripts of their local parish registers and the Federation is placing these transcripts on this website, making a searchable database for many different parishes in England and Wales. Searches are available for various parishes in every county of England and Wales right up to the twentieth century and new additions are being placed onto the website continuously. Currently there are over 67 million records, including certain datasets for Australia. Databases are available for the whole range of parish registers – baptisms, marriages and burials – as well as associated material such as marriage licences, monumental inscriptions and even census records. The website has a section listing what exactly is available per county and a section on new additions to the website.

It is possible to search the website free of charge after registering, but to view the results will require payment. Always remember, however, that as these are transcriptions it is always best to double-check the information with the original parish register, wherever possible.

www.origins.net

This website contains useful databases containing material drawn from parish registers. The section for British origins is probably the most useful as you can search the following:

● *Boyd's Marriage Index:* As described above, over 7 million marriages from every English county are searchable here.

> *Online databases are available for the whole range of parish registers.*

- *Vicar General Index to Marriage Licence Allegations, 1694–1850:* This is a composite index to marriage allegations issued by the Vicar General (the Archbishop of Canterbury). The index is listed by surname only of both parties and identifying the correct record can prove difficult on occasions.
- *Faculty Office Index to Marriage Licence Allegations, 1701–1850:* The Faculty Office was another department of Lambeth Palace that issued marriage licences. This index is more detailed, giving fore-names and surnames of both parties.
- *London burial index, 1538–1872:* This index has been compiled from two sources. First, Boyd's London Burials, an index containing over 240,000 entries, although not including every burial occurring in London and concentrating on male entries. The second source used was London City Burials, compiled by Cliff Webb and containing over 35,000 such entries. The latter source is usually the more detailed, covering both sexes and mostly giving the age at burial.

Overseas Parish Registers

The expansion of the British Empire from the eighteenth century onwards meant that there were pockets of British populations scattered throughout the world. Some of these communities would have their births or baptisms, marriages and deaths or burials recorded in the same fashion as their relatives back home by overseas chaplaincies. These events were recorded in the same way and can be found in a variety of sets of records. Of course, they may also be available online through the IGI and it always worthwhile checking this on the website www.familysearch.org (see above).

The Bishop of London Registers

In 1633 the Bishop of London was appointed as being responsible for Anglican chaplaincies abroad where no local priest had been appointed. Such chaplaincies grew in large numbers in Europe in the seventeenth century, overseen by the Bishop of London. After 1842 the area was spilt into two – Northern and Southern Europe. Northern Europe was still under the remit of the Bishop of London (or the Bishop of Fulham after 1883) and Southern Europe came under the Diocese of Gibraltar.

These Anglican chaplaincies abroad would maintain their own registers and would usually deposit these registers with the Bishop of London, although in an irregular fashion. The Bishop of London maintained registers for these miscellaneous foreign events, and volumes survive from 1816 to 1924, known as the 'International Memoranda'. Indexes to these Memoranda have also been compiled.

The registers themselves can be found at a variety of places. They may be at the Guildhall Library in London or with the Bishop of London's Archives, now part of the collection at the Lambeth Palace Library. The best place to try and locate where records for the country of concern may be is to refer to the book *The British Overseas: a guide to records of their births, baptisms, marriages, deaths and burials available in the United Kingdom* (3rd edition, 1994) published by Guildhall Library. This is a comprehensive country-by-country guide to what records are available for each country and where they may be found. Records of non-statutory registers, many of which relate to births overseas and on board ships, can be found in the collected archives of the Registrar General at The National Archives in series RG 32–36.

Please note, however, that unless indicated otherwise by *The British Overseas*, the majority of registers for Australia, Canada, New Zealand, the United States and the West Indies will still be held in these countries. The records for these territories will be discussed in more detail in Chapter 22. The only large ex-colonial collection of parish registers found in the UK is that for the Indian subcontinent. Parish registers for British communities based there from 1698 to 1968 can be found at the Asia, Africa and Pacific collections at the British Library.

The National Archives' Consular Records

British Embassies located throughout the world also maintained registers for births, marriages and deaths for the country the embassy was located in. However, these embassy registers are incomplete as not every resident of British origin would be registered. These registers can now be found at The National Archives within the appropriate country's records held in the Foreign Office series. *The British Overseas* will indicate what registers survive for what date, along with the relevant reference code. See 'Civil Registration of Britons Overseas' in Chapter 5.

Research Techniques: Using Other Sources

Despite the proliferation of material online, which will only increase over the coming years, it still might be difficult to track down a precise place of burial, baptism date or record of a marriage. Here are a couple of additional places you might consider looking for information.

- *Newspapers:* Sometimes the quickest method of finding where your ancestor may have been baptized, married or buried is to scan through local or national newspapers. Many national newspapers recorded these events, although they tend to include the more notable members of society. *The Times* has digitized its entire newspaper archive from 1785 to 1985 and it can be searched online at many libraries. As it is fully indexed you should be able to search for your ancestor simply by typing his or her name.

  Local newspapers can also give useful information, especially for humbler members of society. Local newspapers began to be published from the early nineteenth century onwards and should be found at the relevant local record office. Most newspapers would have a section for births, marriages and deaths. They can be useful for giving extra information about next-of-kin or stating where the event happened, if not at the local parish church (for example if your ancestor died abroad). Sometimes an obituary may give a full picture of someone's life within the community and information about his immediate family. Obituaries may also be published in newspapers separate from death notices and should also be searched for.

- *Sources in old family papers:* Through the course of your ancestors' lives, documents and papers would be issued to inform others of life events such as births, marriages and deaths. It is always worthwhile searching through your family's private papers to find such documents as they may assist in indicating where to locate the appropriate parish record. Examples of such sources are wedding invitations, funeral cards or old photographs. They would obviously detail where your ancestor was married or buried, and wedding cards in particular would give details of the bride or groom's parents. Old photographs may also have the parish church included in the shot.

  If your ancestors were particularly good correspondents, or had relatives who lived overseas, you may well find key family events mentioned on a regular basis. Sending a Christmas circular letter updating friends and family seems a fairly modern innovation, but in the days before telephone, email or fax, writing a letter was the only way to communicate and it's worth looking for collections of letters in the hope of shedding light on some of these important family events.

Parish Records for Scotland, Ireland and Other Parts

Although parish records were kept for both Scotland and Ireland, they often begin later than their counterparts in England and Wales.

Scotland

Parish registers were first introduced in Scotland in 1553, but only a very small number survive for the sixteenth century, the majority starting from the late seventeenth or early eighteenth centuries, when the Presbyterian Church became the Established Church in Scotland. These were mainly baptism and marriage records as burial records were seldom kept. Additionally, although the registers do not go back that far, the nature of the information is usually more detailed than those of English and Welsh parishes. Baptisms will usually record the mother's maiden name (as taking the husband's name was not always done). An official marriage ceremony was not compulsory in Scotland and often people would just make proclamations of intention to marry. These proclamations contain the same information as official marriage registers and were recorded in the same registers.

Perhaps the greatest advantage of Scottish parish registers is that they are mostly kept centrally, at the General Record Office of Scotland (GROS). The institute houses birth and baptism registers from 1553 to 1854, and marriages and banns for the same date. As these are part of the collection of the GROS, copies of the registers, including images of the actual registers, have also been placed online at www.scotlands people.gov.uk. These can be searched (at a cost) and it is possible to do a large amount of research online, without having to visit GROS in person.

Scottish Non-conformity

Roman Catholicism was disestablished as the official state religion in 1560. Thereafter, due mainly to the influence of John Knox, Presbyterianism began to increase in popularity in Scotland. It became the official Church of Scotland by 1690. There were also groups of non-conformists – Episcopalians, Methodists and Quakers. Parish records for these are now either with the National Archives of Scotland or at local register offices. The National Archives of Scotland also house copies of Roman Catholic registers for Scotland, with the originals still being held at individual parishes.

Ireland

Despite the majority of the population being Roman Catholic, the Protestant Church of Ireland was the official established Church until 1869. It became compulsory to keep registers from 1634, although little survives this far back. Most registers were routinely kept from the mid-eighteenth century and were stored at the Public Record Office after 1870. Unfortunately, they suffered the same fate as Irish census records, more than half being destroyed by fire in the Civil War in 1922. Remaining records are now at the National Archives of Ireland in Dublin, with collections after 1921 relating to Northern Ireland stored at the Public Record Office of Northern Ireland, Belfast.

Non-conformity in Ireland

Although the majority of the Irish population has always been, and still is, Roman Catholic, Ireland suffered a variety of anti-Catholic laws from the eighteenth century onwards, until Catholicism was legalized in 1829. Hence, there was no systematic parish register system kept for fear of prosecution. Nevertheless, some parishes did keep registers before 1829. The cities of Dublin and Cork have information from the mid-eighteenth century. County registers were not kept till some time later. These parish registers, where they exist, survive locally, although copies may be available in the National Library of Ireland.

Channel Islands

Most parish registers for the Channel Islands are still held at the appropriate parish. Jersey Archives does have copies of parish registers for a number of parishes. The Archive also has all the indexes produced by the Channel Islands Family History Societies for all parish churches until 1842, along with transcriptions to all parish registers till 1842 and some non-conformist registers. The Priaulx Library in Guernsey houses parish registers for that island.

The Isle of Man

Parish registers can be found at the appropriate parish church. However, copies dating from the early seventeenth century to 1883 are also available at the Manx National Heritage Library. The Library also holds sets of records for non-conformist registers, but Roman Catholic registers are still mostly held by the individual churches themselves.

> *Parish registers in Ireland suffered the same fate as census records, many being destroyed in 1922.*

CHAPTER 8

Wills and Probate Documents

The last will and testament of a relative can provide a unique insight into their family life, standard of living, circle of friends and even personality. Although very few actual wills survive, many thousands of registers exist which contain copies, and these are available both online and offline. This chapter examines the way wills were registered, where to find the records and a few additional sources you can use to track the whereabouts of a registered will.

> *Wills are the most intimate and personal of all official documents.*

It is often the case that you can learn the most about a person at the end of their life, which is why one of the most important sources of information that a genealogist can unearth is the last will and testament of an ancestor. Wills are the written legal document used to dispose of the property and personal possessions – the 'estate' – of recently deceased individuals, and are helpful to family historians because they can, and often do, provide a lot of detail about the deceased's family, including spouses, children, siblings and even parents. It is therefore possible to tie in many generations through one document. Wills often give information about the occupation of the deceased and where exactly they lived (sometimes the exact house will be mentioned), and in many cases you might find specific provisions about how and where they are to be buried – which in turn can lead you to a specific churchyard, gravestone and burial record.

The amount of property bequeathed is a useful indicator as to the wealth and social status of your ancestor. Indeed, you may be in for a surprise, as the 'last will and testament' of an ancestor may include names of illegitimate offspring or details of other family skeletons that might not be documented elsewhere. As wills are often written in the

HOW TO...

... make the most of wills and probate documents

1. *Probate records are the one last key set of records we can all search for from the medieval period right up to the present day. However, it is worth bearing in mind that if the last piece of information you have for your great-great-grandfather was a census return listing him as a pauper in a workhouse, the likelihood of him having left a will is very slim, although in fairness people did not have to be exceedingly wealthy to write a will.*

2. *Wills can be a mine of genealogical information, giving names of extended family members, explaining relationships between people, and listing their occupations and addresses. If you are lucky enough to find a really detailed will, you could build a small family tree just from the people and relationships it lists, although you should always remember to double-check these details with civil registration documents, parish registers and census records where possible.*

3. *Probate documents are brilliant for investigating family feuds, as some people would make a point of excluding close family members from their inheritance and go to the lengths of explaining why they didn't want that relation to have any part of their estate. You can get an idea of how wealthy your family would have been from the types of items that were left in their wills, and you may even find that inventories of their possessions have been kept.*

4. *In terms of building your family tree, wills are particularly useful for filling in names and details from the twentieth century and for the period before the beginning of civil registration in the early nineteenth century, when there is a lack of census returns and other substantiating material.*

words of your ancestors (even if dictated to a third party) they are the most intimate of official documents, providing invaluable insights into the personality of the deceased – eccentric last requests, or comments about the people they are leaving their possessions to.

Historical Background

The practice of writing wills goes back hundreds of years, far before the start of parish records in 1538, although the earliest wills were generally written by the more privileged members of society who had sufficient money and possessions to pass on. The Romans used wills, and evidence suggests that the practice was continued by the Anglo-Saxons since there are a few surviving wills that date from the ninth century. After the Norman Conquest and throughout the Middle Ages, the system of primogeniture, or descent of property to the first-born son, became entrenched. Under the feudal system, with land being the main asset and source of wealth for much of society, it was usually passed on automatically to the eldest son, without the need for a specific instruction in a will. However, assets other than land (such as gold, money, furniture and other material goods) were termed the 'estate' and would often be divided into portions and left to other family members by the writing of a separate document called a 'testament'; for example, one third of the estate would be left to the deceased's widow to support her, with the remainder divided between any other children.

Over time, people wished to avoid having to give their land automatically to their eldest son and used various means to prevent this from happening, such as passing on the land in their lifetime or placing it in trusts. Eventually the authorities recognized that the need to pass on land via a will was an important development. Therefore the Statute of Wills was passed in 1540, which stipulated that henceforth it would be legal to bequeath land in a will to people other than the eldest son. After 1540 wills and the existing 'testaments' transferring personal possessions were usually amalgamated into one document – hence the phrase 'last will and testament'.

Any man aged over 14 could write a will, whilst a woman over the age of 12 who possessed land or estate in her own right could do so too. There are technical terms used to describe people making a will – a

▼ The will of Caleb Buxton, made in 1793, immediately provides details of where he lived (Smalley, Derbyshire), his status (yeoman) and the fact that his son Joseph recently pre-deceased him.

In the Name of God Amen. This is the last Will and Testament of me Caleb Buxton of the Parish of Smalley in the County of Derby Yeoman. First it is my Will and mind and I do hereby order and direct that all my just debts and Funeral expences shall be fully paid and satisfied so soon as conveniently may be after my decease and subject thereto I do give and dispose of all my Estate (being Personal) in manner following (that is to say) Whereas I have in expectation a Sum of money or other property being the personal Estate of my Son Joseph Buxton lately deceased which by virtue of the Statute of distributions may fall into my hands or in my right into the hands of my Executor hereinafter named, man is referred to as a 'testator' whilst a woman is known as a 'testa-trix'. The Wills Act of 1837 raised the age of writing a will to 21 for both sexes. Those excluded from writing wills were the insane, prisoners and those excommunicated from the Established Church. Additionally, married women could not legally own anything until 1882 when the Married Women's Property Act was passed, so the majority of women writing wills until that point were spinsters and widows. Non-conformists and Roman Catholics could write wills, but – as will be explained below – they would have to be 'proved' in one of the Church of England's ecclesiastical courts to have legal status before 1858, when the process was taken over by the secular courts and religion no longer proved an obstacle to writing legal wills.

The Process of Writing a Will

In general, the wealthier the individual was, the more likely it is that they would have written a will. However, this is by no means true for every person and it is always advisable to look for a will, even if your ancestor was from the humbler sections of society. A will would usually be drafted by lawyers, following instructions from the testator. An executor (or executrix, if female) would also have to be nominated, who would be legally responsible for ensuring the deceased's instructions were duly followed. Before 1858, the executor or executrix would have to attend an ecclesiastical court and legally 'prove' that the will was genuine. This judicial process was known as the granting of 'probate' (probate meaning 'to prove' in Latin) and will registers were created, containing accurate copies of each will for which probate was granted. These will registers are, in most cases, the only documentation that survives as a result of the process of probate, as original wills would have been returned to the executor on completion of probate. Therefore these registers are primarily what you will be looking for in archives. Those wills that survive are likely to be found among private or family papers, some of which might be deposited in county archives along with other estate records. Any wills that were not collected by the executor were likely to have been retained by the relevant ecclesiastical court, and so might have been deposited in a diocesan record office or county archive.

In certain cases there would be no requirement for probate. This occurred when there was no concern that the will might be challenged by disgruntled members of the deceased's family. The family could therefore avoid the cost of going to probate and the will would be settled privately. In these cases no public record of the will would be produced and, if the will survives, it will not be found in the appropriate record office but amongst the family papers or, depending on the time period in question, with the solicitor who drew it up or stored it for safekeeping.

Occasionally, a will may have been dictated orally on the deathbed of the testator. These were termed 'nuncupative wills' and required statements from two witnesses in order for probate to be granted. After 1837 such wills were only valid for military personnel who may have dictated them prior to being killed in battle.

The earliest wills may have been written in Latin, or have some of the preamble in Latin. These early wills also tend to contain a large amount of religious language, although this became less common over time. The more modern wills should be in English. The testator would also be careful to declare himself or herself as 'sound of mind although sick of body', as a will written by anyone deemed to be insane would not be valid.

Probate Inventories

Inventories were detailed lists of the material goods left by the deceased. As such they are extremely useful documents. They can give a precious insight into your ancestor, really bringing him or her to life as a real person, and vividly displaying their wealth and standard of living, as well as their personality and interests. Lists that include such items as jewellery, furniture, silks and linen display a relatively good standard of living and comfort.

Inventories would be compiled shortly after the individual's death by a small number of 'appraisers', who were usually family or friends. They would compile a list for every room, including any livestock that may have been owned. These lists would often appear with surviving original wills or will registers retained by the ecclesiastical courts, as officials insisted on having this documentation during the process of 'proving' a will between the years 1530 and 1782. During this period it was obligatory for executors or administrators to provide this information, although after 1782 inventories were only compiled in matters of dispute.

Intestacy and Letters of Administration

Not every individual left a will dealing with his or her estate. An individual may have died suddenly at a young age; others may simply have chosen not to. Those who did not leave a will or last testament are known as 'intestates'. Even if there was no will, the deceased's estate still had to be duly administered and legal systems were developed to deal with this eventuality, the court issuing 'letters of administration', often referred to by the abbreviation 'admon'. Prior to 1858, the ecclesiastical court would authorize an administrator or two (usually a close relative or, in cases where the individual died owing money, a creditor) to collect the assets of the deceased and, after payment of any debt, distribute them according to the law at the time.

The process of administering the deceased's estate was only required when there was a large amount of property that needed to be transferred to a member of the family not already in possession of it, for example if the deceased had left no children or widow and a more distant relative was the next of kin. Where the intestate only had a small amount to distribute, the family would usually agree amongst one another and consequently no legal record would have been created.

If letters of administration do survive they can provide some useful information to family historians, such as the deceased's date of death and place, residence and the exact relationship of the administrators to the deceased, although they will naturally be less detailed than actual wills.

Changes to Probate in England and Wales in 1858

In order to begin searching for a will – or the record of its probate in a court – it is essential to remember the date 11 January 1858, when a radical reorganization of the system of proving a will was introduced by legislation. Prior to this date, wills would have been proved in a number of different ecclesiastical courts (see below) and therefore stray originals not returned to the family, and the will registers created during the process of probate, may be deposited in various archives – national institutions, specialist repositories, clerical record offices and

> *Letters of administration provide useful information, though they are less detailed than wills.*

county archives. The Probate Act of 1857 ended the authority of the ecclesiastical courts in granting probate and thereafter the process became a civil issue. From 11 January 1858 probate would have been granted in a number of district probate registries under the jurisdiction of the central Court of Probate. Henceforth copies of all registered wills are held centrally at the Principal Probate Registry.

Proving Wills Prior to 1858

As mentioned above, probate prior to 1858 would be granted in an ecclesiastical court of the Church of England, of which there were over 300 operating at various levels of authority. The precise court in which probate would take place depended on where the testator lived, where he owned property and the value of that property. A comprehensive guide to these ecclesiastical courts, their relative hierarchies and jurisdictions can be found in a book by J. Gibson, *Probate Jurisdictions: where to look for wills* (Genealogical Publishing Company, 2002). A very brief summary is given below.

It is also important to stress that, in the majority of cases, what you will be looking for is not the actual will itself, but the will register of the court in which it was proved. Most of the time, the original wills were collected by the executors or administrators, with only those that were left behind being retained by the ecclesiastical court. However, to create a permanent record that probate had occurred, and to prevent fraudulent administration of the terms of the will, the legally proven will was copied into a will register created by the relevant ecclesiastical court. These records survive in far greater numbers, and effectively provide the closest written record to all wills that passed through the probate system.

Archdeaconry Courts

The smallest court that a will would be proved in would be the archdeacon's court. These courts had jurisdictions over a small number of parishes, though the size of the area covered would vary considerably and Gibson's guide provides detailed information on the different boundaries. If an individual's property lay solely within the boundaries of one archdeaconry, probate would be granted in this court. However, if the deceased had property in more then one archdeacon's jurisdiction, probate would be granted in a higher court, described below. Most archdeaconry wills and registers are now held at a county archive or diocesan record office, and should have indexes for their collections.

Diocesan Courts

Any individual leaving property in more than one archdeaconry but still within a single diocese, under the authority of one bishop, would have probate granted in the bishop's Consistory Court. And in certain cases, the appropriate archdeaconry court would not actually take on the task of granting probate (usually because the amount of property involved was higher than normal) and individuals would have to approach the higher bishop's court. These courts were termed 'commissary courts'. Again, wills and will registers proved in these courts should be found in county archives and diocesan record offices, though more comprehensive indexes tend to survive given the increased volume of business handled at this higher level.

Peculiar Parishes

Sometimes a parish would be known as 'peculiar' as, although it physically lay in the jurisdiction of one archdeaconry, it was under the jurisdiction of another. Often such parishes would have small, independent courts responsible for probate. However, an individual would only have probate granted in these courts if he or she only had property within that peculiar parish. If they had property in a peculiar and a non-peculiar parish then probate would still be the responsibility of the relevant archdeaconry or bishop's court. These have been comprehensively listed in the Gibson guide, which states which court was responsible for each peculiar parish.

The Prerogative Courts of Canterbury and York

The Prerogative Courts of Canterbury and York had the highest jurisdiction for any ecclesiastical court, with the Prerogative Court of Canterbury (PCC) being the more senior of the two. Probate would be granted in either of these courts if property was valued at more than £5 (or £10 in London), known as *bona notabilia*, or property lay in the jurisdiction of more than one diocese or bishop's court. If there was property within the jurisdiction of both the archbishops' courts then probate would be granted by the PCC. The jurisdiction of the PCC was the southern part of England; the Prerogative Court of York (PCY),

▲ Probate inventories like this one dated 1487 provide a glimpse into the possessions and wealth of an ancestor when they died.

along with the Consistory Court of York and the Exchequer Court, had jurisdiction over the northern parts of the country.

The PCC was widely seen as the more prestigious court (and some individuals would seek to have probate granted in the PCC even though they were not strictly in its jurisdiction) and was the busier of the two. The PCC sat in Doctors' Commons (or College of Civilians), a society of lawyers with premises in London and regional offices throughout the provinces, with overriding jurisdiction for England and Wales. Over time, more people applied for probate in the PCC as the property valuation qualification became less of a barrier through inflation and increased personal wealth throughout the eighteenth and nineteenth centuries.

Britons who held estates overseas, particularly in the former American colonies and the Caribbean had their wills proved in the PCC.

- *Records of the Prerogative Court of Canterbury:* All probate records for the PCC are held at The National Archives in Kew. The records cover the years 1383 to 1858. There are over one million wills within the series. Additionally, every will proved during the Interregnum period (1653–60) will also be found amongst this series, as all

Possible Difficulties in Locating Wills

Variant or misspelled names in indexes

A very common explanation if you're having difficulty finding the correct record is that the name you are looking for has been spelled differently in an index or database. Surnames often had a number of variant spellings and a will could be indexed using any of the possibilities. Additionally, many indexes have been compiled in the modern age and problems have occurred due to difficulties in reading old handwriting. Hence, mis-transcriptions have become unavoidable when producing the indexes. Remember to search all possible variations of the name.

Probate was granted late

It is important to remember that the date of a will refers to the date probate was granted, *not* the date of death of the deceased. The majority of wills would have been proved within a matter of weeks or months after the death, but this would not always be the case, especially where the will was disputed. Ensure an adequate time frame is used when searching through the indexes; you may need to search several years after the death in particularly difficult cases. There are records available at The National Archives and other archives relating to disputed wills, if you suspect this has happened.

Probate granted in another jurisdiction

Given the complications in probate law prior to 1858 it can be difficult to identify the correct court where probate was granted. This matter is not helped by the fact that people could, and sometimes did, ignore the correct protocols. Executors might decide to seek probate in religious courts were banned and the PCC became a civil court doing the same tasks.

Surviving original wills are located in the reference series PROB 10, whilst will registers can be found in PROB 11. The National Archives has placed all its wills found in series PROB 11 online at www.natioanlarchives.gov.uk/documentsonline. It is possible to search the index free of charge, although payment is required to view and download individual wills. The information is indexed by name, place and date. Like any other online source provided by The National Archives, you can view the information free of charge by visiting the archives. The National Archives also stores the surviving inventories for these wills in another series. However, it is not possible to access these inventories through the Internet, only by visiting in person – though you can search the online catalogue to see if a match appears.

- *Records of the Prerogative Court of York:* Records of probate granted at the Exchequer and Prerogative Court of York can now be found at the Borthwick Institute of Historical Research, part of the University of York. The collection has wills from 1389 to 1858, although few survive before 1630. The Archbishop of York had jurisdiction over all

another court if they themselves lived in a different parish or diocese from the deceased. Alternatively, probate may have been granted in a higher court due to social status or snobbery (as happened with the Prerogative Court of Canterbury) or if the appropriate court was not administering probate. Always remember the existence of 'peculiar' parishes and where probate would have been granted in that case.

Wills not formally proved

Probate was a legal requirement for bequeathing of 'estate' only and not land, up until the 1857 Act. In cases where there was no dispute over the inheritance of the land then no official public record would have been required to enact the will. Although such wills were not commonplace, any surviving documents would only be amongst family papers.

No will made

Do not assume that everyone automatically left a will, even if they were of a higher social status. Although wills were more common than expected, it still remains true across the population as a whole that more people did not leave a will than did; indeed, the further back you go, the less likely it is that a will would have

been made unless your ancestors were more privileged members of society.

Testator died overseas

Whilst many overseas wills were proved in the Prerogative Court of Canterbury, you may need to look elsewhere if your ancestor served in the British civil service in one of the colonies. For example, relatives of deceased servants of the East India Company often used local courts in Calcutta, Madras and Bombay to prove wills, and these records are now in the Pacific, Asia and Africa collections at the British Library and will be examined in more detail in Chapter 23.

> *One of the most difficult parts of locating a will prior to 1858 is identifying where probate was granted.*

of the northern dioceses, and there are far fewer wills and registers proved than by the PCC.

Currently, it is only possible to do a limited search of the PCY collection online, on the Origins website (see below). The website has placed indexes for both the Prerogative and the Exchequer Court of York online for the years 1853 to 1858 and also for 1267 to 1500 in two separate indexes. The first database has over 16,000 names and the next one over 10,000 names. Although you can only view the indexes online, the originals are still held at the Borthwick Institute, and it is possible to order copies of the originals through the website.

Locating a Registered Will Prior to 1858

One of the most difficult parts of locating a will for an ancestor prior to 1858 is trying to identify where probate may have been granted. Unless you are aware of the approximate year of death of your ancestor, his or her residence and a rough idea of their wealth this can be rather difficult, and you may be faced with the necessity of working through all levels of ecclesiastical jurisdiction until you find the will – assuming, that is, your ancestor left a will in the first place. Here are some alternative avenues for you to explore if you have run into difficulties.

www.origins.net

This website has already been mentioned as an online index for the Prerogative Court of York. It also has a number of other indexes for wills collections in England. These include:

- *The Archdeaconry of London Wills Index:* The index covers the years 1700–1807 and contains over 5,000 names. It is an index only and the original records can be found at the London Guildhall Library.
- *Bank of England Wills Extract Index:* This index is for the years for 1717 to 1845 and has over 60,000 names. You can only view the index online, as the original wills are held at the Society of Genealogists, although copies can be ordered online.
- *York Peculiars Probate Index:* This index spans the years 1383 to 1883, containing over 25,000 names for 54 peculiar courts in the

province of York. The originals are held at the Borthwick Institute, although you can order copies online.

- *Prerogative Court of Canterbury Index, 1750–1800:* An alternative index to the PCC wills held at The National Archives, although only for a fifty-year period.

The Society of Genealogists

The Society has a vast range of genealogical sources including collections of local will indexes for many parishes and counties of England, Wales, Scotland and Ireland. The Society is based in central London and anyone is entitled to visit for a small fee or join on an annual basis. For further details, you can visit their website www.sog.org.uk.

Local Wills Indexes Online

Many county record offices and archives are placing their catalogues online, and it is always worth checking the Internet to see if relevant probate material has been made available or indexed. Below is a list of a few of the local record offices that are currently offering this at the time of going to print; more and more local record offices are planning to do this in the near future:

▼ You can now search for all registered PCC wills prior to 1858 via Documents Online, and download images for your records.

- *Gloucester County Council Genealogical Database:* The database can be found on the council's website, www.gloucestershire.gov.uk. It contains all wills proved at Gloucester from 1541 to 1858. It is possible to order copies of the original wills through the website.

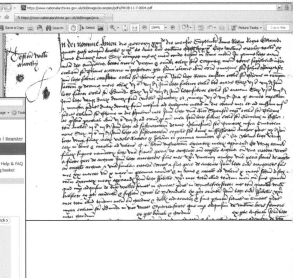

- *Derbyshire Wills Database:* A database of Derbyshire wills can be found at www.wirksworth.org.uk/WILLS.htm. It has been compiled by an employee of the Derbyshire Record Office, Michael Spencer, and covers the years 1525 to 1928. There is an alphabetical list of all wills proved for this period along with other probate documents (including administrations and inventories). However, no images of the wills are available.

- *Cheshire Wills:* The county council of Cheshire has also placed a database of wills proved in Chester (mainly for residents of the Cheshire area) from 1492 to 1940 at http://www.cheshire.gov.uk/Recordoffice/Wills/Home.htm. It contains over 130,000 names in total and you can order a copy of the original will online.

- *London Metropolitan Archives – London signatures:* This database is for wills proved at the Archdeaconry Court of Middlesex from 1609 to 1810. It can be viewed and searched online for free at http://www.cityoflondon.gov.uk/corporation/wills/engine/wills_search.asp. At the moment it contains 10,000 wills. The database also has entries for marriage bonds, and conducting a search on the database will result in hits from both datasets. You need to click on the actual entry to ascertain which dataset the entry applies to. You can also purchase a copy of the image if a relevant entry has been located.

- *Wiltshire Wills Project:* An online database set up by the Wiltshire and Swindon Records Office at http://history.wiltshire.gov.uk/heritage/. It provides an index to all wills proved at the diocese of Salisbury from 1540 to 1858 and contains 105,000 wills and inventories. The index is searchable online and some images of documents can also be viewed. If your entry does not have an online image you can order a copy of the image directly from the record office.

Death Duty Registers

There is another way to research where a will might have been proved, using an associated set of records that were linked to the state's desire to raise taxation. In 1796, death duty on a deceased's estate was introduced, and remained in force until it was replaced by the Inheritance Tax in 1906. As part of administering the tax, registers were created for every estate that came within the financial threshold of the tax, and you can use these to identify which ecclesiastical jurisdiction was used to grant probate, as this is noted in the indexes to the death duty registers. Not everyone would have been liable to pay

CASE STUDY

Jeremy Clarkson

When Jeremy Clarkson was first approached to take part in *Who Do You Think You Are?* he was not convinced that the journey would be particularly interesting. He spotted the surname 'Kilner' lurking amongst the Clarksons. Jeremy knew about the Kilner side of the family, but when the research team showed him the extent of their glass manufacturing empire he became intrigued as to how it had all evaporated so quickly.

Jeremy tracked down a key member of his family, Caleb Kilner, the grandson of the company's founder, John Kilner. Caleb died in 1920, and his will was instrumental in establishing that he was a very wealthy man indeed. His estate was estimated at £142,000 – the equivalent of over £3.5 million in today's money – and he owned stock and property all over the country. Caleb's executors were listed as his son George – who inherited most of his father's wealth – and his son-in-law Harry Smethhurst, who was married to Caleb's daughter Annie – Jeremy's great-grandmother.

George Kilner was clearly not a very good businessman, and by the time he died 20 years later his will revealed that his fortunes had diminished dramatically, being worth just under 10 per cent of his father's fortune. Furthermore, Harry Smethhurst seems to have spent heavily too during his lifetime, earning the nickname 'Flash' – his will also shows that the Kilner money inherited by himself and Annie had been dissipated by the time of his death – 'clogs to clogs in three generations'! Any remaining funds from the Kilner fortune passed to another side of the family after Annie's death; she fell out with her daughter, Gwendoline, and left all her remaining money to her son Tom. Gwendoline was Jeremy's grandmother.

◀ 'Flash Harry' Smethhurst and his wife, Annie, who were rumoured to have spent some of their inherited money on a motor car, around 1901.

death duty; it is estimated that until 1805 the tax was payable on about 25 per cent of all estates. After 1805 until 1815 this increased to 75 per cent, so these tax registers become a far more important source for tracking probate records.

The National Archives has placed online an index for these records for all courts other than the PCC, known as 'country courts'. The index covers the period 1796–1811 and it is free to search, although downloading the document will incur a cost (unless viewed onsite at The National Archives). You can search the indexes of these records online at www.nationalarchives.gov.uk/documentsonline. The entire collection of extant death duty registers between 1796 and 1903 is also available to search online on the commercial website www.findmypast.com. You need to have subscribed to the most expensive package to search these records on this website, though. The Society of Genealogists also has indexes to these registers (although not the actual registers) for 1796 to 1857. Copies of the original indexes, as well as the registers themselves, can be seen free of charge in person at The National Archives, Kew, contained in series IR 26 and IR 27.

The material contained in the death duty registers provides information on the date of death; the value of the deceased's estate; the amount of death duty paid; and details of the administration of the estate under the terms of the bequest made in the will. So they are worth looking at even if you do know the registration district in which the will was proved.

Probate Records Post 1858

The probate system was greatly simplified by the Probate Act of 1857, which stipulated that the process of granting probate would be transferred from the ecclesiastical courts to civil authorities. A a civil Court of Probate was established and became responsible for all probate matters in England and Wales from 11 January 1858. Even if the testator died abroad probate would still be dealt with by this court if there was any property in England or Wales.

Probate registries were created in various districts throughout England and Wales, with a Principal Registry in London. Probate would be granted in any of these local district probate registries, between seven days or six months after the death of the testator. Upon completion of the process, the district probate registry would send a second copy of the registered will to the Principal Probate Registry.

> *The Principal Probate Registery holds copies of all wills for which probate has been granted since 1858.*

Thus the whole probate system became centralized and copies of every registered will made since 1858 can now be found in one place in London.

The Principal Probate Registry is currently situated in Holborn, London, and is open to any member of the public. The Registry houses copies of all wills for which probate has previously been granted, and annual alphabetical indexes to the registered wills are available for inspection in its search room. Looking for a will, or letters of administration, is relatively straightforward. Once you have found an entry in the annual index, you can request a copy of the registered will at the enquiry desk, which will be provided within an hour or posted to any address. A charge will be made for providing a copy, currently £5. As well as the onsite indexes in the search room, a National Probate Calendar exists on microfilm or fiche that can be found in many national and local archives, duplicating the indexed information between 1858 and 1943.

The information in the indexes is quite detailed and informative. They are arranged alphabetically by the surname of the testator, and their last residential address is also provided. In many cases, you will find their date of death, and often where they died, along with the date probate was granted and the relevant probate office; the value of the estate; and, until 1996, who the executor and administrators were (and sometimes including the relationship to the deceased). The indexes will often detail the solicitors who handled the probate process. Indeed, in recent cases you may wish to contact the solicitors' firm to see if they have any extra documentation relating to the handling of the probate process. If you have trouble finding the firm, it might be worth contacting the Law Society, as they can check their lists to see if the firm is still active or not.

Searching for a Will Locally

If you are unable to visit the Registry in London you may be able to search at the appropriate district registry. However, many district registries are now transferring probate documents older than 50 years either to appropriate local record offices, or to the Principal Probate Registry. Don't forget, though, that most archives hold a copy of the National Probate Calendar up to 1943, which might contain the information you need.

Probate Records in Scotland

The laws regarding inheritance and probate in Scotland were slightly different to those of England and Wales. Scotland had strict rules stating that all land had to be given to the eldest son, and in the case of no male children, to the eldest daughter. This remained the case until 1868. Testators were only allowed to bequeath their estate (material goods and chattels) to whom they wished through testaments. Thus there were no such things as 'wills' for this period. Estates of those who died intestate were dealt with by a record known as a 'testament dative', which were similar documents to 'letters of administration' south of the border.

The appropriate church courts had responsibility for dealing with matters of probate in Scotland until 1560. After this the situation changed and responsibility was transferred to non-religious bodies known as commissariat courts. There were many commissariat courts responsible for administering their own specific part of Scotland. These courts all came under the jurisdiction of the Principal Commissariat Court in Edinburgh (a similar system to post-1858 probate administration in England and Wales). In 1824 the Sheriff Courts took over the responsibility for probate. The jurisdiction of these courts broadly coincided with the Scottish counties.

Probate records for Scotland prior to 1824 are now held at the National Archives of Scotland. This archive also has records for the early years of probate records being administered by the Sheriff Courts. You can view this information either by visiting the National Archives of Scotland or going online to view the index at www.scotlandspeople.gov.uk. The website covers the years 1513 to 1901 and has over 611,000 entries in the index. However, it is not possible to search this website without payment. Bear in mind that it was not obligatory for people to leave testaments and it was a less common practice than in England. If you are searching for wills post 1901 then they are likely to be held at the appropriate Sheriff Court.

Probate Records for Ireland

The system of probate administration in Ireland was the same as in England and Wales. The Church of Ireland was responsible for the process through its own ecclesiastical courts, either consistory or diocesan. There were no archdeacon courts and only a tiny number of peculiar courts. The highest ecclesiastical court in the land was the Prerogative Court of Armagh. This system survived until 1858, when it was altered in the same fashion as in England and Wales. A Principal Registry in Dublin was established along with eleven district registries. Copies of probate documents would be submitted by the district registries to the Dublin office.

Unfortunately, a fire at the General Registry Office in the Civil War in 1922 also destroyed probate records housed in Dublin. Original probate records after 1858 should still be with the district registries, and additional copies may have also been sent to the National Archives of Ireland or the Public Record Office of Northern Ireland since 1922. The National Archives of Ireland also has various substitute records for this period. This includes copies of probate documents that were sent from relevant firms of solicitors, records obtained from family papers and also various transcripts that were written prior to the fire in 1922. Although most of the wills did not survive, some indexes did; and they often contain quite detailed information such as the name and address of testators and when probate was granted. The National Archives of Ireland does also hold approximately 10,000 wills that were retrieved from family papers, and these have been fully indexed.

All probate records originating after 1922 will be either in the National Archives of Ireland or in the Public Record Office of Northern Ireland, depending on where they were made.

Records for the Channel Islands

The Ecclesiastical Court of the Dean of Jersey granted probate for all residents until 1949. After that date, it became the responsibility of the Principal Probate Registry in London. In Guernsey probate is still under the jurisdiction of the ecclesiastical authorities, the Ecclesiastical Court of the Bailiwick of Guernsey.

Manx Probate Records

Ecclesiastical courts were responsible for granting probate in the Isle of Man until 1884. The courts would either be the Consistory Court of Sodor and Man or the Archdeaconry Court of the Isle of Man. After 1884 the Manx High Court of Justice became the responsible authority for probate matters.

Areas of Family History

The purpose of this section is to demonstrate the rich and varied ways to add historical context to the lives of your ancestors. Here you can learn about aspects of British and global social history that would have directly affected the lives of your distant relatives, and use the records and resources generated by these events to discover more about them as people – their work, where they came from and their family secrets.

CHAPTER 9

Military Ancestors: The British Army

At some point over the course of the last few centuries, one of your ancestors would have served in the British Army. There are plenty of clues that suggest this military link – an old photograph of a relative in army uniform; some medals passed down through the generations. This chapter shows you how to use these clues to access original records and piece together the army careers of your relatives, as well as gain a greater understanding of what life was like as a soldier.

A Brief History of the British Army

The beginnings of the modern-day British Army can be traced back to 1660 with the restoration of Charles II to the throne, for this was the first time a regular standing army was established. A small number of regiments were created on a permanent basis with the responsibility of guarding the King.

Nevertheless, there had been many wars and conflicts for centuries before 1660 and armies of men would have been raised as and when required, although on a temporary basis, being discharged when the fighting was over. Finding an ancestor who may have served in the army prior to 1660 is more difficult as there was no permanent institution and, therefore, there was no systematic record-keeping system.

The earliest method of raising a group of men to serve in a battle was tied into the feudal system, with the lord of the manor having the responsibility of gathering together such a body. Additionally, all men holding a certain amount of land were required to serve as knights, although the knight system was not used after the fourteenth century.

> *There are many records available to help you trace your ancestor's military career.*

Through the course of time the feudal system came to be replaced by the 'contractual' system, whereby contracts or indentures were used to secure military personnel. Indentures were used between the King and the nobility, and in turn by the nobility with the lesser sections of society right down to peasant level. The start of the English Civil War in 1642 saw the beginnings of a professional military body with the formation of the New Model Army in 1645 by Oliver Cromwell. The New Model Army was disbanded in 1660, shortly after the restoration of Charles II, and it is after 1660 that the origins of the modern British Army can be traced.

The regiments that formed the army in 1660 were recruited from units that were formed during, or even before, the Civil War. Two regiments of Horse Guards and two regiments of Foot Guards were created. Originally, these regiments served as the Army for England and Wales, with Scotland having its own forces. However, after the Act of Union between the two countries in 1707, these Scottish units were united with the English units to form the British Army.

The regiments formed in 1660 are the oldest in the British Army; thereafter the regiment became the basic unit that formed the core of the Army's organizational structure. Army records were organized and retained on a regimental basis, and certain records may still be with the regimental record office. Indeed, it was only in 1920 that an army-wide service number was introduced for each serving soldier.

Each regiment would be under the overall command of a colonel. A regiment could be an infantry, cavalry, artillery or engineer regiment, with various specialist forces, such as the foot guards, also existing from time to time. Originally, regiments would be named after the colonels in charge of the regiment and, as such, the names would change with the change of the colonel. Hence, to make things easier to organize, a numbering system was also introduced in 1694, infantry regiments being named 1st Regiment of Foot, 2nd Regiment of Foot, etc. As the system of naming a regiment after a colonel became increasingly confusing (as more than one colonel could have the same surname) it was ordered in 1751 that regiments also take on official titles, such as the 4th Regiment of Foot being given the additional title of 'King's Own'.

▲ Men of the British Eighth Army are greeted by General Montgomery as they march through the narrow streets of Reggio in the early stages of the Allied invasion of Italy.

Further changes were introduced in 1782, with regiments being attached to geographical parts of the country. Hence, the 5th Regiment of Foot became attached to Northumberland. This was done in part to increase recruitment, although regiments continued to recruit outside their given territories when required. The naming of regiments was changed once more in 1881. From that date, numbers were no longer used. Instead all regiments were given territorial titles, usually county names, although some continued to use their old number unofficially. This system of organization remained largely intact until and during the First World War. Service records for soldiers serving after 1922 have not yet been released to the general public, and changes occurring in army organization in the twentieth century do not require a detailed analysis for genealogical purposes.

The Organization and Hierarchy of the British Army

As mentioned, the basic army unit was the regiment. These regiments would be grouped into four general categories of troops:

1. *Cavalry:* These were the mounted horse regiments of troops. The majority were disbanded or merged in 1922, the horse being replaced by the tank.
2. *Infantry:* These were the basic foot soldiers who served on the line at times of conflict. Infantry regiments have traditionally formed the bulk of the army, and therefore this is where you are most likely to find your ancestor.
3. *Artillery:* Artillery regiments were specially trained to use cannon, mortars and other heavy guns.
4. *Engineers:* Their historical role includes survey work, mining, construction projects and support services to front-line troops.

Specialist regiments have also existed, such as the two foot guard regiments originally raised to guard the monarch in 1660, the Grenadier Guards and the Coldstream Guards. There are other non-regular army elements such as the Militia, and relevant records are discussed later in this chapter.

As regiments were subject to many name changes, mergers with other regiments, or even disbanding at various points in time, it is worthwhile checking the correct name for your ancestor's regiment during the particular period of interest. You can do this online on the website www.regiments.org. Alternatively, consult the following comprehensive guide commonly found in archives and libraries, written by Arthur Swinson, called *A Register of the Regiments and Corps of the British Army: the ancestry of the regiments and corps of the Regular Establishment.*

Although the terms 'regiment' and 'battalion' are often used interchangeably, the difference in their meanings was fixed by 1660. Hereafter 'regiment' would be for the administrative organization, whereas 'battalion' was the term used for the actual fighting unit. Each battalion would be divided into ten units, known as companies. In theory each company would have 100 men, thus a battalion would comprise 1,000 men, but this was not always the case. An individual soldier (or 'other rank') would be recruited into any company of a particular battalion.

The diagram to the right shows how a battalion fitted into the wider hierarchy of the Army.

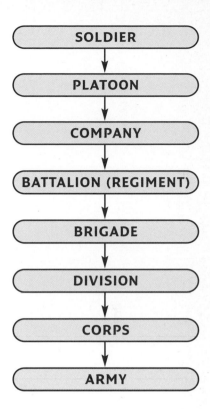

Army Ranks

To begin searching for your ancestor amongst the surviving records it is essential to know whether he was a commissioned officer or other rank, as the records have been organized according to this distinction (unless you are searching the period prior to 1660). A commissioned officer could be any of the following ranks: lieutenant, captain, major, colonel, brigadier or general. Your ancestor would have been an 'other rank' if he had held any of the following ranks: private, lance corporal, corporal, sergeant or sergeant major. Once you have this information you can begin the search for your ancestor's records. The majority of these records are now stored at The National Archives (TNA).

Service Records Prior to 1660

As there was no regular army prior to 1660, there are no comprehensive service records. Additionally, what does survive contains little genealogical information.

The earliest lists of soldiers, recruited in the feudal period, may survive in TNA department series for Chancery or Exchequer records (specifically C 54, C 64, C 65 and E 101). In 1285 the Statute of

Winchester required all males aged between 15 and 60 to equip themselves with armoury that they could afford and they would be duly assessed. By the fifteenth century such assessments were being made by the appropriate local official along with the Lord Lieutenant of the county (also known as Commissioners of the Array). The earliest such 'muster lists' survive from 1522; other miscellaneous lists would also have been forwarded to the Exchequer and Privy Council and form part of their records. A comprehensive guide to these records can by found in *Tudor and Stuart Muster Rolls: A Directory of Holdings in the British Isles* by J. Gibson and A. Dell (Federation of Family History Societies, 1991), including information on where they are held.

If your ancestor served during the English Civil War or Interregnum period after it, you may be able to find some evidence for this. Although there were no specific service records, there may be other records containing relevant information, mostly for officers only, though. The best place to start at The National Archives is to look through the Calendars of State Papers. Other published sources are also useful, such as:

- Edward Peacock's *Army Lists of the Roundheads and Cavaliers* (listing all officers of both sides in 1642)
- W. H. Black (ed.), *Docquets of Letters Patent 1642-6* (1837) (containing additional information on officers of the Royalist camp)
- P. R. Newman, 'The Royalist Officer Corps, 1642–1660', *Historical Journal*, 26 (1983) (ditto)
- R. R. Temple, 'The Original Officer List of the New Model Army', *Bulletin of the Institute of Historical Research* (contains relevant sources for the New Model Army)
- I. Gentles, *The New Model Army* (a history of the New Model Army)

You may find further information in The National Archives State Papers, series SP 28 (the relevant lists have not been calendared) or the Exchequer series.

Individual Army Records Post 1660

Individual army records since 1660 have been organized into those of officers and other ranks, and finding records for each group will be discussed separately. Another important defining moment for the family historian is the onset of the First World War, as records for officers and soldiers serving before this conflict are stored in a different place to those serving after it.

Broadly speaking, most people search in two types of records: service records and pension records.

Officer Records 1660–1913

The officer classes came almost exclusively from the more privileged sections of society for this period, the majority also having rural gentry backgrounds. Indeed, it was a fundamental prerequisite to be a gentleman before an individual was even considered for an army commission. Until 1871, the way most people became army officers was through the purchase of their commission. Nor was promotion based on any meritocracy; promotions were also bought, officers seldom being promoted through ability. It would not be unusual for wealthy young boys (there was no minimum joining age) to have rapid promotions at the expense of more talented, if less wealthy, officers. The problems of this state of affairs became apparent in the Crimean War (1853–56), and by 1871 it was no longer possible to purchase commissions.

Army Lists

As the Army did not have a central record-keeping system until the First World War, the surviving records come from documents created by the relevant regiment. The majority of the records are now at The National Archives, but you may also find additional information at the regimental museum and it is worthwhile contacting these institutes too. Perhaps the best place to begin searching for an officer is by using the Army Lists.

Army Lists were begun in 1702 and listed every officer serving in the Army. After 1740 it became routine to publish these annually (in times of conflict they may have been published more frequently than this). If you know the approximate date your ancestor served in the Army you can trace his entire career, from initial commission to subsequent promotions and eventual retirement, including details of which regiment he belonged to. The lists are somewhat basic, however, and do not expect to find any extensive career details or any genealogical information.

You can find Army Lists in various institutes. A full set from 1759 is in The National Archives. Additionally, the original manuscripts of the

> *There are two main types of records: service and pension records.*

Army Lists from 1702 to 1752 can be located in TNA series WO 64. The British Library also has a complete set from 1754. The National Army Museum, the Imperial War Museum, regimental museums and other large reference libraries also have collections, although not always complete collections.

Another unofficial publication, *Hart's Army Lists*, should also be consulted if your ancestor's army service fell in the relevant time period. Hart's lists cover the years 1839 to 1915 and give more details than are found in the Army Lists, including short biographies of the officers and career details. *Hart's Army Lists* were published on a quarterly basis and cover the regular British Army and officers of the British Army in India. An incomplete set of *Hart's Army Lists* is located at The National Archives on open shelves. The complete collection can be found in series WO 211, along with the working notes Hart used to compile his lists for 1839 to 1873.

Service Records of Officers

As there was no army-wide service record-keeping system until 1914, service records will be found in many different record series. There is no one single document providing the entire service career for an individual. Instead records were kept separately, by both the regiment and the War Office, although the War Office did not start collecting this information until the early nineteenth century. Both these sets of records are now held at The National Archives in series WO 25 (for War Office records) and WO 76 (regimental records). Additionally, there is an incomplete name index to these collections (also at The National Archives). As it is incomplete, searches should still be made in both series, even if the name of your ancestor does not appear in the index.

War Office Records in WO 25

As stated, the War Office records do not contain complete service histories for every officer in the British Army. Rather, they comprise an assessment of officers at various points in the nineteenth century. Five surveys were taken at different dates during the century, officers being required to send returns on their statements of service at that point in time. As the officers themselves were responsible for giving this information, there are gaps in the data as not every officer completed the returns in full. Each survey can now be found in WO 25 and some surveys contain more information then others. The details for each survey are listed here.

1809–10 (WO 25/744–748)

The survey is arranged alphabetically and contains only details of service history.

1828 (WO 25/749–779)

This survey is more detailed and contains information about service completed before 1828. It is arranged alphabetically and military history is provided along with dates of commission. Further information relating to marriage dates and birth dates of children is also included.

1829 (WO 25/780–805)

Organized on a regimental basis, this survey only refers to officers active at the time. It provides the same information found in the second series. In addition the wives of soldiers are indexed separately; their maiden names are provided along with their date and place of birth and, occasionally, their date of death.

1847 (WO 25/808–823)

The survey is arranged alphabetically and contains information very similar to that found in the second survey.

1870–72 (WO 25/824–870)

The survey covers some returns outside those dates and is arranged by the year the returns were compiled and thereafter by regiment.

If your ancestor was an officer in the Royal Engineers, their service records can be located in WO 25/3913–3919, covering the years 1796 to 1922.

Regimental Records WO 76
This series contains the records made by individual regiments which were subsequently transferred to the War Office. As such they are organized on a regimental basis and the entire series covers the years 1764 through to 1913. Of course there will not be information for every regiment from 1764 and the level of information for each regiment also varies, although generally more detailed records were kept through the course of the nineteenth century. To begin searching these records refer to the name index (mentioned above). However, if the name is not found, it is advisable to search anyway. Identify where

the records for your ancestor's regiment are within the series and search through.

There are certain regiments that do not have their records within either WO 76 or WO 25, but in other series. Below is a list of such regiments:

- Royal Garrison Regiment Officer service records (1901–05 only): WO 19
- Gloucester Regiment service records (1792–1866): WO 67/24–27
- Royal Artillery officer lists (1727–51): WO 54/684, 701

A pay list of officers (from 1803 to 1871) can also be found in WO 54/946.

Records for Granting of Commissions

Apart from the service records themselves, documents relating to the granting of commissions can provide additional information. The purchase of commissions was instituted in the reign of Charles II and was not abolished till 1871. It was possible to buy a commission up to the rank of colonel, although the Commander-in-Chief could award free commissions. An officer was awarded their rank by these royal commissions and the date of these commissions can be located in numerous published sources (the Army List, the *London Gazette*, *The Times*, etc).

Perhaps the most interesting documents regarding the granting of commissions are the actual applications for purchasing or selling such commissions. These are also held at The National Archives, in series WO 31, and cover the years 1793 to 1870. The applications contain a lot of personal information, such as baptismal dates. They often have letters of recommendation attached, which can give further details about the character of the candidate and information about his life (such as education and employment details) and his parentage.

Other records of relevance concerning the system of commissioning officers include:

- Warrants issuing commissions from 1679 to 1782, in SP 44/164–418. After 1855 they can be found in HO 51.
- The awarding of commissions was recorded in commission books by the Secretary of State for War. The books cover the years 1660 to 1803 and can be found in WO 25/1–121.
- Succession books recorded the promotions and transfers of officers. They are organized by regiment and date. The regimental records

can be found in WO 25/209-220 for 1754-1808, and another set of records in WO 25/221-229 for 1773-1807.

The Royal Artillery and Royal Engineers regiments have their commission records in series WO 54, for the years 1740-1855.

Half-pay Records

As pensions were not introduced until 1871 a system of half-pay was used instead. Begun in 1641 to compensate officers in reduced or disbanded regiments, it soon became a method of retaining officers as they still held on to their commissions whilst on half pay. Officers on half pay can be found in the Army List. There are also original records in WO 23, although little genealogical information is provided in this series.

The series PMG 4 is more genealogically relevant, the ledgers recording payments of half pay from 1737 to 1921. The series is alphabetically indexed after 1841 and it is possible to find the date of death and, after 1837, addresses of the recipients. The more modern ledgers also give date of birth details.

Further Records Providing Personal Information

In order to join any government service an individual had to provide birth or baptismal details. Hence, collections of these baptismal records have survived in the following series (which both have indexes):

- WO 32/8903-8920: 1777 to 1868
- WO 42: 1755 to 1908. This series also includes marriage, death or burial and birth/baptismal certificates for children

Widows' pension records may also be useful. Widows were entitled to pensions far earlier than retired officers, in 1708.

- WO 25/3995 and 3069-3072 are registers for bounty paid to widows between 1755 and 1856. It is fully indexed
- WO 25 has numerous pieces relating to applications for pensions and may include details of remarriage. You will also find further information in PMG 10 and PMG 11

Lastly, it may be worthwhile searching the records for pensions paid to children and other dependent relatives. Such pensions were introduced from 1720, and were paid out of the Compassionate Fund and the Royal Bounty. Registers recording who received this money and the amount

To join a government service a person had to give birth or baptismal details.

they were paid, from 1779 to 1894, can be found in WO 24/771-803 and WO 23/114-123. Further ledgers can also be viewed in PMG 10 for 1840 to 1916.

Records for Other Ranks

The majority of the British Army was composed of non-commissioned officers (sergeant major, colour sergeant, sergeant, corporal, lance corporal) and rank-and-file soldiers – privates – who were collectively known as 'other ranks'. Before 1806, enlistment was for life; from 1807, discharge was permitted after 21 years' service, and after 1817 soldiers could seek an early discharge after 14 years. Rates of pension were calculated according to length of service, and are discussed later. For most periods in history these men were recruited voluntarily, with conscription being used only on a few occasions. These men usually came from the poorest sections of society, often being labourers endeavouring to escape poverty, even though the pay was very little. Additionally, a disproportionate number were recruited from Scotland and Ireland.

Employment conditions for the other ranks were far from ideal. The majority of the barracks, which housed the soldiers, were cramped and often had poor sanitary conditions. The quality and quantity of food was also inadequate, although rations for beer were generous, leading to the reputation of drunkenness of many of the soldiers. Not every soldier was automatically entitled to marry, and housing for spouses was also limited. These poor conditions were addressed following the failures of the Crimean War in the 1850s. Reforms were introduced in the 1870s in order to improve the living conditions of the soldiers, hoping that this would in turn improve their fighting ability.

Service Records
The National Archives holds the vast majority of service records for other ranks, found in a variety of places.

Soldiers' Documents, 1760-1913
If your ancestor was a soldier during this period, the easiest place to begin searching for any surviving documentation is within the 'soldiers' documents' series of WO 97. Although they are often referred to as 'service records' the documents were compiled for pension purposes and are not actual service records. They were created when a soldier was discharged and awarded a pension by the Royal Hospital at

Chelsea. The Royal Hospital was created in 1682 to house soldiers who had been injured whilst on active service. Pensioners were termed either 'in-house' or 'out-house' pensioners, depending on whether they took up residence at the Hospital or received a pension at their private residence. Most surviving individual soldiers' documents were created in the administration of this pension rather than as specific service records. As such the records are not complete for every serving soldier but, until 1883, only for those discharged to pension. Nor will there be any record for any soldier dying on active service. Additionally, soldiers who were discharged from the Army by purchasing their way out of their remaining period of service would not be included. From 1883 to 1913 the series is fuller and most soldiers should be within these records.

If you do find your ancestor's record amongst the series, there is usually a good amount of detail to be gleaned from it. Most documents include:

- Full name
- Age and place of birth
- Physical description
- Trade or occupation before joining the Army
- Details of next of kin after 1883

The series have been compiled using the following documents, although not every document would have survived for each soldier:

- Discharge forms issued on the day the soldier was discharged from the Army
- Attestation forms, compiled the day the soldier officially enlisted into the Army
- Proceedings of the regimental board, including a record of service, detailing the service career of the soldier
- Any supporting documentation relating to discharge
- Past service questionnaires if other documentation for the soldier's service did not survive
- Affidavits to declare that the soldier was not receiving funds from any other public body

Records for the early period are less complete, but should contain discharge papers at least. The more modern documents usually contain at least the discharge form and attestation papers.

The actual documents themselves are arranged differently according to different time periods. The dates refer to date of discharge not date of attestation. The series is broken down as below:

- **1760–1854: WO 97/1–1271** The records are arranged on a regimental basis, and then alphabetically by the surname of the soldier. This set of records has been fully transcribed by name and it is possible to do a simple name search in The National Archives online catalogue.
- **1855–72: WO 97/1272–1721** These are organized on a similar basis to the collection above, by regiment then surname. However, there is no online index to date and it is only feasible to search within the records if the regiment is known.
- **1873–82: WO 97/1722–2171** The records for soldiers are organized into cavalry, artillery, infantry and miscellaneous troops and then by surname.
- **1883–1900: WO 97/2172–4231** The documents are arranged by surname only.
- **1901–13: WO 97/4232–6322** As above, a simple surname arrangement.

Other Records at The National Archives

If you are still unable to find a record relating to your ancestor, even though you suspect he was awarded a pension, there are other options worth trying:

- **WO 121: 1787–1813** A similar collection to WO 97, containing papers for those discharged and awarded a Chelsea out-pension. This series can also be searched by name online in The National Archives catalogue.

    Records from WO 121/239–257 relate to soldiers discharged without pension from 1884 to 1887.
- **WO 400: 1799–1920** These contain the soldiers' documents for the Household Cavalry.
- **WO 116: 1715–1913** The series contains pension books for those discharged due to medical reasons or disability.
- **WO 117: 1823–1913** Similar to WO 116 but this series contains pension books for those discharged after completing a period of 'long service'.
- **WO 120: 1715–1857** A series of the Chelsea regimental registers of pensioners. These mainly concern soldiers already in receipt of pension who may be required to serve again.
- **WO 69: 1803–63** This series contains service records for soldiers of the Royal Horse Artillery. They are arranged by service number, which can be ascertained from the indexes found in WO 69/779–782, 801–839.
- **Kilmainham Hospital pension registers** This hospital was the forerunner to the Chelsea Hospital, being opened in 1679 just outside Dublin to administer pensions to Irish soldiers. The records for this hospital are kept separately in WO 118 (1704–1922) and WO 119 (1783–1822). The records are arranged by date of admission to pension.

There are also two series for mis-filed documents and they should be checked if you are unable to find any record in the main series. They are also organized by discharge date, by surname only:

- 1843–99: WO 97/6355–6383
- 1900–1913: WO 97/6323–6354

Other Relevant Records for Soldiers

The War Office created other sets of records apart from pension records that can also be used to trace a soldier. These can be useful as you may be able to find a soldier regardless of whether or not he received a pension.

Description Books

As suggested by the title, these books give the physical attributes of the soldier concerned. They also provide details as to his age, parish of birth and occupation. They form two main series:

- **WO 25/266–688 (1778–1878)** These are the description books for the regiments, with the majority only surviving for the early nineteenth century.
- **WO 67 (1768–1913)** This series is for the regimental depot description books. However, books only survive for a few regiments (as detailed in the online catalogue).

Pay Lists and Muster Rolls

This is the main set of records to search for your ancestor if you are unable to find any service history. Each regiment compiled on a monthly or quarterly basis the names of each and every serving officer and soldier in that month or quarter and where they were stationed at that particular time. Additionally, they list when a soldier first enlisted and when he was discharged. They may also contain birthplace details of soldiers. From 1868 to 1883 the Musters would also have a marriage roll listing wives and children of serving soldiers. It is therefore possible to trace the career of any individual soldier, his start date, any promotions, where he served and his final discharge date. However, this can only be done if you know the regiment your ancestor served in and an approximate time period.

The first muster rolls and pay lists date from approximately 1730, although not every regiment kept a muster this far back. They are in a number of series arranged by regiment and chronologically:

- **WO 12: 1732–1878** This is the main series containing Musters for most regiments.
- **WO 10, WO 69, WO 54: 1708–1878** These series contain the Musters for the Royal Artillery.
- **WO 11, WO 54: 1816–78** The Musters are separate for Royal Corps of Sappers and Miners until their merger in 1856.
- **WO 13, WO 68: 1780–1878** This series includes Musters for Militia and Volunteer forces.
- **WO 16: 1878–98** After 1878 the majority of Musters can be found in this series. Bear in mind that the Army was reorganized (thanks to the Cardwell Reforms) in 1881, when many regiments changed names or were amalgamated. Your ancestor's regiment may be known under a different name after 1881 and would have been filed under that name (this can be researched using the Army Lists). The last records are for 1898 as the War Office did not keep Musters after that time.

Army Records for the First World War

The start of the First World War had a large impact on the organization of the Army. It involved an enormous increase in numbers serving for both officer and other ranks, and almost every resident of the country was involved, in either a civilian or a military capacity. The demand for manpower was such that conscription had to be introduced in March 1916, after the early enthusiasm for volunteering gave way to the reality of the horrors of modern warfare. In total approximately 7 million soldiers served in the British Army during the conflict and approximately half of these numbers were conscripts. Most men served until 1919, as the armistice of 11 November 1918 was originally a truce and not an official end to the conflict. Hence soldiers did not begin to be discharged until 1919, when it was guaranteed that there would not be a resumption of hostilities.

Unfortunately, there is no certainty that you will find the service record for your ancestor. A large proportion of service records were destroyed during bombing by the Germans in 1940, during the Second World War. The survival rate depends on whether you are searching for an officer's service record or that of an other rank. It is estimated that approximately 65 per cent of records for the latter were destroyed in the bombing. Additionally, some of the documentation that does survive has been subjected to fire and water damage and may be diffi-cult to read. This will be described in more detail below.

Officer Service Records

During the start of the conflict officers still came from the privileged and gentlemen classes. However, as the war progressed, the large numbers of casualties led to a shortfall in the numbers of officers. Thus promotion from the other ranks became more commonplace and members of the lower middle classes and even some working-class men joined the officer classes.

> *You can trace individual officers through the Army Lists.*

The surviving records for officers serving in this period are at The National Archives. Of course it is still possible to trace individual offi-cers through the Army Lists, although for security reasons the Lists were less detailed than in peacetime. Additionally, the rapid promo-tions given to some soldiers would not always be recorded in the Army Lists. Unfortunately, the main service records were also destroyed by bombing in 1940 and only the supplementary series is now available. The supplementary series was made up from the 'correspondence' file, which each officer had in addition to their service record. These files vary in length depending upon the length of service for each officer. In total The National Archives holds over 217,000 records and the major-ity can be found in two separate series:

- **WO 339** This series is for officers who were serving in the Regular Army prior to the onset of the conflict. It also contains the files for those who were given a temporary commission during the war, along with those commissioned into the Special Reserve of officers. As such it contains the majority of records, approximately 140,000 officer files.
- **WO 374** The smaller of the two series, it contains files for those commissioned into the Territorial Army and contains approximately 77,000 individual files.

A far smaller series, WO 138, holds the files of the most famous and notable officers of the First World War, such as Prince George, Duke of Cambridge, and the famous poet Wilfred Owen.

These series have been indexed and can be searched online in The National Archives catalogue. However, the indexes only give the surname, first initial and rank and, unless the surname is very uncom-mon, you are likely to have numerous entries for the surname. Hence, you will have to use WO 338 to identify the correct entry. WO 338 is an index to the records held in WO 339, but is not available online and will have to be consulted onsite.

Service Records for Other Ranks

Service records for other ranks are also to be found in two series held at The National Archives:

- **WO 363** This is the main series of records. The records were the ones that were subject to enemy bombing in 1940 and, therefore, are sometimes referred to as the 'burnt documents'. They originally contained records for all soldiers who served between 1914 and 1920 and were either killed in action or were demobilized after the end of the conflict. Due to the bombing in 1940 only a small percentage of the original records survive (approximately 20–25 per cent in total). The amount of damage done to each file varies, some pages being only slightly affected whereas other pages are almost illegible.
- **WO 364** These records are known as the 'unburnt documents' and represent a sample of pension records awarded to soldiers after discharge (either for length of service for those soldiers signing up before 1914, or for those injured during the conflict and awarded a disability pension). They represent only about a 10 per cent sample of all such awards, and consequently the collection is significantly smaller than WO 363.
- **WO 400** The series for service records for the Household Cavalry including for the First World War. Service records for Foot Guards are also amongst this series but are not currently at The National Archives.

The length of each individual service file varies considerably depending on how much they were damaged or the amount of correspondence each file generated. WO 363 and WO 364 have been microfilmed and it is not possible to view the original records. Each series is arranged in strict alphabetical order and not by regiment. However, depending on how much information you know about the soldier you are interested in, you may need the service number and regiment to identify the correct file. These can be found using the Medal Index Cards (see below). Additionally, to find out the activities of the regiment during the conflict you will need to consult the relevant war diary (see below).

Ian Hislop

Ian Hislop's family is steeped in military history, and he has always taken a keen interest in the subject, ever since he produced a project at school on the Boer War.

Both Ian's grandfathers served in the First World War. Having asked around the family, Ian discovered that his paternal grandfather, David Murdoch Hislop, served in the 9th Battalion, Highland Light Infantry, and was posted to the front line in 1918. On 29 September he saw action at the battle of Targelle Ravine, with many of his fellow soldiers losing their lives as his regiment suffered heavy losses on one single morning of fighting. After two weeks they finally broke through the German front line, hastening the enemy retreat which finally brought about the Armstice on 11 November. Ian was able to follow in his grandfather's footsteps by locating the unit war diary for the regiment at the National Archives in record series WO 95, and then looking at trench maps in series WO 297 that showed precisely where they were stationed on given days of the campaign. Sadly, though, David Hislop's service record did not survive.

Of perhaps greater interest was Ian's maternal grandfather, William Beddows, an army sergeant during the First World War who, being in his forties, trained new recruits for front-line action rather than serve himself. Nevertheless, he had a distinguished career and saw plenty of action. Beddows enlisted into the army in 1895, and within a few years was caught up in the Boer War, fighting five major campaigns including the notorious battle of Spion Kop, when 1,300 British soldiers were killed advancing towards Boer positions under heavy shell-fire. Much of this information was known to Ian's mother, but Ian was able to research in more detail about the campaign. Although Beddows did not keep a diary, records from other soldiers, located at institutions such as the Imperial War Museum and National Army Museum, reveal in vivid detail the horror

of the battle, when Beddows and his comrades were forced to use the bodies of their fallen friends as shelter from the Boer snipers.

Yet the strangest coincidence lay within his paternal family tree. Further genealogical research linked Ian with a distant relative named Murdoch Matheson, who was born on the island of Uig. Some of his medals were found within Ian's family, which gave crucial information about his regiment and some of the campaigns he served in from the clasps. On checking The National Archives online catalogue, his military service record was located. This confirmed that he enlisted in the 78th Foot Regiment in 1794; in turn, this information led to an examination of Muster Lists in record series WO 12, which revealed that Matheson also travelled to South Africa, landing at Cape Town nearly a century before Ian's grandfather went there to fight the Boers. The same Muster Lists and associated pay books proved that as part of a long career in the army Matheson travelled the world before coming back to Uig to settle down and raise a family. His discharge to pension survives in series WO 97, where his discharge papers reveal that he finally left the army in 1813 when he was 45 years of age – important biographical information to enable a search for parish records and other family material in Scotland.

The website www.ancestry.co.uk is placing the entire collection of WO 364 online, which subscribers to the site will be available to search.

Medal Rolls

▲ Military decorations and medals can provide information about the campaign an individual served in, as well as specific acts of gallantry.

▼ First World War campaign medal index card for Clement Attlee, showing that he received the Victory and British Medals, as well as the 1915 Star for service on the front line from June 1915 onwards.

Many people researching their ancestors' service records may have a medal from that conflict. Indeed every soldier of the First World War serving overseas was awarded two campaign medals – the British War Medal and the Victory Medal. Additional medals were given to individuals who served on particular fronts at particular times, such as the 1914 Star, or who sustained injuries through the conflict, such as the Silver War Badge.

The awards of these medals were recorded in a medal roll and an index card was also created to find a soldier's entry in the appropriate medal roll. These records form the closest surviving documentation for an Army Roll indexing every individual. The Medal Index Cards have been digitized and copies of these images are available on The National Archives website on Documents Online. It is possible to search the Index free of charge, although downloading an image will incur a charge. Often this is the only record that survives for a serving soldier, and the Index Cards can be used to ascertain the regiment and service number of your ancestor. Basic details relating to the soldier's period of service are also given, such as date of enlistment and discharge along with field of conflict. Separate medals, known as gallantry medals, were also issued to individuals who were rewarded for individual acts of heroism. These were indexed by separate cards, also available to search on Documents Online.

Medal rolls for earlier conflicts can be found at The National Archives. There are research guides available online at www.national archives.gov.uk to help you find the relevant record series.

Other Relevant Records

1.
Commonwealth War Graves Commission

This commission was created in order to remember the war dead of the First World War (and later wars). The commission has placed the Debt of Honour Register online on www.cwgc.org. This register lists all casualties along with when they died and where they were buried. Sometimes you may also find information about relatives. This register can be searched free of cost.

2.
War Diaries

These diaries are useful for those interested in the operational history of the First World War. Each battalion recorded their movements, activities and fighting on a daily basis. However, they are not personal recollections of individuals fighting in the war; these recollections are now at the Imperial War Museum. The war diaries can be found at The National Archives in series WO 95. Some have been digitized and placed online at Documents Online, others are on microfilm and some are still in original paper form. Some copies can also be found at the appropriate regimental museum, although they will be the same copies as found at The National Archives.

3.
Pension Records

The National Archives has various series of pension records. The main series is to be found in PIN 26, although only a 2 per cent sample now survives. The files relate to personnel applying for pensions even if they were not awarded one. It has been indexed by name on the catalogue and can be searched by name.

4.
Personal War Diaries

Although the majority of Army records are held at The National Archives, you can also find certain useful information at the Imperial War Museum. The Museum holds a 'documents collection' containing private papers and journals of serving officers and soldiers in the First World War (and other conflicts). It also has a number of records for war poets along with numerous other personal archives. Further information can be found on their website, at http://collections.iwm.org.uk/server.php?show=nav.oog002.

Service Records Post First World War

Surviving service records are only in the public domain for those men who were discharged shortly after the end of the First World War in 1918. If the individual served into the 1920s his service record is still retained by the Ministry of Defence. The dates the records are retained from vary depending on whether your ancestor was an officer or an other rank. If an officer served after 31 March 1922 his records will still be retained, and the same for soldiers serving after the end of 1920. Next of kin can apply to the Ministry of Defence directly at the following address:

> Ministry of Defence's Army Personnel Centre
> Historic Disclosures
> Mailpoint 400
> Kentigern House
> 65 Brown Street
> Glasgow G2 8EX
> Email: apc_historical_disclosures@btconnect.com

The British Indian Army

All of the above information has related to the regular British Army. However, the British also had a military force specifically for maintaining British rule in India, and there are separate archives relating to this force. The British Indian Army had officers of European origin and was under the control of the Viceroy. It was formed in 1859 after the Indian Rebellion of 1857, a successor to the army maintained by the East India Company (EIC). The trading company maintained forces in three regions in India (Bengal, Madras and Bombay) from the mid-eighteenth century onwards. These armies were known as 'Presidency Armies'. Each regiment of these three armies had European officers; however, they had other ranks of European or 'Native' origin. After the failed Rebellion of 1857, the East India Company was abolished and India was governed directly by the British Crown. Henceforth, European officers staffed the Bengal, Madras and Bombay armies until these sub-armies were united into one Indian Army in 1889. Additionally, the other ranks would solely be of Indian origin.

The British Library now holds the majority of records relating to the British colonization of India, including military records. They form part of the Library's Asia, Pacific and African collection.

British Indian Army Officer Records

As with the regular British Army, the best place to begin tracing the career of an officer of the Indian Army is through the Indian Army Lists. They were published from 1759 for the Bombay and Madras Presidencies and 1781 for Bengal. In 1889 a universal India Army List was published due to the amalgamation of all three forces. The British Library has the complete collection, and an incomplete collection can also be found at The National Archives.

Individual service records were only kept from 1900 onwards until 1950. They are held in the British Library in series IOL L/MIL/14 although they are subject to 75-year closure rules (from the date the individual entered the Army). Further information can be found in the information leaflets on the website of the British Library.

▲ The 9th Bengal Cavalry of the British Indian Army during the Sudan war of the late 19th century.

Other Rank Records

The British Library also has recruitment records, muster lists and other registers for European men serving as soldiers within the Indian forces. More information can be found in the appropriate information leaflet in the British Library. The National Archives also has certain lists and registers originally created by the East India Company.

Militia Records

The militia was a group of locally raised volunteer armed forces and can be traced back to Anglo-Saxon times. Records for such units survive from the Tudor period until the end of the Civil War; militia units were next raised in 1757 and continued to exist until 1907 with the passing of the Territorial and Reserve Forces Act, which formed the Territorial Force (which became known as the Territorial Army in 1921). The majority of surviving material concerning the militia can be found at The National Archives or local record offices.

Tudor and Stuart Militia Records

Militia units had been raised for local defence for centuries. The practice was codified in 1285 when the Statute of Winchester was passed and all males aged between 15 and 60 were expected to arm themselves in case they were called for service in the militia. Local Commissioners of Array or the Lord Lieutenants of the county were given the responsibility in the sixteenth century of ensuring these groups were raised adequately and it is their muster records that form the present-day archive for militia units during the Tudor and Stuart periods. These local authorities would forward their muster rolls to the Exchequer or the Privy Council and most of these rolls can be found at The National Archives.

The muster rolls are organized by county, hundred and then parish. The rolls list all male residents who were eligible to be called to arms if required. The earliest known rolls survive from 1522 and the amount of information given on a roll would vary. Some may provide individual ages, occupations and/or income (as individual wealth would determine what arms the men would be expected to provide). Other rolls may be comparatively brief and simply list the total number of eligible men in the parish. As mentioned, the majority of these rolls are at the National Archives (in the Exchequer series and the State Paper series), although some may also be found at local record offices and at The British Library. The easiest method of locating which muster rolls survive for a particular locality and where they are kept is by referring to Gibson and Dell's *Tudor and Stuart Muster Rolls – A Directory of Holdings in the British Isles* (Federation of Family History Societies, 1991).

Militia Records 1757–1914

In 1757 the Militia Act reintroduced militia regiments into each county of England and Wales. These regiments would serve only in Britain and Ireland and not abroad. Each parish had to provide a number of suitable males to serve but, as there were too few volunteers, a form of conscription was introduced. However, this system proved very unpopular and was abolished in 1831, with militia units continuing with volunteers thereafter.

Militia Lists 1757–1831

The surviving records of this conscription process are of great genealogical relevance as local Justices of the Peace or county Lord Lieutenants had to provide annual lists of all men aged between 18 and

45 and a separate list of those chosen to serve. These lists are to be found in local archives and some may give details of each man, his occupation and age and his marital status (and sometimes children too) for each parish. Unfortunately, lists of every parish have not survived. It is best to consult Gibson and Medlycott's *Militia Lists and Musters, 1757-1876* (Federation of Family History Societies, 2000) to ascertain what has survived for each locality.

Records at The National Archives

The National Archives hold the main service records of men who served under the various militia forces from 1769 onwards. The militia was organized on a similar basis to the regular Army, with officers and other ranks.

- **Militia officer records:** Service records for militia officers can be found in the regular Army officer records in series WO 25 and WO 76. Additionally, WO 68 has the records of militia regiments including returns of officers' services. Commissions can be found in HO 50 (1782-1840) and HO 51 (1758-1855). Relevant published sources include *Officers of the Several Regiments and Corps of the Militia* and the *Militia Lists*. The *London Gazette* also published details of appointments for militia officers.
- **Other rank records:** Relevant records for other ranks can be found in the following National Archives series:
  - ➤ Attestation papers can be found in WO 96 (1806-1915). These records provide a service record of each individual and may also give birth details. The papers are arranged by the regular army regiment the militia regiment was attached to, and then alphabetically. WO 97/1091-1112 contains the attestation papers of local militia regiments for the years 1769 to 1854. These have been catalogued by individual name and it is possible to search by name online in The National Archives catalogue.
  - ➤ The main series of records for the militia can be found in WO 68. This series (along with officer records) includes enrolment lists, description books and casualty details. The casualty records may also give details of marriages and children.
  - ➤ Muster and pay lists for the militia can be found in WO 13 (1780-1878).
  - ➤ Records of militia men who were entitled to pensions can be found in the registers in WO 23 (1821-29). Admissions into Chelsea Hospital are located in WO 116 and WO 117.

> *Militia lists may give details of occupation, age, marital status and children.*

➤ Militia men did not generally receive campaign medals but were awarded the Militia Long Service and Good Conduct Medal. The medal roll can be located in WO 102/22.

Volunteer Organizations

Along with information on earlier militia records, information on other volunteer units – and the records that survive for them at The National Archives – can be found in W. Spencer's *Records of the Militia and Volunteer Forces 1757–1945* (Public Records Office, 1997). Volunteer organizations include:

● Volunteers – raised in 1794, these were separate to the organized militia units but were disbanded in 1816, only to be revived in 1859 as the Volunteer Force. The cavalry equivalents of the Volunteers were the yeomanry and imperial yeomanry. Most of the relevant documentation, including musters, pay lists and officers' commissions are held at The National Archives, and are summarized in a research guide online at www.nationalarchives.gov.uk.

● Territorial Force – formed in 1907 when the Volunteer Force was merged with the Yeomanry, with the Territorial Army created in 1921. Documentation is held in relevant regimental museums, although officers' service records during the First World War will be held at The National Archives in series WO 374 (see above).

● Home Guard (Local Defence Volunteers) – service records are closed to the public for 75 years, and can only be obtained by those who served, or their next of kin, by writing to the Army Personnel Centre, Glasgow. There are some records at The National Archives, such as copies of the Home Guard Lists, histories of some regiments in series WO 199 and unit war diaries for the Second World War in WO 166, which provide details of operations and movements.

Tracing a Campaign
Battalion War Diaries

For researchers interested in discovering more about the military history of the First and Second World Wars, and the part each unit played in fighting, the best place to turn to is the war diaries (mentioned above). They were kept by individual battalions during the First World War, from the years 1914 to 1922. A junior officer would be responsible for recording the daily movements of the battalion, the operations of conflicts they were involved with and other relevant information (including lists of casualties, although names of officers would only be mentioned and other ranks would mostly be listed by total number killed only). As such they are an invaluable detailed source for understanding the military conflict in a very microscopic fashion. They can be of great use if your ancestor's service papers do not survive, as the war diaries can act as a substitute service history.

Copies of battalion war diaries were sent to the War Office at the time, and these are kept at The National Archives (series WO 95). However, individual units also retained copies and these can be found in the appropriate regimental museum (where other records for regiments may also be found).

War diaries were also kept during the Second World War and are also retained at The National Archives. They are arranged in a number of series in the WO class arranged by the Order of Battle and then by Command.

Operational Records

It is also possible to trace the history of a campaign or military operation from a wide variety of papers, reports and observations filed at the War Office and other government departments. Most of these official papers are now deposited at The National Archives, and help can be found via research guides online at www.nationalarchives.gov.uk.

Muster Rolls

Another useful source for tracing campaigns is the muster rolls (described above, page 165). As well as tracing an individual soldier's career they are also useful in finding out the details of the campaigns each battalion was involved in. The lists would state exactly where each battalion was stationed and also give details of any casualties or deaths that occurred during the quarter.

Published Regimental Histories

Most regiments have a proud tradition, and the oldest were created many centuries ago. As such individual regiments often have specific histories written about them which will often detail which campaigns the regiment was involved in and the contribution made by that

▲ A statement of service in the British Army for William Gallas, accompanied by his Waterloo medal for taking part in the battle in 1815.

regiment. They may also contain biographical details of eminent soldiers of that regiment. These histories are often published and can be viewed at The National Archives, The Imperial War Museum, The National Army Museum and other large reference libraries. It is also possible to purchase certain histories, if required, particularly from the regimental museum concerned. A good guide to the range of regimental histories can be found in *A Bibliography of Regimental Histories of the British Army* compiled by Arthur S. White (London Stamp Exchange, 1988). Additionally, individual regimental museums should also contain this information.

War Histories

There have also been numerous publications on the many campaigns and wars the British Army has been engaged in through its history. Many will have detailed information on specific battles along with helpful footnotes leading you to primary documents if you wish to research these. These books can be found in the major archives and libraries relating to army history, and in larger reference libraries. Many bookshops will also stock the more popular histories. If applicable, it is also worthwhile researching conflicts through contemporary sources such as newspapers to provide information that may not have been included in later histories.

Regimental Museums

Along with specific histories, many regiments also have museums preserving aspects of each regiment's history. There are 136 military museums scattered throughout the United Kingdom at present. Many of these museums will have regimental records relating to the campaigns they were involved in. They may also have other paraphernalia relating to their campaigns, such as photographs, medals and uniforms worn at the time. Other useful research collections include regimental newspapers and journals. Certain museums may also have donations from ex-soldiers deposited with their holdings. It is possible to find a list of these museums and their contact details at the following website, www.armymuseums.org.uk.

Private Accounts

Many soldiers would write private journals about their experiences during battle. These provide a very personal and unique insight into the campaign in which they were involved. If your ancestor was in the Army such a diary may be found amongst his personal papers. Otherwise, as mentioned above, the Imperial War Museum holds papers that were donated by individual soldiers.

> *Private diaries give a unique and very personal insight into a campaign.*

Life in the Army
National Army Museum

Along with the many regimental museums dedicated to military history, the National Army Museum in Chelsea, London (www. national-army-museum.ac.uk), is the central location for the history of the British Army as a whole. The Museum has significant archives and collections relating to over 500 years of military history. This includes over half a million images from the 1840s onwards. Other items for anyone interested in researching military history include a large collection of army medals and uniforms and over 43,000 printed books including journals, regimental histories and biographies of well-known soldiers. The Museum also has a collection of the different types of weapons used in various conflicts through the ages. There is also a collection of fine and decorative art including paintings, sculptures and ceramics obtained by the British Army through its history. The Museum has been publishing a quarterly journal relating to British Army history since 1921 through the Society of Army Historical Research. These many collections held at the Museum make it indispensable for anyone wishing to gain a fuller picture of the type of life in the Army that your ancestor would have had.

Imperial War Museum

This is another museum dedicated to preserving the history of life in the Army, specifically relating to conflicts from the First World War onwards. It was established in 1917 to record the story of the First World War and the contribution made in that conflict by soldiers from the British Empire. The main museum is based in Lambeth, London. However, there are other branches of the museum at five other sites throughout the country. Along with the personal journals mentioned above, there are collections of medals, firearms, significant film and video archives, and the sound archive has many interviews with ex-soldiers and historic radio broadcasts. The Museum has placed a lot of information online, which at www.iwmcollections.org.uk. As the Museum is related to the history of conflict during the twentieth century it is particularly useful for anyone researching an ancestor who fought during this time.

War Dead

As well as the Commonwealth War Graves Commission, described above, there are other ways to find out more about your ancestors who fell in battle.

Office of National Statistics' Lists

The Army kept separate lists of births, marriages and deaths relating to serving soldiers. Births were recorded from 1761 to 1994, marriages from 1818 to 1994 and deaths from 1796 to 1994. These registers are now found at the General Record Office and the indexes can be searched online at www.findmypast.com. The registers are not available to view, but if an entry is found in the index it is possible to order the appropriate certificate to obtain the details of death.

Memorials

Many cities, towns and villages have erected war memorials to honour their dead. These became especially prominent after the First World War due to the huge numbers of casualties every village suffered. They usually have details for local residents and can give clues to when a soldier served and where or when he died. It is estimated that there are approximately 70,000 such memorials throughout the United Kingdom. The Imperial War Museum has worked in partnership with English Heritage to create a centralized database of these memorials, and the United Kingdom National Inventory of War Memorials has

▼ The Cenotaph in Whitehall, London. Commemoration of the war dead after the First World War also took place via memorials that were erected in villages, schools and places of work.

placed this database of the 55,000 war memorials that have been recorded so far online. The database can be searched free of charge at www.ukniwm.org.uk. The database covers memorials from many wars, starting in the tenth century. It is not possible to search the database by name as the memorials themselves have been listed individually. However, if a memorial was erected to commemorate an individual soldier then that has been listed by his name. Many family history societies have name indexes of war memorials that can be used.

Additionally, there is a database for those who were commemorated for the First World War, and this can be searched by name on the Channel 4 website Lost Generation. The site is dedicated to the history of the First World War, following on from the series of the same name. It is possible to search by the individual soldier's name on the following page free of charge: www.channel4.com/history/microsites/L/lostgeneration/search/person.html.

Obituaries

It is often worthwhile spending time researching any obituaries that may have been written after a soldier died. This is particularly true of high-ranking officers as local newspapers and, sometimes, national newspapers too would publish obituaries. Obituaries can give useful summaries of the deceased's career and can give clues for further research into original sources. The newspaper archive is now part of the British Library's collection, based in Colindale, north-west London. It is possible to search *The Times* online archive through most local libraries.

Visiting Battlefields

If you are particularly interested in the military history of a conflict then visiting the battlefields can be a very enlightening process. There are specific tour groups for those interested in visiting the more popular battlefield sites from the First and Second World Wars. Details of many of those operating such tours can be found online. It is also possible to visit these sites without using a tour guide, but it would be necessary to have a detailed guidebook to make the visit worthwhile. There are also tours for those interested in visiting famous UK and Irish battlefields along with those in Northern France and Belgium. Touring a battlefield is a very good way to help visualize the actual fighting your ancestor would have been involved in.

CHAPTER 10

Military Ancestors: The Royal Navy

Alongside service in the army, the other main military destination for many of our ancestors, particularly those living along Britain's hundreds of miles of coastline, was the Royal Navy. This chapter explains how you can locate relevant service papers, where they survive, as well as find out more about life on the ocean wave from official material such as ships' logs, admiralty correspondence and pension records, as well as in museums, archives and research institutions around the country.

A Brief History of the Royal Navy

It was in the reign of King Alfred (871–901) that the first known naval fleet in this country was established to defend against Danish invaders. Hence, King Alfred is credited as the founder of the Navy. About a century later, Edward the Confessor established the institution of the Cinque Ports, key ports on the south coast where merchant vessels could be refitted for military purposes in defending against Norse attacks. As this conversion process was relatively easy, there was no specific requirement for a separate military navy, although King John commissioned the construction of a large military fleet in 1212 to attack France. The year 1340 marked the first occurrence of a major naval engagement at the Battle of Sluys off the coast of Flanders, though previous large-scale skirmishes had frequently taken place for over a century beforehand. From an institutional point of view, though, the origins of the navy as a recognized 'department of state' come with the

> Henry VIII is credited as the 'Father of the Navy'.

appointment in 1391 of the Earl of Rutland as the first ever Lord High Admiral.

The origins of the modern Navy are to be found in the Tudor period. The first dry dock was constructed in Portsmouth by Henry VII, who personally owned a fleet of seven ships. His son Henry VIII built a large number of fighting ships including the *Henry Grace a Dieu*, the largest warship at the time in 1514. Henry VIII also established the Navy Board in 1546 as well as the Office of Admiralty. He is thus credited as the 'Father of the Navy'.

The Tudor monarchs were well aware of the benefits of exploratory travel and the potential spoils of the New World and were keen to develop a suitable naval force to undertake such voyages. Of course it was during the Tudor period that the Navy faced one of its most famous conflicts against the Spanish Armada in 1588. The Navy did not at this time have a permanent staff, either officers or ratings. Instead men were recruited to serve when required.

▲Ratings training on board a Royal Navy ship, 1904.

The reign of Charles II led to further developments in the history of the Navy; it was given the official title of the 'Royal Navy' in 1660, and the Royal Society of London was also established during the reign of Charles II to help further knowledge of seafaring and scientific knowledge. The Royal Navy became an increasingly important fighting force in the wars against the Dutch during the mid-seventeenth century. Developments in fighting techniques also meant that converting merchant ships to military vessels was no longer an effective option, and a unique military Navy became essential. It was Samuel Pepys who instituted the Naval Discipline Act in 1661, bringing in strict discipline rules and codes of conduct marking the beginnings of a professional naval force.

The growth of the British Empire during the eighteenth and nineteenth centuries placed new demands upon the Royal Navy. Dockyards were built in various parts of the world, ensuring the fleet remained rapidly mobile. Scientific innovations in seafaring such as sophisticated means of navigation, accurate charts and powerful weaponry were embraced by the Royal Navy, helping it to dominate the seas. Many buildings were constructed at the turn of the eighteenth century to administer the Navy, such as the dockyards in Portsmouth, Chatham, etc., the Royal Hospital in Greenwich and the

Admiralty building in Whitehall (in 1699). This era was marked by numerous conflicts with various European powers, and the Royal Navy established its reputation as the world's foremost maritime power after the Battle of Trafalgar in 1805. The growth of the Navy in this period led to a large increase in the numbers employed in the force, both ratings and officers, up to the end of Napoleonic Wars. However, once this conflict was over, the Royal Navy did not face any large-scale war until the First World War. The primary role of the Navy changed to policing and protecting trade routes and the numbers employed decreased considerably.

The living and working conditions for ratings were far from comfortable. Living space on the ships was very cramped, food was far below adequate and pay was minimal. The work itself was extremely physical, exhausting and also hazardous, with injuries not being uncommon. The risk of disease was also ever present for the average rating.

Unsurprisingly, officers fared better than did ratings whilst serving on ships. Although officers had to pass a lieutenant's exam to be considered as an officer, promotion afterwards would mostly be by 'selection'. This ensured nepotistic practices, as officers would be more likely to be promoted if they came from influential families. They would be mostly from privileged backgrounds and considered gentlemen.

The first steamships began to be used in the nineteenth century; their running depended upon a more specialized workforce of engineers and electricians. The Admiralty established training schools to equip its men with the required skills and the Royal Navy became an increasingly professional force in the twentieth century.

Individual Navy Service Records

The National Archives holds the majority of records for the Royal Navy, mostly in its Admiralty series (ADM). Additional information may also be found amongst the archives of the National Maritime Museum in Greenwich. Similar to the Army records, service records are divided between naval officers and ratings. The main series for officer records begin in 1660. Records for ratings start much later, in 1853, although it is possible to research the career of a rating prior to that date if details of the ships the individual served in are known. Unlike for the Army, the First World War does not mark any new opportunities for searching out service records as many series continued after 1914. However, Royal Navy records are more complete than Army records.

Officer Records Pre 1660

Prior to 1660 records of individuals serving in the Navy were not kept on a systematic basis. Any surviving documentation will be found amongst the appropriate State, Chancery and Exchequer series held at The National Archives.

Officer Records Post 1660

Broadly speaking the Royal Navy employed two types of officers: commissioned officers and warrant officers.

The main ranks of commissioned officers ranged from sub-lieutenant, through lieutenant, captain, commander, commodore and rear-admiral to admiral at the very top. They were recruited from 1660 onwards through the grant of a royal commission after an examination had been passed.

Warrant officers included masters, engineers, surgeons, boatswains, carpenters and other ranks that were involved in the practical aspects of running a ship. These individuals held their rank by a warrant and were also subject to examination. Warrant officers became more likely to be promoted to commissioned officers during the latter half of the nineteenth century and into the twentieth.

Records for these individuals may be found in a myriad of sources.

▲ Naval service records can provide biographical information such as date and place of birth, name of father and name of spouse, as well as a list of ships on which someone served.

Published Sources

For the earliest periods it may be easier to research an individual by consulting the published sources listed below. All of these sources are available for consultation at The National Archives, the National Maritime Museum, the British Library and other large reference libraries. You may be able to view certain publications online.

- *The Commissioned Sea Officers of the Royal Navy, 1660–1815* by Syrett and DiNardo: This three-volume book lists all the commissioned officers serving in the Royal Navy for the above years, along with their rank and the dates they served in each rank. The book is now available to search online at www.ancestry.co.uk.
- *Naval Biographical Dictionary* by William R. O'Bryne (1849): This publication lists all serving officers who were alive at the time of publication, from the rank of lieutenant to admiral. Each entry varies in detail depending on each officer. This book has also been digitized and can be searched online as part of the Ancestry website.

▲ Early ships' musters only provide basic information, but can be used to trace someone's movements from ship to ship and to compile a career record.

- *Biographia Navalis* by John Charnock (1797): Charnock's book is a comprehensive survey of all naval officers serving between 1660 and 1797, from the rank of captain to admiral.
- *Lives of the British Admirals* by Dr John Campbell: A list of all admirals serving up to 1817.
- *Royal Naval Biography: or, Memoirs of the services of all the flag-officers, superannuated rear-admirals, retired-captains, post-captains, and commanders, whose names appeared on the Admiralty list of sea officers at the commencement of the present year, or who have since been promoted; illustrated by a series of historical and explanatory notes … With copious addenda* by John Marshall (1823–25).
- *The Navy List*: This publication is based on the same principles as the Army List. It was started in 1796 as *Steel's Navy List* and was published quarterly from 1814 onwards as *The Navy List*. Similar to the Army List it details all officers of the Royal Navy from the rank of lieutenant onwards along with the date of each promotion. After 1810 it records each naval ship and which officers were serving on each. The National Archive series ADM 177 has confidential Navy Lists published during the two world wars.
- An unofficial Navy List was also published from 1841 to 1856, as the *New Navy List*. It is particularly useful as along with all the information provided in the official lists it gives biographies of officers.

Original Service Records

The National Archives holds numerous series of records that contain service history information and, depending on the date your officer ancestor served and the rank he held, you may have to consult more than one of these series. Below is a list of the types of records held by The National Archives and the information each series contains.

Register of Officers' Returns

There are three main series that contain service register histories after the Royal Navy started a central record-keeping system from the mid-eighteenth century onwards:

1.

ADM 196: Officers' Service Records (Series III), 1756–1966

This is the main series of service records compiled by the Admiralty. Although the dates begin from 1756 the majority of records are from between 1840 and 1920. However some records may well go back to the mid-eighteenth century. The series has been microfilmed and there is an index to the series in The National Archives, although it is not complete. If your ancestor is not mentioned in the card index, he may still be in the series. There are numerous indexes within the series itself that can also be consulted. The registers record the details of the individual's career (including promotions and ships served on), birth, residence and marriage details.

The records cover all commissioned officers entering the Royal Navy until May 1917, and warrant officers till 1931. If your ancestor entered the service after these two dates the records will still be retained by the Ministry of Defence. They are only available to next of kin and can be requested from two different departments, dependent on whether the officer concerned is alive or dead:

The Directorate of Personnel Support (Navy) – for deceased personnel
Navy Search TNT Archive Services
Tetron Point, William Nadin Way
Swadlincote, Derbyshire DE11 0BB
Tel: 01283 227910
Fax: 01283 227942
Email: navysearchpgrc@tnt.co.uk

Data Protection Cell (Navy) – for personnel still alive
Building 1/152
HM Naval Base Portsmouth
Victory View, Portsmouth PO1 3PX
Tel: 02392 727381
Fax: 02392 725829

2.

ADM 29: Royal Navy, Royal Marines, Coastguard and related services: Officers' and Ratings' Service Records (Series II), 1802–1919

Although this series is mainly for ratings it also includes warrant officers amongst its records. The records were produced for individuals seeking pensions or medals, as they had to provide a record of service. Hence the Navy Pay Office produced these certificates of service. These are also available on microfilm.

3.

ADM 9: Survey Returns of Officers' Services, 1817–48 and ADM 11: Officers' Service Records (Series I), 1741–1903

The information found in ADM 9 and ADM 11 constitutes the various surveys of officers that occurred in the nineteenth century. One survey was carried out in 1817, shortly after the end of the Napoleonic Wars. As hostilities had now ceased the Navy found itself vastly overstaffed and had to decide which officers to retain in peacetime. This was done by sending out circular letters to individual officers and asking them to fill out their details of service. A similar survey was also carried out in 1846.

The records do not include every serving officer as not every officer received the letters, nor did every officer choose to fill in the required information. A large number of returns were also lost. Nevertheless, the surveys that do survive can be found in ADM 9 or ADM 11 (and in ADM 6 for warrant officers). Indexes can be found in ADM 10/1–7.

Naval Officer Pay Registers

The Navy Pay Office was responsible for compiling pay registers for officers. The information contained in these registers was used to compile certificates of service and for pension purposes. As such they contained a full, but basic, record of each officer's service with the Royal Navy. There were registers for officers on full pay or half pay (for officers who were paid a retainer to keep them on reserve or a type of pension). The registers were kept from 1668 until 1920.

- Full pay registers can be found in ADM 24 (1795–1872) and ADM 22 (1847–74)
- Half-pay registers are in ADM 18 (1668–89), ADM 25 (1693–1836), ADM 23 (1867–1900) and PMG 15 (1836–1920, containing a name index for each volume)

Succession Books

It is also possible to trace an officer's career using succession books. These books detail which individual held a particular position on each individual ship during the years each volume covered. The books would also state the previous ships each officer served on for both warrant and commissioned officers. The books are indexed by ship and by name. They cover the years 1673 to 1849 and can be found in ADM 6, ADM 7, ADM 11 and ADM 106.

Passing Certificates

Another useful source for researching naval officers is the surviving passing certificates. As the Navy required a skilled workforce to function properly, officers would have to sit a number of examinations in the initial stages of their careers to demonstrate competence. The certificates are held in a number of series depending on the rank of the officer. The certificates start from 1660 and go up to the early twentieth century.

- The largest collection of certificates is that of lieutenants' passing certificates, which cover the years 1691 to 1902. As the rank of lieutenant was the most junior of the officer ranks, most officers ascending through the Navy would have started as this rank. The certificates contain details of service to the date of examination and may also include copies of baptismal records. They can be found in The National Archives series ADM 107 (1691–1832), ADM 6 (1744–1819) and ADM 13 (1854–1902). There is an index to the majority of these certificates compiled by Bruno Pappalardo, called *Royal Naval Lieutenants: Passing Certificates 1691–1902*.

- Masters were personnel who had specialized knowledge on the ship they were stationed on and usually spent lengthy periods on one ship only. Their passing certificates can be found in ADM 106 (1660–1830) and ADM 13 (for the second half of the nineteenth century).
- Gunners were responsible for the artillery or guns on a ship. Their passing certificates can be found in ADM 6 (1731–1812, not complete) and ADM 13 (1856–67).
- Pursers (later named paymasters) were involved with the financial aspects of the ship, paying seamen and also ensuring supplies were adequate in the ship stores. Their certificates can be found in ADM 6 (1813–20) and ADM 13 (1851–89).
- Boatswains' responsibility was for the sails of the ship and also for summoning other seamen to their duties. Their certificates are held in ADM 6 (1810–13) and ADM 13 (1851–87).
- Surgeons' passing certificates are held in ADM 106 for the eighteenth century.

> *Succession books and passing certificates can be used to trace an officer's career.*

Confidential Reports

These reports were compiled from commanding officers' comments on the suitability of officers seeking promotion. They can be very enlightening as comments made by the commanding officers were often very frank and honest. They were kept from 1884 to 1943 and can be found in ADM 196. They are arranged by rank and by date of promotion, although some volumes also contain name indexes.

Naval Records for Ratings

The 'other ranks' personnel of the Royal Navy were known as ratings: the non-officer seamen on the ship. There were numerous different ratings within the Royal Navy through its long history and the line between ratings and warrant officers would not always be clearly drawn. Until 1853 ratings were not recruited by the Royal Navy on a permanent basis. Instead, individuals would sign up for one single commission on a particular ship. Nor was the system of recruitment on a strictly voluntary basis, with the system of impressments (commonly known as 'press-ganging') not uncommon, especially in times of war. Men were often conscripted or impressed without consent or prior knowledge; the system started in 1664 but was particularly popular in the eighteenth and early nineteenth centuries. It was usually sailors of the merchant navy who were sought by recruiters as they had experience of sailing. Impressments were not used after 1814 as the ending of the Napoleonic Wars meant the Navy no longer required so many men.

In 1853 the system of continuous service was introduced and ratings signed up to serve a fixed period within the Royal Navy. After 1853 it is relatively straightforward to trace the career of a rating, as the Navy kept an individual service record for each man. However, prior to 1853 finding records can be more complicated and a variety of sources may be needed.

Naval Rating Records Prior to 1853

Searching for a rating in this period involves turning to muster books and pay books, as no service records for ratings were kept. However, this is only feasible if you know which ship your ancestor served on and the approximate date. Muster books were kept for each ship listing each person aboard. Alongside the name of each rating are birth details and when he joined the ship. Sometimes the muster will also note the previous ship the sailor joined from and which ship he was discharged to. If all this information is given it is possible to trace an individual's career from first ship to final discharge. Muster books would be kept on an eight-week basis and a separate general muster would be kept annually. Muster books can be found in The National Archives, in series ADM 36–40, ADM 41, ADM 115 and ADM 117, covering the years 1667 to 1878. The ratings are not listed in alphabetical order, and unless there is an index, searching for a particular name can be a somewhat lengthy process.

Pay books can also be used to trace individual ratings. They were kept from 1691 to 1856 and are found in ADM 31 to ADM 35. The information is very similar to that found in muster books. However, these books usually have more indexes then those found in muster books, as the pay lists included 'alphabets' (whereby names would be indexed by first initial of surname only) from 1765 onwards, far earlier than musters did.

There are other options that can also be used when tracing ratings. The series ADM 29 has been discussed above relating to warrant officer records. It also contains service history information for ratings as certificates of service were compiled for both groups in ADM 29/1–96. As mentioned, the records in this series were created for seamen seeking pensions, medals or gratuities to detail their service within the Royal Navy. For ratings they were compiled from the pay books

> *Muster books were kept for each ship, listing every*

mentioned above. The series begins in 1802 although an individual's service may predate that year as the date relates to when the certificate of service was issued. Some of the series has been fully transcribed and can be searched online by name on The National Archives catalogue.

Naval Ratings' Service Records Post 1853

In 1853 a system of 'continuous service' was introduced and ratings were now guaranteed work for a number of years, rather than signing up on a ship-by-ship basis. These service records are held at The National Archives, until 1923. Each rating has an individual service record stating which ships he served on along with personal details (such as birth details and residence).

Men (over 18 years old) would sign up initially for a period of ten years. Boys younger than 18 could also join with parental consent, and so could those already in the Royal Navy in 1853 (signing up for a period of seven years). Ratings were also entitled to sign up again after their initial term for a further ten years, thereby entitling them to naval pensions. Each recruit to the new system would be given a 'continuous service' number and their service history would be entered into a register on an ongoing basis. The registers were organized by the given service number. These registers can now be found in The National Archives series ADM 139 from 1853 to 1872. Indexes to retrieve the continuous service number and find the appropriate record are also contained amongst series ADM 139. ADM 139 is currently available to view only on microfilm but there are plans to digitize the series and

◄ British sailors watching the surrender of the German fleet at Scapa Flow in the Orkneys, the principal British naval base, on 21 November 1918.

make it available online as is the case with the later series, ADM 188 (see below).

The ratings' registration system was subject to certain alterations as a new numbering system was introduced from 1 January 1873, and thereafter the service records are held in another series, ADM 188. ADM 188 begins in 1873 and ends in 1923. These dates relate to when a rating was enlisted into the Royal Navy and, therefore, service records may contain information for those still active in the Royal Navy after 1923, up to 1928.

The numbering system was subject to further alteration in 1894. Previously, each rating was given a number on a sequential basis, regardless of what job he held. From 1 January 1894, specific number sequences were assigned for different types of ratings. For example, stokers would have any service number from 276001 to 313000. The numbering system was further modified in 1 January 1908 to avoid any overlap in service numbers. Henceforth service numbers were also prefixed with a letter. The nature of the information in the earlier service records is similar to that found in ADM 139, containing birth details and also physical attributes. The later records after 1892 will provide additional details such as remarks on the individual's character and more details as to their physical appearance.

As mentioned, the series ADM 188 is now available online on The National Archives website at www.nationalarchives.gov.uk/documentsonline. It is possible to search the entire series by name, birth details and also service number. However, due to the changes in the numbering system it is perhaps easier not to search by service number only. As always, the index can be searched online free of charge, with a cost being incurred to view the actual document.

Naval Pension Records

There was no centralized pension system for either officers or ratings until the nineteenth century. Prior to that a variety of different systems were in place, depending on whether your ancestor was an officer or a rating. Four main institutes were responsible for administering pensions: the Navy Pay Office, the Chatham Chest, Greenwich Hospital and the Charity for the Payment of Pensions to the Widows of Sea Officers.

Officers' Pension Records

As there was no official retirement or pension system in the Royal Navy between the seventeenth century and the mid-nineteenth century, officers were either placed on half pay or continued to work regardless of their ability to carry out their roles effectively any more.

The Admiralty made certain provisions at certain points, awarding pensions to specific officers. For example, thirty of the most senior lieutenants were awarded pensions in 1737. By and large, however, many officers were promoted and then retired on a better rate of half pay on the understanding they would no longer be required for service. This system continued until a more effective method of retiring officers was introduced from 1836 onwards. Prior to that, specific pensions were only awarded on an individual basis for deserving cases.

The Admiralty also awarded pensions to officer's widows. Wives of officers who died on active service were entitled to a pension awarded from the Compassionate Fund from 1673 onwards (extended to next of kin in 1809). This was further extended to warrant officers' widows in 1830.

The Chatham Chest, founded in 1581, was used to award pensions to warrant officers. In 1814 the Chatham Chest was merged with the management of the Greenwich Hospital and pensions were now awarded by Greenwich Hospital only. Another independent body (responsible for paying pensions to officers' widows) was established in 1732, known as the Charity for the Payment of Pensions to the Widows of Sea Officers. The funds were raised from public funds and through a salary contribution from officers. Surviving records are now found in ADM 6.

After 1836 Greenwich Hospital became the only institute involved with pension payments, when the Admiralty took over the running of the service (see below for further details). Henceforth records can be found in either the Admiralty series or in the Paymaster General records (PMG). For further details please refer to *Tracing Your Naval Ancestors* by Bruno Pappalardo.

Ratings' Pension Records

It was only after 1859 that pensions were awarded to ratings as a matter of right. Prior to that there was no guarantee that a rating would be entitled to one.

The Chatham Chest, mentioned above, awarded pensions to ratings as well as to warrant officers. The records are held at The National Archives, in series ADM 82, covering the years 1653 to 1799.

The Royal Hospital in Greenwich was the main institute responsible for administering pensions. It was established in 1694 by a royal statute of William and Mary for the relief and support of seamen and to

CASE STUDY

Robert Lindsay

Robert Lindsay had long suspected that his maternal grandfather, Raymond Dunmore, had been actively involved in the Royal Navy during the First World War. Family rumours had been passed down that he had served on board the *HMS Prince of Wales* during the major naval engagement of the war, the Battle of Jutland on 31 May – 1 June 1916. Indeed, it was claimed that he had been 'blown up' twice during his career!

Given some basic information about Raymond, such as his date and place of birth (from his birth certificate), it was possible to investigate these stories by obtaining his Royal Naval service record from The National Archives. Today, these records (in record series ADM 188) are available to search on Documents Online via the website www.nationalarchives.gov.uk and can be downloaded for £3.50, but indexes books are available onsite, where the documents can be viewed for free.

The information in his service record showed all the ships he served on from enlistment to discharge, including *HMS Prince of Wales*. But further research at The National Archives in the official Captain's Logs in record series ADM 53 confirmed the movements of the vessel, and showed that it played no part in the Battle of Jutland. The truth was even more alarming for

▶ Robert's grandfather, Raymond Dunmore, as a young man, in naval uniform.

Robert's family, as the documents revealed where he had actually been stationed.

At the time that Raymond Dunmore was on board, the *HMS Prince of Wales* was involved in the Dardanelles campaign, an attempt to weaken the Ottoman Empire's grip on the region and open up access once more for Allied shipping and troop movements. Although originally conceived as a naval campaign, thousands of troops – the majority from Australia and New Zealand, the ANZACS – were to be landed on the Gallipoli peninsula with an overall aim to capture Constantinople.

According to the ship's logs, and official printed histories of the campaign, *HMS Prince of Wales* was one of three vessels that launched 48 landing craft on 25 April 1915 as part of an attempt to secure a beachhead at Anzac Cove. It is possible that Raymond was part of these landing parties, making sure that the four steamboats, each taking three rowboats, launched from *HMS Prince of Wales* successfully made the beach. If so, this is the moment when he was most likely 'blown up', due to the heavy Turkish defensive shelling of the boats as they made their way to shore. Many hundreds of soldiers and seamen lost their lives during these frantic attempts to leave the water, and it is perhaps unsurprising that Raymond chose not to elaborate too much to his family about his experiences during this phase of the war.

support their wives and children. Similar to the Chelsea Hospital for army pensioners, Greenwich Hospital had both in-pensioners living on the premises and out-pensioners. Both ratings and officers were looked after by the Hospital. There is a series of records for the admission of in-pensioners between 1790 and 1865 in ADM 73. These registers provide detailed information on the individuals who stayed at the hospital. The majority of ratings awarded pensions by the Royal Hospital did not live in the Hospital but in their own accommodation as out-pensioners. Records for out-pensions from 1781 to 1859 can be found in ADM 6, ADM 73 and ADM 22.

After the introduction of continuous service, ratings became entitled to a pension by right, after serving a period of twenty years. Unfortunately, few records survive for pensions awarded in this period.

Payments were also made to widows and other dependants from the Royal Bounty from 1675 to 1822. A lump sum would be made after the death of an officer or rating whilst on active service, and applications for such awards can be found in ADM 16 and ADM 106.

Royal Naval Reserve and Auxiliary Forces

Apart from the main body of the Royal Navy, various other smaller forces were also created to support the work of the main navy. As at times the merchant navy and Royal Navy would be closely connected, organizations were established that made it possible for the two bodies to interlap when required.

The Sea Fencibles

The earliest auxiliary naval force established was the Sea Fencibles. They were in existence during the height of the Napoleonic conflict, between 1798 and 1810. They were a local defence unit staffed by fishermen and boatmen to guard against the threat of invasion. The National Archives retains a variety of records for this force in ADM 28, including pay lists and musters along with details of a charity for widows.

Royal Naval Reserve

The Royal Naval Reserve was created thanks to the recommendations of the 1858 Royal Commission on the Defence of the United Kingdom. The Commission was established in response to threats of a French invasion and amid questions of whether the United Kingdom had adequate defences. The Commission closely examined all aspects of the country's military defence and made a number of recommendations. One of these recommendations was to form the Royal Naval Reserve (RNR), staffed from merchant seamen who had experience of working on deep-sea ships, and who would be required to serve in the Royal Navy in case of any emergency. The RNR had commissioned ranks along with ordinary seamen. The service records for this force are also at The National Archives.

- Officers' records can be found in ADM 240. The series covers the years 1862 and 1920 but also lists honorary officers serving up to 1960. The Navy List also includes RNR officers from 1862 onwards.
- Ratings' records from 1860 to 1913 are held in BT 164 (although only a selection of service records was kept and records do not survive for every individual). From 1914 to approximately 1921 the records can be found in BT 377, organized by service number. BT 377 contains copies of service records of ratings who served up till 1958; the originals of these records can be viewed at the Fleet Air Arm Museum (see page 199).

Royal Naval Volunteer Reserve

The Royal Naval Volunteer Reserve (RNVR) was established in 1903 and was to be staffed by volunteers from all sectors of society apart from merchant seamen and fishermen (who were expected to join the RNR). Service records can also be found at The National Archives.

- RNVR officer records are in series ADM 337 for the years 1903 to 1919. There is a card index available for this period and details can also be found in the appropriate Navy Lists. Service records after this period are still with the Ministry of Defence.
- RNVR ratings records can also be found in ADM 337 for 1903 to 1919. The records are organized by service number, which can be ascertained from the medal rolls in ADM 171, if the rating served during the First World War.

Further records can also be found in the Fleet Air Arm Museum. In 1958 the RNVR was merged with the RNR.

Royal Naval Division

The Royal Naval Division (RND) was established in 1914 from surplus recruits from the RNR. It served as a division of the Army and not a naval service during the First World War and was disbanded in April 1919. Service records for both ratings and officers are at The National Archives in ADM 339, arranged in three sequences:

1. Ratings alive at the end of the conflict
2. Ratings killed on active service
3. All officers (officer details are also in the Navy Lists)

As the RND was actually part of the Army, any medals awarded can be found by searching The National Archives website under the army campaign medals section on Documents Online.

Women's Royal Naval Service

Women were first recruited into the armed forces by the Royal Navy in 1917 to help fill the shortfall of male sailors due to the heavy losses being suffered during the war years. Thus the Women's Royal Naval Service (WRNS) was formed in November 1917. Women were recruited to work in a variety of roles, from cooks and clerks to storekeepers and electricians. Women were not expected to serve in ships, but worked onshore, and the organization was disbanded by 1919 (although it was re-formed in 1939). Service records for WRNS personnel during the First World War are also with The National Archives. Officers' papers can be found in ADM 318 (the entire series has been catalogued by name and can be searched online) and ADM 321, and ratings' records in ADM 336.

Service records for women serving after 1919 are still with the Ministry of Defence.

Additionally, the Royal Naval Museum has a large collection relating to the history of the WRNS. Further information can be found online at http://www.royalnavalmuseum.org/info_sheets_WRNS.htm.

> *Women were recruited into the WRNS in 1917 to fill a variety of onshore roles.*

Royal Naval Dockyards

The growth of the British Empire in the eighteenth and nineteenth centuries increased the international presence of the Royal Navy. Ships would be in all parts of the world and dockyards were needed to service these ships throughout the globe, not just the UK. As such naval dockyards needed employees to enable them to function and these employees were likely either to come from the Royal Navy or to join the Royal Navy after working at a dockyard. Often working in a dockyard was a good option for retired ratings and officers.

   Records of the dockyards are split between the National Maritime Museum and The National Archives. The National Maritime Museum's collections relate in general to the administration of the dockyards. More detailed information can be found on their website at http://www.nmm.ac.uk/server/show/conWebDoc.581. The National Archives holds staff records for the dockyards, the main series being ADM 42 (Yard and Pay books from 1660 to 1857). Additionally, ADM 106 contains a number of description books (giving physical characteristics) for employees from 1748 to 1830. The entire collection, including specific references for individual dockyards, is described in detail on the website at www.nationalarchives.gov.uk/catalogue/rdleaflet.asp?sLeafletID=50.

Medal Rolls

The Royal Navy issued a variety of campaign medals from the late eighteenth century onwards, the first one being the Naval General Service Medal. Medal rolls themselves give only basic information for each individual receiving such a medal. The rolls can be found at The National Archives, in series ADM 171, organized by conflict and ship. Prior to 1914 there is no specific index for the series.

   The rolls are perhaps most useful for the First World War period, as for this conflict they are organized by force (RNR, RNVR, etc.) and then by surname. Hence, it is often a useful shortcut to finding the service number of your ancestor, enabling you to then search for his service record. Gallantry awards during this period are also in ADM 171, and it may be possible to find further references to such awards in the *London Gazette* or *The Times* newspaper.

Medal rolls are a useful shortcut to find your ancestor's service number.

Admiralty Records

The National Archives series ADM 1 is the Admiralty collection of correspondence and papers of the entire Admiralty department from 1660 to 1976. As such it is a vast series, containing a whole range of documentation relating to the Royal Navy. It can be searched using the index in ADM 12. ADM 1 can be extremely useful for anyone researching a Royal Naval employee, giving all sorts of detailed information. However, finding a reference to an individual can take some time and there is no guarantee of finding anything.

Additional information on ship movements can be found via logs in various ADM series. A full summary of operational records can be found via four research guides on The National Archives website, listed under 'Royal Navy: Operational Records' and divided into relevant chronological periods.

Searching in Other Archives

Although The National Archives holds the majority of service records for the Royal Navy, there are other institutions that also have significant collections covering the history of the Royal Navy, and you might find them worth contacting or visiting.

- The National Maritime Museum, Greenwich, London, www.nmm. ac.uk: The Museum holds a large collection on the history of seafaring and naval history in general, including some Royal Naval operation records and archives of the Royal Naval Hospital in Greenwich. There are a number of information leaflets on their website providing further details.
- The Fleet Air Arm Museum, Yeovil, Somerset, www.fleetairarm. com: The Museum has been given a large number of service documents from the Ministry of Defence for various branches of the Royal Navy, RNR and RNVR amongst others. The nature of these collections is described in detail on their website at www. fleetairarm.com/pages/research/archivep1.htm.
- The Royal Naval Museum, Portsmouth, www.royalnaval museum.org: The Museum has collections on all aspects of naval history, both primary and secondary sources. Further information can be found on their website.

CHAPTER 11

Military Ancestors: The Royal Marines

Although the Army and Navy were traditionally seen as the two major branches of the armed forces, the Royal Marines also recruited large numbers of personnel and therefore a search of the surviving documentation can provide details of an elusive ancestor who saw military service, but cannot be traced in other records. This chapter explains how the Royal Marines were organized, and outlines the various ways you can search for information in enlistment and discharge papers, pension awards and other official documents.

A Brief History of the Royal Marines

> *Marines were originally to act as soldiers aboard Royal Navy ships.*

The origins of the Royal Marines are to be found in the seventeenth century, when the first marine regiment was raised in 1664 as part of the mobilization for the Second Dutch War. The regiment was originally raised with the intention that the men were to act as soldiers aboard Royal Navy ships. This regiment was titled 'The Duke of York and Albany's Maritime Regiment of Foot', after the title given to James (later King James II), the brother of Charles II, although it soon became known as the Admiral's Regiment as James was also Lord High Admiral. At its inception it was not a permanent force, being disbanded in 1689, after the deposition of James II in the Glorious Revolution.

Subsequent regiments were raised in wartime and disbanded in peacetime until 1755, when a permanent force was created in response

◄ The infantry of the Navy. During strenuous exercises in 1941, Royal Marines leap from a landing barge armed with Bren guns, rifles and anti-tank rifles.

to fears of war. Although the previous Marine regiments had come under the authority of the Army, henceforth the Marines would be part of the Admiralty. His Majesty's Marine Forces were formed on 5 April 1755, comprised of fifty companies in three divisions – Chatham, Portsmouth and Plymouth. A fourth division was opened in Woolwich in 1805, although this was closed later in 1869. Each division would have a depot centre in its town and people would be recruited at each of these depots. The depots recorded information in a similar fashion to that of the Army's foot regiments, while Marine bodies at sea had normal ships' records.

The number of regiments grew throughout the latter eighteenth century as they were used increasingly in various international conflicts of that period, the American War of Independence being one such example. In 1802 the force was given the official title of 'Royal Marines' by George III. From 1804 onwards separate artillery companies were also formed and attached to the Royal Marine Divisions (the Royal Regiment of Artillery no longer being responsible for artillery functions for the Royal Marines), and a new separate Marine unit of the Royal Regiment of Artillery was formed in 1859.

At that point the distinction between the 'Red Marines' and the 'Blue Marines' began, as this new artillery regiment had a blue-coloured uniform (as did the Royal Regiment of Artillery) whereas the rest of the Marines (known from 1855 as the Royal Marine Light Infantry) had a red-coloured uniform. This distinction lasted until the infantry and artillery units were merged in 1923 to form the Corps of Royal Marines.

Throughout the eighteenth and nineteenth centuries, Marine regiments served mostly on Royal Navy ships. Their functions changed somewhat in the twentieth century with the onset of the First World War as they now fought on land. During the Second World War they became a specialized unit, named the Royal Marine 'Commandos', responsible for raids on enemy coasts from 1942 onwards. This has become the main function of the Marines to the present day.

Service Records for Officers

Officers who served in the Royal Marines did not have to purchase their commissions as men were appointed without payment. The officer ranks comprised many junior posts but very few senior posts, and opportunities for promotion were very limited. Hence, many officers came from less prominent and influential families, often from the poorer gentry who did not have the funds to buy commissions in the Regular Army.

▼ Attestation papers give important biographical information, such as the age of the applicant and where they came from.

Records Prior to 1793

Service records do not survive for officers before 1793. However, there are a variety of other sources that relate to Royal Marine officers at The National Archives. Original sources include records of commissions and appointments. These can be found in archive series ADM 6/405, between 1703 and 1713. Commissions from when the permanent Marine regiment was established in 1755 are in ADM 6/406. The latter has information up to 1814. Please note, however, that the information is very basic, with nothing of genealogical relevance, and neither series has an index. ADM 96 also has some details of officers' records from 1690 to 1740.

Some published sources may also be helpful.

- Marine Officer Lists were published from 1755 onwards and The National Archives has collections from 1757 to 1860 in ADM 118/230-236 and ADM 192 for 1760-1886.
- The Army List contains details of officers on full pay and half pay from 1740 onwards.

Service Records 1793 to 1925

As mentioned above, the Army Lists may contain information on Royal Marine officers after 1740. Additionally, the Navy List and *Hart's Army List* also include details from 1797 and 1840 onwards, respectively.

The National Archives has additional series that contain information for this period too.

- There is a register of commissions in ADM 201/8, covering the years 1849 to 1859.
- The main body of service records are found in ADM 196. The records are only complete for the years 1837 to 1925, although some records do contain information going back to 1793. The records should contain service histories and may occasionally give details of birth and parentage. (Service records start in ADM 196/58-65, 83 and from ADM 97-112.)
- There is an index for records up to 1883 in ADM 313/110. After this date, indexes are provided within the individual volumes.
- Service records for Royal Marine Artillery officers are held in ADM 196/66, covering the years 1798 to 1855.

It may be worthwhile consulting these additional records to find more information on your ancestor:

- Pay lists and records are in ADM 96 and ADM 6. The latter has full- and half-pay registers for 1789 to 1793 and 1824 to 1829 in ADM 6/410-413.
- Marine officers were also surveyed on the same basis as those of the Royal Navy in 1822 and 1831. These records are now held in ADM 6 and may give individuals' ages.

> *Information about Marine officers might be found in either the Army Lists or the Navy List.*

▼ Soldiers of the 2nd battalion, 124th Infantry wade ashore from landing craft at Morotai Island, Indonesia, in the final stages of the Second World War.

USEFUL INFO

Marine officers' service records for those who were recruited after 1925 are still with the Ministry of Defence and are available to next of kin only. They can be contacted at:

DPS(N)2
Building 1/152
Victory View
PP36
Her Majesty's Naval Base
Portsmouth
PO1 3PX

- ADM 63/27-30 contains confidential books with letters on the work of officers from 1868 to 1889.

As with officer ranks of the other forces, researching in published sources such as *The Times* and *The Dictionary of National Biography*, along with other biographical sources, may turn up interesting information. If your ancestor served during the First World War his service details up to 1916 may be found in *The Naval Who's Who*.

The Royal Marines have been publishing their own journal since 1892, called *The Globe and Laurel*. These are now archived at the Royal Marines Museum (see below, page 207) and may also contain relevant information.

Warrant Officers

The Royal Marines also had a body of warrant officers, many of whom later became commissioned officers. Only a limited number of service records for these individuals are known to have survived. They can be found in ADM 196, covering the years 1873 to 1920.

Records for Other Ranks

In contrast, a large amount of documentation relating to the careers of other ranks survives for the Royal Marines. Information can be found in service records, muster books and pay lists. Service records include attestation forms, description books (detailing the physical attributes of new recruits) and service registers. These records are arranged by the appropriate Division (Portsmouth, Plymouth, Chatham or Woolwich) and therefore it is essential to know which division the soldier served in before searching for his service records.

Service Records for Other Ranks

Once you have identified the correct Division it is possible to begin searching for the service records within the Divisions. There are three main sets of records to consult: attestation forms in ADM 157, physical description books in ADM 158 and service registers in ADM 159.

If you are researching an 'other rank' who enlisted after 1925, the records are still retained by the Ministry of Defence. They can be obtained by contacting them directly at the address given above.

1.
ADM 157: Attestation forms

The series covers the years 1790 to 1925, comprising the forms completed by young recruits upon enlistment. They are loose documents containing similar information found in the Soldiers' Documents series of WO 97 (see Chapter 9), including birth details, any previous occupation, physical characteristics and details of service with the Marines. Although they are attestation forms they are largely organized by date of discharge per Division. There are a few exceptions to this rule: the records of the Chatham Division are all organized by enlistment date until 1883. Further details of how the series is organized can be found in the appropriate research guide, available at www.nationalarchives.gov.uk/catalogue/RdLeaflet.asp?sLeafletID=56.

There is also a card index to the early part of the series in The National Archives. The index covers the records up to 1883 and there is a card for each individual Marine in the first 659 pieces of the series ADM 157. The cards will direct you to the relevant piece for that Marine and also detail the Division he belonged to.

2.
ADM 158: Description books

This is the second main series relating to other ranks' service records. They begin around 1750 and the series finishes in 1940. The series contains books detailing the physical characteristics of the new enlistees. Additionally, you will find where the individual enlisted, his age and place of birth, previous employment and details of injury or if killed on service. There is no service information. The books are arranged by Division, then company and date of enlistment, then by the first letter of surname (not strictly alphabetically).

3.
ADM 159: Service registers

The service registers are from 1842 to 1936, allowing for a 75-year closure rule. Along with the same biographical information provided in the other main series, the records provide a full service history of each Marine and may include comments regarding the individual's service. In 1884 a new numbering system was introduced for the Royal Marines and was applied retrospectively. Hence, the numbering system begins at different times for each Division:

- Portsmouth begins in 1843
- Royal Marine Artillery from 1859
- Chatham from 1842
- Plymouth from 1856
- As the Woolwich Division was disbanded in 1869 the records of marines serving in this Division will be with the Divisions the men were transferred to

The records in ADM 159 are organized according to this numbering system within each Division. If you are not aware of the service number it can be obtained by using the indexes found in the series ADM 313. ADM 159 also contains service records for the Royal Marine Band (from 1903 to 1918, in ADM 159/103-112) and the Medical Unit for the First World War (in ADM 159/209-210).

Discharge Books

Discharge Books were also kept by the three main Divisions, detailing when each Marine was discharged. They can be found in the following series:

- Chatham: ADM 183
- Plymouth: ADM 184
- Portsmouth: ADM 185

Finding your Ancestor's Marines Division

Most individuals would be in the same Division throughout their service with the Royal Marines. The oldest Divisions were Portsmouth, Plymouth and Chatham, while Woolwich was a Division between 1805 and 1869. It is possible to ascertain which Division your ancestor would have been in through the following means:

• Most would have been recruited from the nearest division to where they lived. Hence, if you know where your ancestor was living, identify the nearest Division to the locality as it is very likely he would have been recruited from there.

• If your ancestor was awarded any medals they would have been recorded in the Medal Roll found in ADM 171. These rolls would also provide the Division and service number.

• Garth Thomas has written a detailed guide to Royal Marine records, called *The Records of the Royal Marines*, which has a table that can be used to find the correct Division (if you know your ancestor's company number and when he served).

• Knowing which ship your ancestor served on can also provide this information. By using the Navy List you can discover the home port of the ship. Until 1947 those serving on ships would belong to the same division as the home port of the ship (i.e. if the home port of the ship was Chatham then the Marine would belong to the Chatham Division).

• One part of the service records (attestation forms) is found in The National Archives series ADM 157. There is a partial card index to this series which gives the appropriate Division.

Medals Awarded to Royal Marines

The vast majority of medals awarded to Royal Marines will be found in the same National Archives series as for the Royal Navy, in ADM 171. As mentioned, these rolls can be used as a useful method of obtaining the correct division for individual Marines. Prior to the First World War the rolls are organized by serving ship. However, medals issued for the First World War are organized alphabetically, within Royal Marine Officer and Other Rank series.

Pension and Casualty Records

Pensions for Royal Marines were administered in the same way as those for Royal Naval officers, the Greenwich Hospital being used for both forces. As such it is possible to find relevant service records in a number of different sources:

- ADM 201/22-23 lists Marine officers being awarded Greenwich pensions in alphabetical registers from 1862 to 1908.
- Information for widows of Marine officers who were awarded pensions can be found in ADM 196/523, PMG 16, PMG 20 and PMG 72. The earliest records begin in 1712 and details can be found for widows up until 1926.

The Admiralty also recorded casualties suffered by Marines whilst on service. These can be found in ADM 242/7-10 for 1893 to 1956. This series contains detailed information about each individual including birth details, burial and next of kin. ADM 104 also has numerous registers of Marines who were injured or died on service, from 1854 to 1941. The Commonwealth War Graves Commission also records Marine casualties from the First World War onwards.

Royal Marine Genealogical Information

The Royal Marines kept good genealogical records; specifically registers of births, marriages and deaths were maintained by each Division. These registers were kept as Marines were entitled to certain benefits if they had children whilst on active service. Each Division kept records for different time periods. Chatham has the most complete collection, covering 1830 to 1913 in ADM 183/114-120. Plymouth Division's records start in 1862 and go on to 1920 and can be found in ADM 184/43-54. Portsmouth and Woolwich Divisions only have registers for marriages, in ADM 185/69 (for 1869-81) and ADM 81/23-25 (for 1822-69) respectively. The Royal Marine Artillery's registers are also complete, ranging from 1810 to 1853 (in ADM 193) and 1866 to 1921 (in ADM 6/437).

SUMMARY

The vast majority of records relating to service in the Royal Marines are held at The National Archives. However, the Royal Marine Museum in Southsea, Hampshire, may also be worth visiting. Along with the entire archive of The Globe and Laurel *the Museum has a large collection of medals and other items relating to Royal Marine history. The Library has collections of Navy and Army Lists and journals such as the* Naval Chronicle. *The Museum also has a large document collection including diaries of prominent Royal Marine servicemen, prisoner of war papers and operational unit diaries and reports. Further details of the Museum can be found on its website at www.royalmarinesmuseum.co.uk.*

CHAPTER 12

Military Ancestors: The Royal Air Force

Very much the junior branch of the armed forces, the Royal Air Force has only been in existence since 1918. This chapter provides information about how to trace the service histories of its individuals, including the various branches of the Army and Navy from which the RAF was formed, as well as general research into the operational activities of the RAF in both war and peace, and where to look for further details of planes, RAF bases and general aviation history.

A Brief History of the RAF

The RAF played a decisive role in securing Britain's safety during both world wars.

The Royal Air Force was officially created on 1 April 1918, but this was not Britain's first military air force. On 13 May 1912 the Royal Flying Corps (RFC) was formed by royal warrant in response to Germany's rapidly increasing air fleet. The development of aeronautical technology had only begun a decade earlier in America, and initially its potential as a resource for the British army was not recognized by the government. Britain was more concerned about German naval competition, until in 1911 it was realized that the Royal Engineers' newly formed Air Battalion, which had grown out of the Balloon Section formed in 1890, could not match the number of aircraft and experienced pilots possessed by other European countries.

The Royal Flying Corps was set up under the command of the War Office and was made up of four main components – the Naval Wing, the Military Wing, the Central Flying School (CFS) based on Salisbury

Plain, and the Royal Aircraft Factory at Farnborough. The principal function of the RFC was to provide unarmed reconnaissance support to the army and navy rather than as a defence and attacking force in its own right. The Military Wing was responsible for building an aerial reconnaissance force designed to support army ground operations. The Naval Wing, however, took a more proactive approach to the use of aircraft in warfare and began to explore the possibilities of using planes during long-range attacks. The Navy had used seaplanes since 1911 to help protect naval ships against submarine attack, and the growing threat of German airship development (and sightings of unidentified aircraft off the British coast) led to the RFC's Naval Wing being separated from the rest of the Corps in July 1914 to form the Royal Naval Air Service (RNAS) under the command of the Admiralty.

▲ British pilots wait to take off from an RAF fighter command station in England during the Battle of Britain, January 1940.

When the First World War broke out in August 1914, Britain's air force was still an under-equipped, fragile resource relying on primitive aeroplanes and observation balloons. There was an initial reluctance to employ aircraft in the war, as the cavalry were considered more reliable for reconnaissance missions, but once the fighting in France became mainly trench-based the advantages of using aircraft over cavalry were clear. The planes could cover a wider area for observation and the information they provided the army proved to be highly accurate. A year after the outbreak of war the Germans developed fighter aircraft built with specially designed machine guns, and Zeppelin airships began bombing London in 1915, which forced the British to make technological advances in their aircraft's capabilities and led to a largely airborne war by 1917.

The RNAS was principally responsible for home defence while the RFC was based on the Western Front in France, but as the war progressed it became apparent that the functions of the RFC and RNAS overlapped. In 1917 General Jan Smuts recommended in a special report that the RFC and RNAS be amalgamated under an Air Ministry that was independent of both the War Office and the Admiralty. The Royal Air Force was established as a result of this report, and amongst much controversy it came into being just eight months before the end of the First World War at a time when the German offensive was at its height. Despite its difficult beginnings, the RAF played a significant supporting role helping the Allies to win the war. By the time the Second World War started, aircraft had changed the face of warfare and the RAF was vital in securing victory.

Most men in the Air Force were not pilots. The majority worked on the ground as engineers and mechanics, and women were employed as cooks, telephonists, drivers, intelligence agents and administrators as well as mechanics. Air Force records do not give much genealogical information, simply stating the name, usually a date and place of birth and details of next of kin, which help you to identify your ancestor's records. However, the records enable you to build up a picture of your ancestor's life, to understand their experiences and to place their career within the wider context of the history of the RAF.

Officers' Service Records
RFC Officers' Records

Service papers for officers who joined the RFC between 1914 and March 1918 were forwarded to the RAF and are held with RAF officers' papers for 1918 to 1919 in The National Archives series AIR 76. These are kept on microfilm and are arranged alphabetically. Consult the index to AIR 76 to establish which reel contains the surname you are looking for.

If an RFC officer died or was discharged prior to the formation of the RAF in 1918 then his service record is likely to still be kept with War Office papers in WO 339, an alphabetical index to which is held on microfilm in WO 374 (see Chapter 9 for further information on these series). There are biographies for some officers of the RFC and RAF from the First World War period kept in AIR 1 that can be searched by keyword in The National Archives online catalogue.

RNAS Officers' Records

Officers who joined the RNAS between 1914 and March 1918 had their records kept with the Admiralty, and can be found in the Registers of Officers' Service for the Royal Naval Air Service in ADM 273. The records in ADM 273 need to be ordered as original documents and are organized numerically by service number, but each volume has an alphabetical index and there is a combined alphabetical card index for ADM 273

USEFUL INFO

Prior to March 1919 it is possible to trace the careers of officers of the RFC and RNAS in the Army Lists and Navy Lists.

among the Finding Aids at The National Archives for you to establish whether your ancestor has a service record among these documents.

RAF Officers' Records

All officers who joined the RAF from April 1918 and who were discharged before 1920 should have service papers held in AIR 76 as described above.

Air Force Lists were published from March 1919 onwards. The information they contain and frequency of publication vary over the years, but generally they list the names of officers, their rank and squadron, and the entry for each station and squadron lists the aircraft and the date they were posted to a unit, so you may use these to trace an officer's career over time. Each book has an index so that an officer's name can be found easily.

Air Force Lists from 1919 until the present day are on open access on shelves at The National Archives and Society of Genealogists, or you can check with your local library and archive to see if they have copies. Confidential Air Force Lists for September 1939 to December 1954 can be ordered as original documents at The National Archives from AIR 10/3814–3840, AIR 10/5237, AIR 10/5256, AIR 10/5413–AIR 10/5422, AIR 10/5581 and AIR 10/5582.

Payments made to officers who were invalided out of service between 1917 and 1920 are kept with the Ministry of Pensions records in PMG 42, and records of pensions paid to the next of kin of officers who died between 1916 and 1920 are in PMG 44. Special grants and allowances paid to officers and dependants between 1916 and 1920 are in PMG 43.

Airmen's Service Records
RFC Airmen's Records

The biographies of the first men to join the RFC between 1912 and August 1914 (service numbers 1 to 1,400) along with many photographs can be found in *A Contemptible Little Flying Corps* by Webb and McInnes.

If an airman of the RFC died or was discharged before April 1918, then his service papers should be held with those of other ranks of the army in WO 363 and WO 364. These records are kept on microfilm at The National Archives and arranged alphabetically by surname. Those documents held in WO 364, created from pension claims, can now be accessed online from the subscription-based website www.ancestry.co.uk, although most local libraries have a subscription to the site that visitors can use for free. See Chapter 9 for more information about locating documents in WO 363 and WO 364.

RNAS Airmen's Records

If an airman served in the RNAS prior to April 1918 then his record of service up until 31 March 1918 will be found in the Registers of Seamen's Services held in ADM 188, which can be searched from the Documents Online search engine found on The National Archives website. A search can be conducted for name and official number. Each entry in the register should give a name, date of birth, the name of the ship or shore establishment where the individual served and a short account of their service and appointments. The RAF kept service records for men who served after 31 March 1918.

Other RAF Records

Accident Reports

Information about accidents that occurred during operations is usually found in the Squadron's Operational Record Books in AIR 27 (see below). Some correspondence relating to crashes and casualties is kept in AIR 1 and there are reports for crashes that took place between April 1916 and November 1918 held in AIR 1/843/204/5/369 to AIR 1/860/204/5/427, 914–916, and 960–969.

There are also Accident Record Cards, Casualty Files and Aircraft Record Cards held by the RAF for non-operational accidents. If you would like the RAF to conduct a search of these cards for you, then write to them at:

Air Historical Branch
Building 266
Royal Air Force
Bentley Priory
Stanmore
Middlesex HA7 3HH

The RAF Museum at Hendon has copies of the Aircraft Record Cards and you can request a search of those by writing to

The Department of Research and
*   Information Services*
Grahame Park Way
Hendon
London NW9 5LL

Records of Prisoners of War

If you have an RFC, RNAS or RAF ancestor who was held as a Prisoner of War (PoW) during the First World War then it is easy to find evidence of this if he was an officer simply by checking the *List of British Officers Taken Prisoner in the Various Theatres of War between August 1914 and November 1918* by Cox and Co., a copy of which is held in the Library at The National Archives in Kew. This gives information about the regiment, theatre of war, name and rank of the officer, the date he was captured, when and where he was held, and the date he was released or his date and place of death if he died while a prisoner. If the officer survived you should find a report about the circumstances of his capture in his service record.

There is not a similar published list for airmen who were captured as PoWs, but it is worth looking at *Researching British and*

RAF Airmen's Records

A muster list was taken of all the airmen who joined the RAF on its first day on 1 April 1918, and the original document is kept in AIR 1/819/204/4/1316 with supplements in AIR 10/232-237. Service records of the first 329,000 men who served in the RAF and men who joined the RFC but continued service with the RAF after 1 April 1918 are held as original documents in AIR 79. They are arranged numerically by service number but there is an alphabetical index held on microfilm in AIR 78. If your ancestor had a service number higher than 329,000 or if his service number was between 1 and 329,000 but he continued to serve during the Second World War, then his records will still be with the RAF. Ex-service personnel can request a copy of their service history to be sent to them by writing to ACOS (Manning) at the address given above (page 211), detailing their name, date of birth and service number if known. If the service person has died, their next of kin can request a

Commonwealth Prisoners of War: World War One by Alan Bowgen, held behind the Research Enquiries Desk at The National Archives to see which records might be able to help you. Series AIR 1 holds some records about RAF, RFC and RNAS Prisoners of War, and there are additional registers in ADM 12 for RNAS servicemen who were captured during the First World War. Service records of ordinary RFC, RNAS and RAF personnel or their Medal Index Card should contain some notes about their capture.

The International Committee of the Red Cross (ICRC) in Geneva holds a list of some known Prisoners of War for both the First and Second World Wars and will respond to written queries at an hourly rate if enough information is supplied in writing. The address is:

International Council of the Red Cross
Archives Division
19 Avenue de la Paix
CH-1202, Geneva
Switzerland

The names of RAF Prisoners of War captured by the Germans during the Second World War may be found in an alphabetical list compiled towards the end of 1944 held in AIR 20/2336. There are related files in AIR 40, including escape and evasion reports kept in AIR 40/1545-1552, and AIR 14/353-361 contains reports of escaped RAF Prisoners of War and nominal lists of reported RAF prisoners. There is a nominal list of British and Commonwealth PoWs held during the Second World War in WO 392/1-26 with specific sections dedicated to the Air Forces held in WO 392/8, WO 392/18 and

WO392/21. (For more detailed information about finding records for Second World War PoWs have a look at The National Archives' Military Information Research Guide 20 on the online Research Guide alphabetical index under 'Prisoners of War, British, 1939-1953'.)

Courts Martial Records

If you believe your RAF ancestor was tried by a court martial then have a look in the RAF Courts Martial Registers in AIR 21, where the name, rank, the nature of the offence, the place of the trial and the sentence given to each prisoner between 1918 and 1965 are recorded. The RAF Courts Martial Proceedings records for district, general and field courts martial trials both for officers and airmen from 1941 to 1994 are held in AIR 18, but there is a closure period of 30 years on the later documents.

> *Most men in the Air Force were not pilots. The majority worked on the ground as engineers and mechanics.*

copy at a fee of £30 by sending a letter to the same address giving the service person's name, date of birth, service number and any known information about their career with evidence of their death certificate.

Births, Marriages and Deaths in Service

Records of births and marriages for RFC personnel and their families should be found among indexes to GRO Army Births 1881–1965, GRO Army Marriages 1881–1955 and GRO Army Marriages within British Lines 1914–25, while the Army Chaplains' returns for 1796 to 1955 will contain records for the RAF from 1920.

The deaths of service personnel during the two World Wars are well recorded. The Commonwealth War Graves Commission (CWGC) has compiled a Debt of Honour Register from its First and Second World War cemetery and memorial records. The 1.7 million names contained on this can be searched for free on www.cwgc.org. Entries on this database can give as much information as the name, age, nationality, regiment, rank, date and place of death and burial or memorial as well as the names of next of kin and their address.

If you cannot find an entry on the Debt of Honour database for an RAF ancestor who died during the Second World War then it is worth searching the indexes to GRO War Deaths, RAF All Ranks 1939–48. Indexes to the deaths of RFC personnel during the First World War may be found among the GRO War Deaths, Army Officers 1914–21 and GRO War Deaths, Army Other Ranks 1914–21, while those of RNAS service personnel may be found among the GRO War Deaths, Navy All Ranks 1914–21. The RAF Museum at Hendon also has a casualty card index for RFC and RNAS personnel during the First World War.

Deaths on service of RAF personnel since the wars may be found in the indexes to Air Deaths and Air Deaths – Missing Persons between 1947 and 1965, held by the GRO for all deaths occurring on aeroplanes registered in the United Kingdom.

All of the birth, marriage and death indexes for service personnel and their families recorded by the General Register Office are held at The National Archives in Kew. If you find a likely entry in the indexes for your ancestor you will need to order a duplicate copy of the certificate from the GRO at www.gro.gov.uk/gro/content/certificates.

The National Archives of Scotland have recently catalogued the wills

CASE STUDY

Julian Clary

Julian Clary didn't know much about his paternal grandfather, Jack Clary, as he died in 1951, several years before Julian was born. Julian's father, Peter, had vague memories of Jack, though mainly associated with visiting him whilst he was admitted to a psychiatric hospital in the 1930s. This was a line of research that Julian followed up, discovering that Jack voluntarily admitted himself to Napsbury hospital in Hertfordshire between 1926 and 1938. Although there appeared to be no superficial connection with his stay in hospital, it emerged that Jack has spent time during the First World War as chief mechanic for the Bristol fighters of 48 Squadron.

Having obtained this information from within the family, Julian was able to follow up with some detailed research at the Imperial War Museum, Duxford, and The National Archives, Kew. At Duxford, he was able to find out more about working conditions, as well as the distinguished history of the squadron which was formed as part of the former Royal Flying Corps in Netheravon on 15 April 1916, and moved to France in March 1917 as the first Bristol Fighter Squadron.

Julian learned that his grandfather's daily routine in France would have been tough, back-breaking work, living under constant threat from enemy air attacks and with the daily pressure of ensuring the aircraft were serviced and mechanically sound, ready for action at any time. In short, the pilots' lives

depended on the skill of Jack and his team.

Furthermore, by examining the service records for the RAF at The National Archives in record series AIR 79, and official records of the squadron in record series AIR 1, Julian was able to work out the precise history of Jack's involvement with the 48 Squadron; his

▲ Julian and his Aunt Doreen look through his grandfather's photo album from his days with the Royal Flying Corps.

movements around France; and indeed associated combat reports that the airmen filed on return to base. By reading these contemporary accounts from official sources, it was possible to step ninety years back in time and gain a real sense of what it must have been like to work on these early planes on the front line – and what it was like to face such danger on a daily basis.

◄ Jack Clary, Julian's grandfather, during his RAF service in the First World War.

The Women's Royal Air Force and Women's Auxiliary Air Force

There are no surviving records of service for officers of the WRAF or WAAF, but records of service for airwomen for the First World War period can be found in alphabetical order in AIR 80, kept on microfilm at The National Archives. (There is an index in AIR 78.) Women who did not have domestic responsibilities were sometimes sent to help out in France and Germany between 1919 and 1920 and were known as 'mobiles' because they could live on camp and be posted anywhere, while 'immobile' airwomen would have lived at home and worked part time. An airwoman's record of service will indicate whether or not she was mobile, along with her age, address, marital status, names of any dependants, and information about any promotions.

The WRAF was formed at the same time as the RAF in 1918. It was disbanded in 1920, but re-formed at the start of the Second World War under the name the Women's Auxiliary Air Force (WAAF). It reverted to the title Women's Royal Air Force in 1949, and became a fully integrated part of the RAF in 1994. Records of service for women who served in the Second World War need to be requested from the RAF by writing to ACOS (Manning) at the address on page 211.

To learn more about the WAAF and WRAF read a copy of *Women in Air Force Blue: The Story of Women in the Royal Air Force from 1918 to the Present Day* by Beryl E. Escott. Some women also worked for the RAF as nurses from January 1919, in the Royal Air Force Nursing Service, which became Princess Mary's Royal Air Force Nursing Service in 1923. All records of service for these women are still with the RAF.

▲ Women played a vital role in the RAF, particularly during the Second World War.

of 61 RAF airmen and officers from 1939 to 1950 and some RFC airmen from the First World War, which can be searched via the NAS online catalogue: http://www.nas.gov.uk/guides/soldiersWills.asp.

Medals and Awards

Campaign medals are awarded to members of the armed services for their part in a particular war or battle. There is no First World War campaign medal roll for the RAF in the AIR series. Most men who were in the RAF after 1 April 1918 would have come from the RFC, RNAS or another part of the armed forces, so their Medal Index Card stating which medals they were entitled to will be found in the relevant series for their earlier service. If your ancestor only saw service with the RAF during the war then they are unlikely to have a Medal Index Card, but any entitlement should be detailed on an individual's record of service.

Medal rolls for those awarded to men of the RFC prior to the outbreak of the First World War are held in WO 100. First World War campaign medal rolls for the RFC are also kept (with those of the army) in WO 329. They are arranged by battalion but there are also alphabetical Campaign

Medal Index Cards kept in WO 372 that can be searched by name and rank via The National Archives website, at Documents Online. The indexes themselves give very little information about the person the medal was awarded to, but abbreviations in the index will indicate that they served in the RFC, their service number, rank, the first theatre of war they served in and the type of medal they were awarded. This can then be used to establish which campaign the medal was awarded for.

First World War medal rolls for the RNAS are held in microform in ADM 171. These are arranged alphabetically, but are split up according to rank. The medal roll for officers is found in ADM 171/89–91, while that for ratings is in series ADM 171/94–119. Again, the entry gives you the name, service number, rank and abbreviations indicating which medal was awarded, but also states where the medal was sent.

The National Archives only holds medal rolls for one campaign fought since the First World War, known as the African General Service Medal, which was given to around 200 RAF servicemen who were involved in operations in Somaliland at the beginning of 1920. This medal roll is kept in series AIR 2/2267–2270. Any other campaign medal rolls issued to the RAF since then are kept with the Ministry of Defence, but information about them can be requested by writing to:

> CS Sec 1d
> Room F93, Building 256
> HQ RAF PTC
> RAF Innsworth
> Gloucester GL3 1EZ

Gallantry medals were created specifically for RAF service personnel in 1918, namely the Air Force Cross, the Air Force Medal, the Distinguished Flying Cross and Distinguished Flying Medal. Prior to this RFC and RNAS servicemen would have received army and navy honours.

Medals awarded for gallantry and meritorious service were announced in the *London Gazette* and some are accompanied by a citation of the reason for the award. Digital copies of the gazette are now available from www.gazettes-online.co.uk. Recommendations for some gallantry awards issued to the RFC and RAF for the First World War are held in AIR 1. Recommendations for gallantry medals issued to RNAS servicemen prior to 1918 are in ADM 1, ADM 116 and ADM 137. Recommendations for the Second World War and post-1946 are in AIR 2, and AIR 30 also contains records of gallantry medal recommendations for RAF personnel.

Operational Record Books (ORBs)

ORBs are made up of Summary of Events forms and Detail of Work Carried Out forms, otherwise known as Forms 540 and 541, along with appendices containing operational orders, relevant reports and telegraphed messages. They are worth tracking down if you would like to discover a little bit more about your ancestor's experiences during their time with the Air Force, particularly for the Second World War period when service records are difficult to access.

A search for an Operational Record Book requires definite knowledge of which Squadron your ancestor was with and for what dates, as ORBs do not tend to state many names and give no biographical details about those men they do mention. However, they will fill you in on the movement of that Squadron each day and their location and activities. This can be of especial use if you have been told an anecdote about an incident that happened while your ancestor was on duty and you would like to find out more about the facts.

From 1914, the air forces were divided into Squadrons and each Squadron had subdivisions known as Flights. In 1936 the RAF was divided into four main Commands – Bomber, Training, Fighter and Coastal Commands, which were subdivided into Groups, or collectives of Squadrons, and the Groups were divided further into Wings.

◆ Squadron ORBs for the period 1911 to 1977 can be found in AIR 27. They have both daily and monthly summaries and list details about the aircraft, crew, weapons and any casualties.

▲ Detailed descriptions of flying missions can be found in Operation Record Books at The National Archives.

Finding Out More

For a detailed history of the RFC, RNAS and RAF take a look at the History pages on the www.raf.mod.uk website, where there is an abundance of photographic archive material as well.

The RAF museum in Hendon, North London, has its own library and an archive mainly made up of documents donated by private individuals, ranging from personal diaries, letters and memoirs to service papers, log books and operational records. The museum's exhibits and collections include medals, uniforms, photographs, film footage and audio recordings, as well as aircraft, engines, weapons and vehicles. Visit www.rafmuseum.org to find out more and have a look at the RAF Online Exhibitions, or search the archive's records via the National Register of Archives at www.nationalarchives.gov.uk/nra. The RAF also has a museum at Cosford in Shropshire that does not have an archive but does exhibit many aircraft and other large artefacts relating to the history of the Royal Air Force.

If you had ancestors who fought in the RAF or its predecessors during a conflict, then you may be interested in visiting the Imperial War Museum at Duxford in Cambridgeshire. Visit http://duxford.iwm.org.uk

◆ Squadron Combat Reports held in AIR 50 cover the Second World War period and these can be used to supplement information found in the relevant Squadron's ORBs.

◆ There are some miscellaneous units' ORBs covering 1912 to 1973 in AIR 29. The ORBs of smaller units were kept with those of the larger units to which they were attached.

◆ ORBs for Stations to which your ancestor may have been posted are kept in AIR 28 for the period 1913 to 1966.

◆ Groups ORBs for 1914 to 1970 can be found in AIR 25.

◆ ORBs for the Wings from 1920 until 1964 are in AIR 26.

◆ Commands' ORBs for 1920 to 1973 are in AIR 24.

There are also some logbooks for airships kept in AIR 3 for the First World War period and Aircrews' Flying Log Books for 1915 until 1983 are held in AIR 4. The South African Air Force ORBs from 1937 to 1947 are kept in AIR 54. Squadron Diaries of the Fleet Air Arm for 1939–57 contain similar information to RAF Operational Record Books.

There is a detailed guide to finding operational records for the RAF at The National Archives, which can be located online by going to the Research Guides page on their website and scrolling down the alphabetical index to 'Royal Air Force: Operational Records'.

Operational Record Books for the most important event for the RAF during the Second World War, the Battle of Britain, have been transcribed into Daily Reports and made available for free online from www.raf.mod.uk/bob1940/bobhome .html. There are unit histories of all the Squadrons that took part, a list of the Stations they were based at and information about the periods they were resident there. If you had an ancestor who was involved in the Battle of Britain then this website is certainly worth a visit.

to find out more about the museum's upcoming events or browse the Imperial War Museum website's 'War in the Air' collection at www.iwmcollections.org.uk/inair, which includes photographs and audio recordings and descriptions of film footage, documents and arte-facts held at the museum.

The Royal Air Forces Association (RAFA) was established in 1929 so that RAF personnel could maintain contact with one another after the First World War, and it has a list of useful links on its website at www.rafa.org.uk. You can post a message in the *Airmail* magazine run by RAFA to try to find any living ex-service personnel who may have recollections of working with your relative by writing to:

Royal Air Forces Association
Central Headquarters
43 Grove Park Road
London W4 3RU

If you would like to contact an association dedicated to a specific squadron or station that your ancestor worked with then have a look at the Royal Air Forces Register of Associations (RAFRA) website at www.associations.rafinfo.org.uk. These associations may be able to help you learn more about your ancestor and their time with the RAF.

USEFUL INFO

Some suggestions for further reading:

• Air Force Records for Family Historians *by W. Spencer (Public Record Office Publications, 2000)*

• RAF Records in the PRO *by S. Fowler, P. Elliott, R. Conyers Nesbit and C. Goulter (Public Record Office Publications, 1994)*

• The Squadrons of the Royal Air Force and Commonwealth, 1918–1988 *by J. J. Halley (Air Britain (Historians), revised edition, 1988)*

CHAPTER 13

Occupations: The Merchant Navy

As a nation surrounded by the sea, Britons have a long association with maritime professions, perhaps no more so than the Merchant Navy. Fortunately, a vast array of material survives that allows us to trace the movements of our ancestors around the world, and this chapter shows you all the steps you'll need to take, as well as highlighting some of the potential pitfalls involved.

> " Until 1835 finding records for ordinary seamen is complex – up until 1747 masters didn't have to keep crew lists. "

The Merchant Navy was a private organization that was not regulated by government until the nineteenth century. In 1835 merchant shipping became the responsibility of the Register Office of Merchant Shipping (RGSS), regulated by the Board of Trade. It is from this date that the major collections of Merchant Navy records begin, especially for individual seamen. Prior to that date some records do exist, but locating an individual is significantly harder. The RGSS, later known as the Registry of Shipping and Seamen (RSS), has been the administering body for the Merchant Navy since 1835. It is currently based in Cardiff and still retains the most recent Merchant Navy records (since 1996).

Records for individuals who served in the Merchant Navy are found in a number of archives. Again, the majority are held at The National Archives. However, there are also important collections in the Maritime Museum in Greenwich, London, and the Guildhall Library of London and some local repositories. The British Library holds the collection of the East India Company's merchant marine. The Memorial University of Canada also holds information that may be of relevance. Additionally, the General Record Office has indexes recording births, marriages and deaths of British nationals on British ships.

Service Records Prior to 1835

The Merchant Navy, like the Royal Navy, distinguished between masters and captains, who were responsible for the entire ship, and ordinary seamen, who worked on the ships. Until 1835 finding records for ordinary seamen is a complex task with little guarantee of success. Indeed, up until 1747 there was no requirement for ships' masters to keep any muster rolls or crew lists. Surviving records to that date simply detail the owner of the ship and its master.

Crew Lists

Crew lists were only kept in a small minority of cases prior to 1747, when ships were involved in maritime disputes. Records of these maritime disputes may record crew lists, although this was not always the case. These records are now held at The National Archives, in series HCA (High Court of Admiralty) and DEL (Records of the High Court of Delegates). Such lists as do survive in these series are not indexed and the chances of locating your ancestor are rather limited.

In 1747 it became a requirement for all masters or owners of merchant ships to record the individuals employed on their ships on various muster rolls and crew lists. Along with the seaman or officer's name they would usually record home residence, name of last ship served on and the date the sailor was engaged on and discharged from the ship. These are now held in The National Archives series BT 98. Unfortunately, the survival rate of these lists is not complete. The lists would be stored at the arrival or departure port of the ship and until 1800 lists only survive for Dartmouth, Liverpool and Shields (now North and South Shields).

▼ The Port of London in 1831, showing Customs House in the background.

The latter is the only port that has lists starting from 1747. The number of lists increases greatly after 1800. However, they are arranged annually by name of port and there are currently no indexes. Hence, it is a very lengthy task to search for an individual unless you are aware of the relevant port and the ship on which they sailed.

Other than crew lists, there are some alternative sources that may give information about seamen in this period.

- *The Trinity House petitions:* The Corporation of Trinity House was formed in the sixteenth century as a guild to help mariners and their families in times of poverty or sickness through charitable funds. By the nineteenth century the organization was supporting many mariners through almshouses or awarding pensions. Mariners or their families had to submit petitions in order to receive assistance from Trinity House. These petitions survive from 1787 to 1854 and are stored at the Guildhall Library. They contain useful genealogical information and also career service details. The Society of Genealogists has copies of these petitions, numbering around 8,000, and an index to these records can be searched on the Origins website, at www.origins.net. The Guildhall Library also has registers of almspeople and out-pensioners being assisted by Trinity House, ranging from 1729 to 1946.
- *Apprenticeship records:* The system of apprenticeship was widespread amongst numerous trades and professions from the seventeenth century onwards, merchant shipping being no exception. Some pauper boys were forcibly indentured to serve as apprentices on ships by parishes from 1704 onwards. If records survive for this practice they will be held with local record offices. Taxes were placed on the entire apprenticeship system for each type of profession from 1710 to 1811. These records are held in The National Archives series IR 1 and there are indexes for boys (from 1710 to 1764) and masters from (1710 to 1762).
- *The Marine Society:* This society, founded in 1756 and still in existence today, was another charitable organization training individuals for employment in the Merchant or Royal Navy. Their records can now be found at the National Maritime Museum. The records include registers of boys who were sent to sea from the Society from 1756 to 1958. Some of these registers may be indexed and contain genealogical information such as the name of the parents. There are separate registers for adult males who came through the Society, although these are not indexed.

Merchant Navy Records From 1835

The bulk of the records for merchant seamen start from this period. In 1835 the Merchant Shipping Act stipulated that all crew lists should be stored centrally. Hence a large amount of documentation was produced to comply with this regulation. These include registers of seamen (compiled from the crew lists with the aim of providing additional personnel for the Royal Navy in times of conflict), officers' certificates and more modern crew lists and apprenticeship records.

Apprenticeship Records, 1824–1953

Document collections for the more modern apprenticeship records begin in 1824 due to the passing of the Merchant Seamen Act of 1823. The Act made it compulsory for all ships over a certain tonnage to include indentured apprentices. Lists of indentured apprentices would be made by local officials and, after 1835, these lists would be sent to the central authorities in London. After 1849 these apprenticeships became voluntary and started to be used as a means of training officers. As mentioned above, indexes for these records survive in BT 150, although only a sample survives for the actual records. Some local registers also survive and can be found in the CUST series at The National Archives.

Crew Lists and Agreements

Along with the many other regulations brought in with the Merchant Shipping Act of 1835 was a new method of keeping crew lists. From 1835 to 1860 the records are held at The National Archives in series BT 98. Records up to 1857 are arranged by port and then alphabetically by ship's name. Between 1857 and 1860 the lists are organized by the official ship number. Each seaman serving on the ship is noted in these lists as written agreement had to be sought from these seamen about conditions of service. Each individual's entry should include the name, birth details, rank (known as quality), previous ship and date of joining and leaving the current ship. However, there are no indexes from 1835 to 1857 and it is only feasible to view these records by finding out ship details from seamen's registers (see below).

The later crew lists (from 1861 to 1976) are found in a number of

USEFUL INFO

The official ship number can be found by consulting Lloyd's Register of Shipping *or the* Mercantile Navy List.

archives, and the survival rate is generally much higher then for the earlier ones.

- *The National Archives:* They have a random 10 per cent sample of crew lists from 1861 to 1938 and then from 1951 to 1972. They can be found in a few different series: BT 99, BT 100, BT 144 and BT 165. The crew lists during the Second World War are held separately, in BT 381, BT 380 and BT 387. There is a card index to all ships' logbooks and crew lists received by the RGSS between 1939 and 1950 in BT 385. The index is alphabetically arranged by ship name and is the simplest way of ascertaining where records of the ship of interest will be found. After 1972 only a 2 per cent sample has been retained. The remainder have been destroyed up to 1994, apart from the lists retained by the National Maritime Museum (see below). All lists after 1994 are still with the RSS in Cardiff.
- *The National Maritime Museum:* The Museum holds the remaining 90 per cent of the crew lists for the years 1861 and 1862. It also has the remainder for every year ending in 5 from 1865 onwards (apart from 1945). The Museum also holds card indexes to the ships' crew lists similar to those found at The National Archives for its collections.
- *Local Archives and County Record Offices:* Certain local repositories may also store crew lists for the years 1863 to 1913 for their local ports. The website www.crewlist.org.uk can provide further details regarding which lists such archives may store.
- *The University of Newfoundland, Canada:* The Maritime History Group in the University of Newfoundland, Canada, also retains significant holdings relating to crew lists and agreements. The Group has the remaining crew lists not found elsewhere for the years 1863 to 1938 and then from 1951 to 1976. The Group has also produced three indexes by ship number for their earlier collection:
  - ➤ An index for its crew lists from 1863 to 1913
  - ➤ An index for its crew lists from 1913 to 1938
  - ➤ An index of crew lists deposited with local repositories and archives
  All three indexes can be found at the Guildhall Library, and The National Archives also has a copy of the first index.

Service Records of Merchant Navy Personnel

As mentioned above, the Merchant Navy, like the Royal Navy, distinguished between officer personnel and ordinary seamen. As such there are different records depending on which rank your ancestor held.

▼ The Crew List Index Project aims to bring together information about British seafarers between 1861 and 1913 (when there was no central register) from crew lists and agreements.

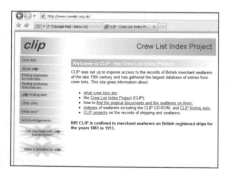

Officers' Service Records

A specific system of registering officers' service in the Merchant Navy was begun in 1845. Masters and mates of ships were recognized as the officer class of the Merchant Navy. If your ancestor served before this time there is no guarantee of finding any relevant documentation; the best route is to trace any vessels that he may have served on, as records for these vessels may mention your ancestor. From 1854 you may find information in any of the sources here.

If you can't find information on an ancestor who you suspect was a captain, master or mate, then it is worth looking in the records listed below for ordinary seamen, particularly in the period after 1913.

1.

Alphabetical Register of Masters

From 1845 to 1854 the RGSS produced such a register by using the crew lists. They are now held in BT 115.

2.

Other sources of officers' service information

- BT 336 contains a register of changes of ships' masters between 1894 and 1948, arranged by ship number.
- The National Maritime Museum keeps records of the applications for certificates until 1928. These would often include details of service up to the date of application. No applications survive after that date.
- Further information can be found in *Lloyd's Captains' Registers* post 1913. Extra information for masters was extracted from the record of certificates issued to foreign-going masters and sent to Lloyd's. Lloyd's began collecting this information from 1869 in manuscript form and these registers can be used to supplement information found in records after 1913.

3.

Certificates of competency and service, 1845–1969

A voluntary examination system was introduced by the Board of Trade in 1845 for men serving on foreign-serving ships. Records of those passing the exams are held in BT 143/1 (this piece also includes a name index) and BT 6/218-219. *Lloyd's Register of Shipping*'s appendix contains an alphabetical list of those passing such exams during these years. The *London Gazette* (available online) also published the names of those men passing such exams.

The system was made compulsory in 1850 and officers were only entitled to certificates if they passed the exams (therefore showing themselves to be competent) or could prove a long period of service with the Navy. Many registers were compiled recording the issuing of certificates that are now held at The National Archives. These registers should include name, date and place of birth, the date the certificate was issued and rank of the individual. Details of which ships the individual served on are also included until 1888. The certificates themselves are in series BT 122 to BT 126 (BT 128 has colonial certificates). An index for them can be found in BT 127.

From the late nineteenth century onwards there were separate series:

- *Engineer certificates* are held in BT 139–142 (colonial registers in BT 140) from 1862 onwards. The indexes can be found in BT 141.
- *Certificates for skippers and mates of fishing vessels* are in BT 129 and BT 130 from 1882 onwards, with the index in BT 138.
- *Certificates for cooks* were introduced in 1908 and indexes to them are in BT 319 for 1913 to 1956. The National Maritime Museum holds the actual registers for 1915 to 1958.

From 1910 the series' indexes were combined to form one single index to all the different types of certificates issued. It is now held in BT 352.

Records of Seamen's Service

Records began to be kept for serving seamen of the Merchant Navy from 1835 onwards. However, this is not a continuous sequence as there were times when no records were kept even after that date. As mentioned, the process of registration was begun in 1835 in order to have a reserve force for the Royal Navy in times of conflict. In total there are five registers of seamen that are open to members of the public. They start in 1835 and the last series ends in 1972 although there are gaps in those dates. Additionally, there is a sixth series from 1973 but it is currently held by the RSS in Cardiff and only available to either the individual or, in cases of decease, the next of kin. The RSS can be contacted at the following address:

> Registry of Shipping and Seamen
> Maritime and Coastguard Agency
> Anchor Court
> Keen Road
> Cardiff
> CF24 5JW

The first three series were kept from 1835 to 1857, compiled from crew lists as required by the Merchant Shipping Act of 1835.

1.

First Register of Seamen, 1835–36

This series can be found in The National Archives series BT 120. The records are organized alphabetically and provide age and birth details, which ship the individual was serving on and his status.

2.

Second Register of Seamen, 1835–44

These records can be found in BT 112. The series is further divided into two parts. Part 1 is from 1835 to February 1840, the sub-series being arranged alphabetically by the first two letters of the seaman's surname (there is an index to this part in BT 119/1). Part 2 is from December 1841 to 1844 and is arranged entirely alphabetically. There appears to be a gap from March 1840 to November 1841.

The first gap in the series appeared in 1845 when the registration system was replaced by a ticketing system for seamen who were leaving the country. This system lasted until 1853 and each individual had to have a register ticket upon departure. As these tickets were then

3.

Third Register of Seamen, 1853–57

In 1853 the ticket system was replaced by a new series of registers which ran till 1857. These records can be found in BT 116, men being listed alphabetically, and birth and career details are given for each seaman.

retained by individual seamen most have not survived. However, there is an alphabetical index to seamen in BT 114. The index provides their place of birth, and the ticket numbers refer to the ticket registers in series BT 113. This latter series provides additional details of each seaman including physical description and career details.

Between 1857 and 1913 there was no system of registering individual seamen as the authorities believed the crew lists were sufficient information. During this period it is only possible to research a merchant seaman if the name of a ship is also known in order to access the crew lists.

The original cards of the Fourth Register of Seamen are stored at Southampton. As many of these cards contain photographs it may be worthwhile viewing these originals:

Southampton City Archive
Southampton City Council
South Block Basement
Civic Centre
Southampton SO14 7LY
Tel +44 (0)23 8083 2251

www.southampton.gov.uk/leisure/history/archives/default.asp

4.

Fourth Register of Seamen, 1913–41

This series, also referred to as the Central Indexed Register, contains records of seamen going up to 1941 and includes approximately 1.25 million cards. Unfortunately, the cards for 1913 to 1918 were destroyed. The originals of the surviving cards are currently stored at Southampton Archives Office. However, The National Archives has microfilm copies of the cards in a variety of series, some organized numerically and others alphabetically:

- **BT 351** This series is an index to all seamen awarded the Mercantile Marine Award for participating during the First World War. As such, they can provide some information about seamen during 1914-18, the period where no other cards survive.
- **BT 350** This series contains the cards originally found in RSS card series CR 10. It is an index begun in 1918 to the cards from 1918 to 1921 and may contain photographs of the seamen.
- **BT 349** The series stores the card series originally known as CR 1. The records are arranged alphabetically by surname, provide birth details, a career summary and the discharge number. Photographs may also be included. It is essential to note down the discharge number as the other relevant National Archive series are organized by this number.
- **BT 348** Further career details are given in this series, originally known as the CR 2 collection. The cards are organized by discharge number.
- **BT 364** This is a combined index of the CR 1, CR 2 and CR 10 cards and is also arranged by discharge number. Most cards are for seamen who served after 1941 and, therefore, it is possible to find the discharge number by using the series for the Fifth Register of Seamen, BT 382 (see overleaf).

Fifth Register of Seamen, 1941–72

The last publicly available Register of Seamen was begun in 1941 in response to the Essential Work (Merchant Navy) Order. The Government wished to ensure that there would always be adequate personnel to staff vessels and hence created a Reserve Pool of seamen. A system of continuous paid employment was introduced to fulfil this requirement and merchant seamen could now have permanent employment. A new registration system was also introduced along with the Order, and all men who had been serving during the past five years had to register with the authorities. The Fifth Register has two main sources that form its records:

1. Seamen's pouches

When seamen were registering they were required to provide their CR 1 and CR 2 cards from previous service. They were placed in an envelope and any subsequent paperwork resulting from service would be placed in this envelope when they were discharged. These collections came to be known as 'seamen's pouches'. Unfortunately, not every pouch survives; it is estimated that there is a 50:50 chance of finding a pouch for a seaman during this period. The surviving pouches can be found in the following series:

- **BT 372** This series contains the majority of pouches including records up to 1973. The series has been fully catalogued by name and can be searched by surname online in The National Archives catalogue.
- **BT 390** This includes pouches for merchant seamen who worked on Royal Navy ships during the Second World War (up to 1946). It is currently being catalogued by individual name, thus making it possible to search the series online on The National Archives website, although at the time of publication this process is not complete.
- **BT 391** This is another series specific to merchant seamen involved in the Second World War, in this case regarding individuals who worked in 'special operations' between June 1944 and May 1945. It has been fully catalogued by surname and can be searched online on The National Archives website.

2. Docket books

Docket books (CRS 10 forms) can be found in The National Archives series BT 382 and give service details of merchant seamen along with birth details and discharge numbers. They are arranged alphabetically by surname, subdivided into eight different parts. The first two parts contain the largest number of records, relating to European and British dependent seamen for 1941 to 1946 and 1946 to 1972 respectively. The majority of records are held within these parts although they may not necessarily be in the right date range. Parts three to six are specific to Asiatic (mainly Indian and Chinese) merchant seamen. Part seven contains records for men who served on ships requisitioned during the Second World War. The last part has records for pension purposes.

Medals Awarded

Merchant Navy personnel have been entitled to medals for service and gallantry from the mid-nineteenth century onwards. Indeed, individuals of the Merchant Navy were often involved in conflict during the two world wars and thus became entitled to the same medals as were awarded to military personnel. Below is a summary of the main types of medals given to merchant seamen.

Gallantry Medals

The main medal awarded for gallantry at sea was the Albert Medal. Medal registers for the Albert Medal can be found in BT 97, covering the years 1866 to 1913. Further medal registers for awards of gallantry (from 1866 to 1986) can be found in the following series: BT 261, BT 339 and MT 9.

Medals Awarded for the First World War

Merchant seamen who were involved in the First World War were entitled to four medals:

- The 1914–1915 Star
- The British War Medal
- The Victory Medal
- The Mercantile Marine Medal

The first three medals were the same medals awarded to branches of the armed forces. As such, awards of these medals will be found in the same series as that of the Royal Navy, ADM 171.

The Mercantile Marine Medal was specifically awarded to individuals who served for a minimum of six months during the First World War and made one or more journeys through a danger zone. BT 351 holds an index to each recipient of the award. Each entry should record the name, birth details, rank and other career details.

Merchant seamen and officers were often awarded other Royal Naval gallantry awards and records can be found in ADM 116, ADM 12 and ADM 137.

If you know that the individual you are researching died in conflict then refer to the Commonwealth War Graves Commission's 'debt of honour' roll, available online at www.cwgc.org (discussed in more detail in Chapter 9).

▼ An entry from the Fifth Register of Seamen, showing biographical details for Donald Main and a list of ships on which he served.

Medals Awarded for the Second World War

The Merchant Navy formed an important part of the fighting force during the Second World War. The Ministry of Shipping took control over the Merchant Navy as soon as the conflict started in 1939.

The main record series relating to seamen who were awarded a service medal for this period is in BT 395. However, medals were not issued automatically; rather the individual had to claim his medal. Hence, if your ancestor did not claim his medal there will be no record of him in this series. (Indeed, it is still possible for individuals to claim their medals, if they can prove their service to the Registry of Shipping and Seamen. As such the series covers the dates 1946 to 2002.) Any of eight different medals were awarded to merchant navy personnel: the War Medal, the Atlantic Star, 1939-1945 Star, Africa Star, Pacific Star, Burma Star, France and Germany Star, and the Italy Star. This series has now been placed on The National Archives website at Documents Online and can be searched free of charge (viewing a record will incur a cost, though).

Gallantry awards for this period have been described above. Additionally, award of such medals would also be recorded in the *London Gazette* and *The Times*, both of which can be searched online.

The Merchant Marine Memorial was built at Tower Hill, London, to commemorate those merchant seamen who died during the world wars. The Trinity House Corporation has the register for this memorial, which can be consulted on their premises at Tower Hill.

The Lloyd's Marine Collection

The London Guildhall Library houses the marine records of the Corporation of Lloyd's of London. The collection includes a number of different sources useful for tracing seamen and shipping in general. They include:

● *The Mercantile Navy List:* This is the official list of British registered merchant ships and has been published annually since 1850 (excluding 1941-46). They list each ship on an individual basis, including its official number, owner and size of vessel. From 1857 to 1864 details of the certificates of masters and mates on the ship were also included.

● *The Lloyd's Register:* Another comprehensive list of vessels, produced from 1775 onwards. The lists were compiled for insurance purposes and give details of each individual vessel's name, number, owner, master and construction details.

● *Lloyd's Captains' Register:* This is the most useful genealogical register, listing holders of masters'

certificates from 1869 to 1948. Each entry also gives the birth details of the individual and a brief career history.

● *Lloyd's List:* This list was begun in 1741 and was a weekly register of ships' arrivals and departures at each particular port, detailing from which port the vessel had come and where it was headed. The name of the master of the ship is also recorded. The Guildhall Library has an index to the ships from 1838 onwards. The National Maritime Museum and the British Library's Newspaper Archive also have copies of this list.

CASE STUDY

Amanda Redman pt 1

Amanda Redman knew very little about her mother Joan's family, the Herringtons, other than rumours that her grandfather – Joan's father – William Herrington was a strict disciplinarian. Yet on talking to Joan a long-forgotten family secret emerged – a missing half-brother, Cyril Herrington, who was the illegitimate daughter of Joan's mother, Agnes, conceived before she married William. Intrigued, Amanda decided to find out more about her elusive ancestor.

No birth certificate could be found for Cyril, but on asking around the family it turned out that he had died in Liverpool, and that he had married and had a child. By painstakingly looking through civil registration material, Amanda tracked down her half-cousin and, on meeting her, gained some important clues about her half-uncle's life – including the fact that he served in the Merchant Navy.

By running a search online and checking The National Archives catalogue, Amanda discovered that a key document survived in record series BT 372 – Cyril's seaman's pouch, which contained personal information about him. On inspecting its contents, she found photographs of him dating from 1936, when he first signed up as a cook, and 1969, at the end of his career, as well as details of his height, next of kin, address and date of birth – listed as 19 April 1917. However, further documentation also indicated that he had, at some point, changed his name from Pillings to Herrington (possibly giving a clue as to his natural father's surname) and that his real date of birth was 1919, not 1917.

Amanda was able to cross-reference the information in his seaman's pouch with another document in the Fifth Register of Seamen in record series BT 382, including a complete list of all the vessels that he had served on from 1941 until his discharge in 1969, including in 1942 an unauthorized stint on an American vessel, the *Gulf Coast*.

Rather remarkably, further investigation into Amanda's family tree led to a maternal 2 x great-grandfather who also earned a living in the Merchant Navy – a man with the rather grand name of James St Ledger, who was born in 1837 in New Ross, County Wexford, Ireland. At the time of the 1881 census he was listed as the master of a vessel called the *Alma*, which was in port at Porthcawl when the enumerators visited. The ship was described as a coasting schooner, with its home port listed as Falmouth, where the rest of James's family lived. However, he was accompanied on board the *Alma* by his son, William Henry St Ledger, an ordinary seaman clearly following in his father's footsteps. Some of the later voyages of the *Alma* were researched by checking the official ship's number against the surviving crew lists and agreements at The National Archives, in record series BT 99.

Births, Marriages and Deaths at Sea

Masters of British ships were required to record all births, marriages and deaths that occurred on their vessels from approximately the mid-nineteenth century onwards. The masters would pass this information on to the Registrar General of Shipping and Seamen who would subsequently send on the information to the appropriate General Record Office for England, Scotland and Ireland. The National Archives also has the records that were originally with the RGSS. Separate registers were kept for seamen and passengers until 1889. After that date a new combined series of registers was introduced. Details about General Record Offices and the system for recording civil registration certificates for births, marriages and deaths can be found in Chapter 5.

Deaths of Seamen

In 1851 the Seaman's Fund Winding-up Act was introduced. The Act required that masters of British ships hand over the personal belongings and wages of seamen who died on their ship to the sailor's next of kin. Records of this were maintained in registers now found in BT 153 (indexed by BT 154). The registers should detail where the seaman was engaged from, his cause of death and his wage details.

Marine Records of the East India Company

The East India Company (EIC), being originally a trading company, was very prominent in marine activities. The company owned and chartered many ships until its abolition in 1858. Many published sources listing the ships owned by the EIC can be consulted either at the National Maritime Museum or at the British Library (where the original records of the EIC are stored). The various registers produced by Lloyd's (mentioned above) also include information regarding EIC ships.

The British Library has the following EIC records that may be useful to look at:

● The *East India Register and Directory*, which contains lists of masters and mates under 'marine establishments', along with Army officers.
● The EIC records sometimes refer to the appointments of some individual marine officers.
● Journals and logs for approximately 10,000 ships (although not every ship's record has survived), from the early seventeenth century to 1856. The India Office records are described in more detail in *India Office Library and Records: A Brief Guide to Biographical Sources* by I.A. Baxter (London, 1979; second edition 1990).

In addition, BT 156 has printed monthly lists of deceased seamen (giving age and birthplace and death details) from 1886 to 1890. BT 157 lists deaths and cause of death of seamen from 1882 to 1888.

Births, Marriages and Deaths of Passengers, 1854–91

All of these three events were legally required to be recorded after the Merchant Shipping Act of 1854. They were extracted from the official ships' logs and can be found in the following sources:

- BT 158 has details of births, marriages and deaths of all passengers at sea from 1854 to 1890. After 1883 there are no marriages recorded; only births and deaths to 1887 (with deaths only from 1888).
- BT 159 has registers of deaths at sea of British nationals from 1875 to 1891.
- BT 160 lists births of British nationals from 1875 to 1891. The registers of BT 159 and BT 160 were formed in response to the Birth and Death Registration Act of 1874, whereby ships' masters had to record births and deaths separately for all British and foreign-registered ships.

Combined Registers of Passengers and Seamen, 1891–1964

After 1891 a new system was introduced whereby births, marriages and deaths of passengers and seamen (marriages and deaths only) were recorded together in a combined register. The majority of these records are in BT 334 and include details of non-British nationals too. There are indexes to births and deaths, indexed by individual surname and by the name of the ship. All entries should include the ship's name, official number and where it was registered.

- *Births:* Births are recorded from 1891 to 1964 (indexes to 1960). The record gives the father's name, rank or occupation, birthplace or nationality and last residence and the mother's name, maiden surname and last residence.
- *Marriages:* Marriages start from 1854 and go up to 1972 and are all found in piece BT 334/117 (also available online).
- *Deaths:* Death registers are available from 1891 to 1964. Additionally, BT 341 included any inquests for deaths at sea (organized by ship name).

HOW TO...

... find online sources for births, marriages and deaths at sea

Some websites provide information on births, marriages and deaths at sea:

1. www.findmypast.com
This website has placed all of the main Registrar General of Shipping and Seamen records available at The National Archives on its website. It is possible to search births from 1854 to 1887, marriages from 1854 to 1883 and deaths from 1854 to 1890. It can be searched free of charge but viewing the record incurs a cost.

2. www.shiplist.com
This website has transcribed every marriage found in BT 334/117 and can be viewed free of charge at www.theshiplist.com/Forms/marriagesatsea.html.

CHAPTER 14

Occupations: The Sea

Aside from joining the Merchant Navy, there were plenty of other lines of work that our seafaring or coastal ancestors chose to embark upon, and they are described in more detail in this chapter – along with ways you can find out more about them in archives, libraries and other institutions.

Coastguards

The history of the Coastguard began with the desire of the authorities to limit and even eliminate the large amount of smuggling prevalent since medieval times when duties were first imposed on various imports and exports. Indeed, whenever a duty was placed on a particular product, such a product would automatically become suitable for smuggling. The Board of Customs was historically responsible for collecting the duties placed on such goods and appointed 'Preventive Officers' in wartime to fight against smuggling, and also for intelligence duties. However, this was not a permanent force.

The modern-day Coastguard service was established in 1822 to counteract smuggling, but its immediate predecessors – the Riding Officers, the Revenue Cruisers and the Preventive Water Guard – were formed earlier in the late seventeenth century as the first peacetime forces to guard the sea and to curtail smuggling. Riding officers were deployed in Kent and Sussex, with over 300 people being employed by the early eighteenth century. Their specific task was to stop any movement of smuggled goods inland, goods that may have escaped the notice of Revenue Cruisers or customs officials, while the Revenue Cruisers originally operated on the Kent and Sussex coasts to prevent any

> *The Coastguard service was established to combat smuggling – only later did lifesaving become its function.*

smugglers arriving on shore. By 1782 Revenue Cruisers comprised 40 armed vessels, with 700 men working on these ships. The two bodies' jurisdiction gradually increased by the end of the eighteenth century to cover most of the British coastline (Scotland had its own forces).

The third force, the 'Preventative Guard' was formed later in 1809. The force also operated on the coastal waters seeking out smugglers. Thus there were three stages at which a smuggler could be prevented from successfully bringing his goods into the country.

▲ Smuggling was a common problem in the 18th century, represented here by an attempt to break open the King's Customs House in Poole.

After the end of the Napoleonic Wars in 1815 the idea of uniting these three branches was first suggested by Captain Joseph McCulloch for his forces in Kent, and in 1816 the Admiralty established the Coast Blockade service. This new force began patrolling the Kent coastline and the service proved very successful. In 1821 a committee suggested that the three services be united for the whole country and that this new body still remain under the overall control of the Board of Customs, with the Admiralty appointing officers for this service. Thus the 'Coastguard' service started operating on 15 January 1822. The Coast Blockade service was amalgamated into the Coastguard service in 1831 and in total the force employed approximately 6,700 men. These men would work on the seas and onshore. Onshore Coastguard stations were established throughout the country, each under the command of a Chief Officer (usually a lieutenant of the Royal Navy).

Although the Board of Customs was responsible for the Coastguard service, the Admiralty controlled its recruitment as it wished to use the service as a reserve force for the Royal Navy. At the end of the Crimean War in 1856 entire control of the service was transferred to the Admiralty, which subdivided the Coastguard into three separate units: the Shore Force; the Permanent Cruiser Force; and the Guard ships.

The functions of the Coastguard grew; along with controlling smuggling the body was responsible for coastline defence and continued to act as a reserve force for the Navy. Additionally, lifesaving and dealing with wrecks became important functions. These responsibilities made the recruitment of personnel with naval skills to the Coastguard increasingly important. In this the service was aided by the fact that many new recruits came from the Bengal Marine, after the East India Company disbanded its Navy.

In 1925 the Royal Navy became responsible for the service's Naval

Signalling Force, the Board of Customs took control of the Coast Prevention Force, and the Board of Trade became responsible for the Coast Watching Force. Overall control of the Coastguard service is now in the hands of the Department of Transport.

Service Records for the Preventative Forces and the Coastguard Services

The majority of the surviving records are to be found in The National Archives. However, as many different departments controlled the service throughout various points in its history, the records are scattered amongst many series.

Preventative Forces' Records

Riding officers and Revenue Cruisers were used both by the Board of Customs and the Board of Excise. The difference was that a Riding officer employed by the Board of Excise would be responsible for collecting duties and ensuring individuals were not evading excise duty inland throughout the entire country. (Excise duty differed from customs duty as it was a tax paid on goods within the country.) It is important to know which Board – Customs or Excise – your ancestor worked for, as their records are stored in different National Archive document series, under CUST. This series is discussed later in this chapter, in the section looking at the careers of Customs and Excise officers.

It may be possible to find information for the earliest recruits, serving in the seventeenth and eighteenth centuries, in the Admiralty and Secretariat Papers (ADM 1) indexed in ADM 12, or in the Treasury Board Papers (T 1). Treasury papers have a published calendar for 1556 to 1742. Otherwise, searching through these sources can be somewhat time-consuming.

As many employees of the preventative forces would have joined the Coastguard service upon its inception, Coastguard service records would also detail the earlier service details of such individuals. These records will be discussed in further details below. Pension records for some men can be found in CUST 40/28 (1818 to 1825).

The Admiralty was jointly responsible for administering the Revenue Cruisers. Hence, some records may be found for these officers amongst the Admiralty series (ADM). As the Admiralty appointed officers and other ranks to the various posts from 1816, the appointments of some officers can be found in ADM 6/56 (1816 to 1831). Additionally, the Navy List included officers for this service from 1814 onwards.

Service Records for the Coastguard

As already mentioned, depending on when your ancestor served in the Coastguard service, the majority of records will be found in either the Admiralty or the Board of Customs' records. Unfortunately, there is no single name index available to assist in your search.

Admiralty records for coastguard employees are in series ADM 175. Depending on which part of the United Kingdom your ancestor served in, records can be found as follows:

- *England:* Registers in ADM 175/74–80 for 1819 to 1866
  Indexes in ADM 175/97–98 for 1823 to 1866
  Nominations in ADM 6/199 for 1831 to 1850 and ADM 175/101 for 1851 to 1856
- *Ireland:* Registers in ADM 175/74, ADM 175/81, ADM 175/99 and ADM 175/100 for 1820 to 1849
- *Scotland:* Registers in ADM 175/74 for 1820 to 1824

ADM 175/102 also has indexes for discharge records for 1858 to 1868. Unfortunately, from 1866 to 1886 there are no surviving records. Thereafter service records are divided between ratings' records and officers' records.

1. Ratings' records

Ratings' records are available from 1900 to 1923. Service records for those dates can be found in ADM 175/82A to ADM 175/84B. These records are in card form and arranged alphabetically.

Further records for 1919 to 1923 are in ADM 175/85–89, indexed by ADM 175/108. In 1919 the Coastguard was significantly reduced in numbers with the end of the First World War. Hence many individuals were discharged and their records are in ADM 175/91–96. ADM 175/96 is indexed by ADM 175/107. Additionally, as many Coastguard personnel were involved in the First World War they would have been entitled to medals. These awards can be found in ADM 171.

2. Officers' records

Records for officers are available from 1886 to 1947. There are indexed service registers in ADM 175/103–107 and ADM 175/109–110.

As many men of the Coastguard had previously been with the Royal Navy it may be possible to find further information on them using the Royal Navy service records (discussed in Chapter 10). Specifically, records of Revenue Cruiser officers (lieutenants, masters and boatswains) being appointed by the Admiralty can be found in ADM 6/56 (for 1816 to 1831) and ADM 2/1127 (for 1822 to 1832). ADM 119 also has quarterly musters of Revenue Cutters for 1824 to 1857.

▲ This coastguard register of nomination for appointments, 1857, can be used to trace the start of a period of service for an individual.

Other Coastguard Records

There are collections of coastguard records (succession books detailing the individuals serving at various stations at various points) arranged by location of ship, type of ship or name of ship. These records also contain personal details of individuals, but it is only feasible to search them if you are aware of either the type of ship your ancestor served on or its location. Below is a summary of the arrangement of these records:

- *Location:* From 1816 to 1866 there are records in ADM 175/1-10, ADM 175/13-19 and ADM 175/22-23 arranged by location, under the British Isles, England, Wales or Scotland. Further details can be obtained by searching in The National Archives catalogue.
- *Ship type:* Records for Cruisers survive from 1822 to 1863 in ADM 175/24-25. Records for Tenders are in ADM 175/26 for 1858 to 1868.
- *Establishment books:* Records for men working on Revenue Cruisers in numerous locations throughout the United Kingdom can be found in ADM 175/24-73. These records are the ships' establishment (crew list) and record books and run from 1816 to 1879. As they are arranged by location you need to know the approximate location where your ancestor was serving to use these books.

As the Board of Customs also had responsibility for the Coastguard service on different occasions it is possible to find records in the CUST series in the National Archives too. Below is a list of the most useful documents:

- CUST 19/52-61 contains records relating to the establishment of Revenue Cruisers from 1827 to 1829
- CUST 29/40-42 has Coastguard minute books from 1833 to 1849
- CUST 38/32-60 has Coastguard statistics
- CUST 39/173 details salary and incidents of the Thames Coastguard, 1828 to 1832

Coastguard Pensions

Pension records for retired Coastguard personnel can be found in a variety of sources:

- **ADM 23** Pensions paid by the Admiralty and civil pensions can be found in this series from 1866 to 1926.
- **PMG 23** This series contains additional records of civil pension payments paid by the Paymaster General from 1855 to 1935.
- **PMG 70** This series includes pensions paid to chief officers of the Coastguard by Greenwich Hospital. Payments began after an Order in Council of 16 February 1866. They cover the years 1870 to 1928.
- **T 2** This series contains the annual lists of individuals who were awarded pensions and superannuations. It was usual practice for the Board of Customs (prior to 1856) and thereafter the Admiralty to notify the Treasury of such awards. The Treasury would publish these in their 'Public Office' registers. Unfortunately, only a tiny fraction of the paperwork referred to in T 2 survives in T 1.
- **CUST 30** This series ('Out-Letter Entry Books: Extra Departmental') contains the actual texts of the Customs' letters to the Treasury, mentioned above. The information should provide details of the individual's career and the circumstances of his retirement.

Miscellaneous Documents

It is possible to find original documents relating to the Coastguard stations themselves, including architectural plans, in The National Archives series WORK 30 (1844 to 1914). CUST 42/66 has a schedule of leases and deeds for Coastguard properties throughout Great Britain in 1857. ADM 7/7-39 contains records of the repair and upkeep of individual stations, arranged in a series of registers for 1828 to 1857.

Other records of relevance include:

- CUST 38, containing annual abstracts of the Coastguard providing a variety of statistical data including salaries, details of expenses and costs involved with Coastguard houses and stations.
- CUST 29/40-42, containing a number of entry books of orders and minuting other such information.
- ADM 114/11, relating to the many issues and instructions concerned with the transfer of control of the Coastguard from the Board of Customs to the Admiralty in 1856.
- Paperwork relating to a myriad of administrative topics relevant to the Coastguard service can be found in ADM 1, ADM 116, MT 9, BT 166 and ADM 120.

USEFUL INFO

- *There was a volunteer Coastguard service established in the 1860s, the volunteer Life Saving Apparatus Companies. In 1911 individuals working in the volunteer force became entitled to the Rocket Life Saving Apparatus Long Service Medal, whose registers are in BT 167/84 (for 1911 to 1935).*

- *BT 167/87–97 contains annual lists of enrolled volunteers from 1920 to 1932. These registers can provide useful genealogical data such as age, birth details and residence.*

Lastly, as HM Coastguard had many stations throughout the country, local archives may also have records for the Coastguard service that operated in its locality. Most often these records are the local station records.

Customs and Excisemen

HM Customs and Excise is one of the oldest government departments still in existence today. Historically, they were two separate departments (the Board of Customs and the Board of Excise) until their amalgamation in 1909. The difference in meaning between the two terms 'customs' and 'excise' lay in the type of good that is taxed. 'Custom' was defined as the 'ancient and rightful custom' of paying duty of any type, be it in kind, regal or ecclesiastical. It later became limited to the payment on certain commercial goods. Excise duties were those taxes placed on goods produced inland (and not imported), most commonly alcoholic beverages, but historically on other goods such as salt and pepper.

A nationwide custom-collecting body originated in the thirteenth century. The Winchester Assize of 1203/4 declared that the custom duties collected in various English ports should be given directly to the Exchequer of the King, and in 1298 government custom officials were appointed to collect duties on behalf of the crown. These individuals were known as 'custodes custumae' and a more famous example of such an official would be Geoffrey Chaucer who was appointed as an official for the port of London in 1378.

In 1643 a Board of Customs was created, staffed by custom collectors. In 1671 a permanent Board of Customs was formally established by Letter Patent. By 1823 all the separate boards for England, Scotland and Ireland were merged to form one single Board of Customs for the whole of the United Kingdom.

The Board of Excise was also formally established in 1643 and by 1660 excise duty on alcohol was confirmed by legislation, even though it was unpopular. Permanent Boards of Excise for the different home nations over the next fifty years were established: Ireland in 1682, England and Wales in 1683 and Scotland in 1707. During the course of time excise duties were placed on more goods other than alcohol, such as paper, salt and soap. During the late eighteenth and early nineteenth centuries increased taxes were placed on a number of goods. This in turn led to an increase of smuggling, which resulted in Customs and Excise departments becoming more efficient at policing this smuggling, and (like the

> *The difference between 'customs' and 'excise' lies in the type of goods that are taxed.*

Customs boards) the Excise departments for the different countries of the United Kingdom were amalgamated into one unified Board of Excise for the entire United Kingdom. In 1849, it was further combined with the Board of Stamps and Taxes and became known as the Board of Inland Revenue.

The two bodies encountered further changes during the twentieth century. The Finance Act of 1908 ordered that the Commissioners of Customs be responsible for the collection of excise duty. Henceforth a new body was created on 1 April 1909, Her Majesty's Customs and Excise, directly responsible to the Chancellor of the Exchequer.

Records of HM Customs and Excise

The records and archives of HM Customs and Excise and its various predecessor bodies are held in different places depending on where your ancestor was employed. As there were separate institutions for Scotland and Ireland up until 1823, records prior to this date will be at the National Archives of Ireland, the Public Record Office of Northern Ireland and the National Archives of Scotland (discussed below). The records for officials for England and Wales can be found at The National Archives.

Custom Officials in England and Wales

The very first Custom officials were appointed in 1294. If you believe your ancestor was a Custom official from the thirteenth century onwards there are two published sources that may be a useful starting point for your search. Both can be found in the library in The National Archives and at the British Library:

- *The English Customs Service 1307–43, A Study of Medieval Administration* by Robert L. Baker (The American Philosophical Society, 1961)
- *An Account of the Commissioners of Customs, Excise, Hearthmoney, and Inland Revenue, 1642–1913* (HMSO, 1913)

The main series of records begin from the late seventeenth century onwards and are to be found mainly in the CUST series in the National Archives, along with a small number in the T series (Treasury documents) and some in the C (Chancery) series.

The following is a list of the key documents worth viewing when tracing a Customs official:

- *Bills and Warrants for the Appointment of Customs Officers:* These are found in C 208 from 1714 to 1797. Every officer employed during that time would have been by warrant. The series is indexed in C 202/267 to C 202/269.
- *Custom Board Minute Books:* These books run from 1734 to 1885 and contain information about individual officials and are in series CUST 28. They should detail initial and subsequent postings of officers along with any promotions awarded to individuals. Each volume should contain an index.
- *Pay and staff lists:* This is a large collection of documents spanning four centuries. They are organized by place and seldom give any personal or genealogical information about the individual but record the post held by each individual, usually on a quarterly basis, along with salary details. As they are organized by place it is important to know where your ancestor served to begin using these documents. The main collections are:
  - ➤ PRO 30/32/15-19: 1673 to 1689
  - ➤ CUST 18: 1675 to 1813
  - ➤ CUST 19: 1814 to 1829
  - ➤ CUST 39: 1671 to 1970
  - ➤ T 42: 1747 to 1847

  Additional records for Scotland can be found in T 43 (1714 to 1829) and CUST 39/104-121 (1860 to 1885). This is continued after 1885 up to 1894 in CUST 39/141-144 but also includes Ireland. Separate information for Ireland is also located in CUST 20 (1684 to 1826) and CUST 39/122-140 (1840 and then 1860 to 1885).
- *Pension records:* These records can be useful as they may contain additional details about family members. The main series can be found in CUST 39/145-151 (1803 to 1922). Separate records for Ireland and Scotland can be found in CUST 39/161-22 (1795 to 1898) and CUST 39/160 respectively. Further records regarding applications for widow's pensions are found in T 1. There is a published calendar for this series covering the records till 1745. Thereafter indexes can be found in T 2, T 4 and T 108.
- *Appointments, pay, discipline and other matters:* Two general series that contain a large number of documents relating to appointments, pay, discipline and other such matters can be found in CUST 119 (1802 to 1926) and CUST 40 (1831 to 1921).
- *Ham's Custom Year Book:* This is a very useful publication covering the nineteenth and twentieth centuries. It was an annual publication listing all the employees of HM Customs and Excise, the

CASE EXAMPLE

Working for HM Customs

Customs pay lists were particularly useful when tracing the career of **John Hurt's** *ancestor,* **William Richard Browne.** *William Richard Browne's son,* **Walter Lord Browne,** *had claimed that his father was 'head of the bond office, H.M. Customs, London' on his marriage announcement in the local newspaper,* The Great Grimsby Gazette and Commercial Advertiser, *in 1857. However, by tracing his career through the original pay lists, it was discovered that this was an exaggeration and that William Richard Browne was only a cocket-writer for HM Customs at the port of London.*

date of their first appointment, where they were stationed and also details of their retirement. As such it is perhaps the easiest method of tracing an ancestor who was part of this profession. The British Library and The National Archives both have copies, from 1875 to 1930.

- *Correspondence:* If you have ascertained where your ancestor worked, you might want to examine the correspondence files between the local Collector of Customs and the Board of Customs based in London. These are to be found at The National Archives in a wide variety of CUST series, arranged geographically by the name of the relevant Collector's district. Correspondence letter books contain copies of all letters sent to the Board in London, including disciplinary matters, promotions, appointments of senior staff and occasional pay lists, as well as incidents that occurred on a daily basis – many of which relate to piracy, as well as more mundane matters such as assessment of customs duties and information on general maritime activity in the port.

> *Letters sent to the Customs Board cover promotions, disciplinary matters and piracy.*

Excise Officers

Records for Excise officers are similar in nature to that of Customs' officials and are mostly held at The National Archives.

- *Pay lists:* There are lists for English Excise officers in T 44 (1705 to 1835) and for Scotland's officers in T 45 (1708 to 1832). Unfortunately, few records survive apart from an index in the National Archives for Scotland due to their destruction in a fire.
- *Board Minute Books:* These can be founding CUST 47, starting in 1695 and finishing in 1867.
- *Pension records:* These are stored in CUST 39/157–159 (1856 to 1922).
- *Entry papers:* This series of records are unique to Excise officers and are held in CUST 116 for England (1820 to 1870) and Scotland (for 1820 to 1829). They are useful documents giving birth details and marital status. They were created following an individual's application and form two parts of a letter: a letter of recommendation and an existing Excise officer's report on the candidate's training. They are searchable online by name in The National Archives catalogue. Records for Ireland can be found in CUST 110 and CUST 119, although these are not currently searchable by name.
- *Staff lists:* The earliest list available is one for 1692 in CUST 109/9. The main series are for the nineteenth and twentieth centuries (1870 to 1937) in CUST 39/225 to 249.

- *Ham's Inland Revenue Year Book:* This is a similar publication to the one for Customs officers (above). However, as the Board of Excise was part of the Inland Revenue until the two Boards' amalgamation in 1910, Excise officials may be noted as Inland Revenue officers until that date.

Customs and Excise Records in Scotland

If your ancestor was working in Scotland there may be additional sources worth consulting at The National Archives of Scotland. They are organized into two series, before and after the Act of Union of 1707.

1.

Customs records

Prior to 1707 records can be found within the series for the Scottish Exchequer in E38 and E71-74. These include customs book, which may detail names of ships' masters and merchants although the primary purpose of these records was to record the duties placed on goods entering Scottish ports and therefore there are no lists of crew included. After 1707 Scottish Customs establishment books list the names and salaries of individuals employed at each port. They can be found for the years 1715 to 1822 in series CE3. CE12 contains the establishment lists of those involved in the collection of salt duty in particular, covering the years 1714 to 1798. Additionally, letter books of the Board of Customs can give information on employees and their families along with personal information on other people not employed by the Board but in contact with the authority, such as merchants, ship-owners and mariners.

2.

Excise records

Many excise records were destroyed in a fire during the nineteenth century. However, the Archive has a biographical card index detailing Excise officers between 1707 and 1830 that can be found in the records that did survive. Additionally, piece GD1/54/10 is a list of all officers in 1743. CE13 contains registers of appointments and removals from 1813 to 1829. Scottish Excise minute books have some details of its staff from 1824 to 1830 and are in CE16. One further list of officers in 1794 is in CE6/19.

Fishermen

If you find from a birth, marriage or death certificate or a census return that your ancestor was a fisherman it may also be possible to find some further documentation in local archives or The National Archives. Prior to the Merchant Shipping (Fishing Boats) Act of 1883, many records of fishermen would be included within the general records of the Merchant Navy (see Chapter 13). This new legislation made it compulsory for skippers to have written agreements with any seaman he carried on his ship as crew from any port in the United Kingdom. Prior to this Act the crew lists would be within the general series of crew lists, in The National Archive series BT 98 and BT 99. From 1884 to 1929 the records can be found in BT 144. However, although similar to the records in BT 98 and 99, these later crew lists do not survive in their entirety and only a 10 per cent sample was retained. Post 1929 these lists are once more included within BT 99, arranged by the official number of each vessel.

The majority of the remaining records can be found in the University of Newfoundland or the National Maritime Museum (for years ending in 5, from 1885 to 1935 and 1955 to 1985). A small number may also be found in local repositories. Appendix 2 of *Records of Merchant Shipping and*

▲ Fishermen faced constant danger from the elements, as well as back-breaking work, to bring their catch to shore.

Fishing Apprentices

Similar to many other occupations and trades, there was an apprentice system for fishing. This system became centralized in 1894 with the Merchant Shipping Act, which extended the apprentice system for general merchant shipping to cover fishing vessels too. Henceforth copies for all apprenticeship indentures were to be forwarded to the Registrar General of Shipping and Seamen. These records can now be found in BT 152, covering the years 1895 to 1935, although only on a five-year basis. This series is indexed by BT 150.

Many local archives may also have information on fishing and apprenticeship records for their locality. If you are aware of where your ancestor was based it is always worth checking to see what such repositories may contain. For example, Hull and Grimsby archives contain crew lists along with apprenticeship records, the area having an important maritime history. Further information on the nature of their holdings can be found at www.hullcc.gov.uk/portal/page?_pageid=221,575789&_dad=portal&_schema=PORTAL.

CASE STUDY

Amanda Redman pt 2

In the previous chapter, Amanda Redman discovered various ways to find out more about her ancestors who had served in the Merchant Navy, in particular her elusive relative Cyril Herrington. However, whilst building her family tree she also found that earlier branches of the Herrington family had a long association with the sea, particularly her 2 x great-grandfather Benjamin Herrington, who – according to census returns – lived with his family in the coastal village of Southwold, Suffolk.

Although little more than the names and occupations of family members, and their addresses, were listed in the census, Amanda was able to undertake further research on Benjamin Herrington at the Lowestoft archive, and it would appear he was something of a hero. Apart from earning a living from the sea as a fisherman, he also became actively involved in the local lifeboat service. In 1854, as Coxswain he was awarded a Silver Medal along with a colleague, William Waters, when they helped rescue nine crew from the ship *Sheraton Grange*, after it had run into difficulties during a winter storm on 29 November 1853.

Indeed, Benjamin Herrington was awarded a second Silver Medal in 1859 for his part in saving lives of eleven crew members from the *Lucinde*. Yet the year was tinged with personal tragedy for Benjamin. Amanda also discovered at the archive, in a list of parish burials, that his wife Ann had died, and was buried in St Edmund parish church, Southwold. Benjamin eventually remarried, and it was his son from this second union – John Clair Herrington – from whom Amanda is descended.

Seamen by Smith, Watts and Watts (PRO Publications, 1998) has details on the information held in local repositories.

These crew lists should provide:

- Name, birth details, salary, length of engagement and total number of the crew
- Owner's and skipper's name and address

Additionally, skippers and mates of fishing vessels would also be required to take the competency examinations introduced for masters and mates for other merchant vessels from 1884 onwards. As stated in Chapter 13, certificates for skippers and mates of fishing vessels from 1882 onwards can be found in The National Archives series BT 129 and BT 130 with the index in BT 138. The index covers the period 1880 to 1917 and should provide the individual's name, date and place of birth. Although the index in BT 138 extends to 1917, after 1910 the system was modified once more. The competency certificates for skippers were combined with competency certificates for masters and mates to form one single series – BT 352 (discussed in Chapter 13).

Dockyards and Shipyards

As seafaring was such an important part of the country's history, the construction and maintenance of ships would have created many jobs and employment opportunities for our ancestors. Many dockyards and shipyards operated at various ports throughout Great Britain and Ireland. The Royal Navy itself had dockyards throughout the world as well as its ones at home. You may be able to find documentation relating to your ancestor's employment in this industry depending on when and where they worked. There is a large amount of material relating to the Royal Naval dockyards in particular.

Royal Naval Dockyards

The first six Royal Naval dockyards sprang up in England in the seventeenth and eighteenth centuries, sited at Deptford, Woolwich, Chatham, Sheerness, Portsmouth and Plymouth. There were additional outposts in England and also overseas. These dockyards were under the control of the Navy Board although officers would be appointed by the Admiralty. Most employees of the Royal Naval dockyards would either have come from the Royal Navy itself or would go on to work for the Royal Navy and it may be possible to find records for both periods of employment.

Records for the naval dockyards are held at The National Archives and the National Maritime Museum. Most of the staff records are with The National Archives, while the National Maritime Museum holds a number of letter books, correspondence papers and plans of individual dockyards. Further information can be found about their holdings in their research guide available online at www.nmm.ac.uk. The museum also has a card index of senior dockyard officials from the mid-seventeenth century to 1832. The National Archives also has some documentation relating to correspondence between the dockyards and the Navy Board in series ADM 106. These records are currently being catalogued by name by a volunteer project and can be searched by name of individual. This is also being done for series ADM 354, which contains the same type of records as held by the National Maritime Museum.

The main staff records held at The National Archives can be found in the following series:

- Yard Pay Books in ADM 42, ADM 32, ADM 36 and ADM 37
- Payments to widows from the Chatham Chest in ADM 82
- Pension records, 1836 to 1928 in PMG 25; for earlier periods, ADM 23
- Apprentice records for dockyard workers in ADM 1, CSC 10 and CSC 6.
- Photographs of work being carried out in ADM 195

Royal Naval dockyards also used the labour of convicts (kept in prison hulks) and slaves during the eighteenth and nineteenth centuries. Many black slaves, and former slaves, were used in naval dockyards in the West Indies, and the latter were paid. Their pay books can be found for Antigua in ADM 42/2114, and for Jamaica in ADM 42/2310. Bermuda's dockyard was constructed by convicts during the nineteenth century and further information can be found at http://www.bermuda-online.org/rnd.htm.

The Royal Naval Museum in Portsmouth has created a website dedicated to the history of the Royal Navy during the twentieth century at www.seayourhistory.org.uk. The site also contains pages relating to the Royal Naval dockyard at Portsmouth and details the many different types of roles people had whilst working in the dockyards and what each role entailed. As such it is a very useful way of understanding the exact nature of the occupation your ancestor was involved in.

Shipyards

There were also a large number of commercial shipyards concentrating on constructing and maintaining commercial ships. The work would have been hard and very physical; the hours would have been long, and workers would have had few rights.

The main industrial ports where the largest shipyards were situated were Southampton, Hull, Liverpool, Glasgow and Belfast, although there were many others in such places as Tyneside and Sunderland. Scotland and Ireland had particularly strong traditions in shipbuilding. The River Clyde in Scotland became the centre of the shipbuilding industry and by the 1870s the area was producing a quarter of the world's ships and 80 per cent of Britain's. At the peak of the industry in the nineteenth and early twentieth centuries the number of shipyards

By the middle of the twentieth century, British shipyards were constructing merchant ships every seven or eight months.

building ships and employing workers would have been in the hundreds. The industry as a whole began to decline in the decades after the Second World War.

If you discover that your ancestor was employed in a shipyard, they would most likely have worked in a one local to where they lived. Surviving staff records for shipyards are not held centrally and may not survive for every shipyard. The Maritime Museum in Liverpool has a number of records for shipbuilding companies in the Merseyside area, which may include individual staff records. More details of their archival collections can be found on their website at www.liverpoolmuseums .org.uk/maritime. The Scottish Maritime Museum (www.scottish maritimemuseum.org.uk) also has similar documentation in its archive in its branch in Irvine. One of the largest and most famous shipyards in Belfast was Harland & Wolff, the company that built the *Titanic* in 1912. Their archives can now be found at the Public Record Office of Northern Ireland. Indeed most surviving shipyard records will be held at the appropriate local repository. If you know the name of the shipyard your ancestor worked at you can search for the exact location of any documentation on the National Register of Archives database at www.nationalarchives.gov.uk/nra.

USEFUL INFO

Suggestions for further reading:

• Coastguard: An Official History of HM Coastguard *by W. Webb (London, 1976)*

• Something to Declare! 1000 years of Customs and Excise *by G. Smith (London, 1980)*

• Family Histories in Scottish Customs Records *by Frances Wilkins (Kidderminster, Wyre Forest Press, 1993)*

• Naval Records for Genealogists *by N. A. M. Rodger (London, 1998)*

• Records of Merchant Shipping and Seamen *by K. Smith, C. T. Watts and M. J. Watts (Public Record Office Publications, 1998)*

CHAPTER 15

Occupations: Mining

One of the most physical lines of work was, and remains, coal mining. With so much of the Industrial Revolution dependent on coal, thousands of men across the country literally carved out a living below ground. Although the industry and the once vibrant mining communities have declined, there are still many places you can visit to find out what life would have been like. This chapter explains what material survives, and how you can use it.

A Brief History of Mining in Britain

Mining has formed a substantial part of the British economy over the past millennium. The Domesday Book mentions mining activity in 12 counties and there are frequent references to mines in documents dating back as early as the twelfth century. The demise of the mining industry, moreover, only really began twenty or so years ago as a consequence of market competition from abroad and from cleaner sources of fuel, and due to technological advances requiring less manpower in the industry.

The type of mining your ancestors would have been employed in depended on where they lived. Scotland, Northern England, the Midlands and Wales were rich in coal deposits; therefore people living in those areas were needed to work the coal mines, while Cornwall and Devon were once famous for copper and tin ore and Kent for its chalk and lime quarries. Migration was commonplace among mineworkers, particularly into the nineteenth century when the demands from industrialization rapidly opened new pits around the British Isles, and

> *Miners' records can be hard to find; workers moved from pit to pit and were paid cash-in-hand.*

▲ A typical colliery in 1913. Thousands of people relied on the pits for work, both above and below ground.

advances in the development of machinery made existing pits much larger. As one pit was run down a new one would be opening elsewhere, which in some cases led to people migrating to the other side of the country in pursuit of work. Tin miners in Cornwall in the 1840s moved to Yorkshire to work in the coal mines at times of recession in the tin industry, while coal miners from Yorkshire moved to new pits in Kent in the early twentieth century in search of a higher income. Learning about how your ancestor's occupation changed during his lifetime may help to explain their movements (and vice versa).

The economic hardship endured by early miners meant that child labour was an integral part of the industry. Though it is difficult for us to imagine now, children as young as five were employed in mines. In the 1840s a Royal Commission investigated child labour and produced a report depicting children working in mines, carrying heavy loads, leading and pulling horses into the pit. These reports and emotive pictures prompted an outpouring of concern and rage directed at mine owners from the educated classes within Britain. Subsequently in 1842 a Mines Act was passed, establishing a Mines Inspectorate to regulate employment conditions in privately owned mines. The act also banned women and children under the age of ten from working underground.

Mining was a highly dangerous occupation and many of those who did not meet an untimely end while at work had to live with a number

Tin and Copper Mining

Tin mining is believed to have started in Cornwall and Devon as early as the first century AD. These two counties provided the majority of the United Kingdom with tin and copper. Up until the sixteenth century Devon was the main source of tin production. Tin was mined at stannaries, where steam engines pumped out the mines, and there were Stannary Courts for Devon and Cornwall that administered justice among the tinners, as they were known. In King John's time the tinners negotiated the right to their own Parliament and Stannary Courts continued to keep law and order over the tinners until the 1896 Stannaries Court Abolition Act. The records of the Court of Stannary are held at Cornwall Record Office in Truro.

From the mid-sixteenth century onwards the Devonshire Stannaries were worth very little and Cornwall became the principal supplier of both tin and copper. Copper mining was at its height in the eighteenth century when Cornish copper demanded a high price, mainly because there was little competition. With the discovery of copper deposits elsewhere in the world in the nineteenth century and Cornish reserves running out, tin started to fuel the mining boom in the south west of England. Tin was found below copper deposits, which simply meant that many former copper mines were mined deeper to get to the tin.

of ailments and diseases as a result of their working conditions. Pneumoconiosis and silicosis, both lung diseases caused by dust (the former also referred to as 'black lung' and the latter as 'Grinder's disease' or 'Potter's rot'), bronchitis and rheumatism were common among miners, as were lesser irritations like 'pit knee'. If you find a mining ancestor who died at a fairly young age it might be interesting to order his death certificate and find out whether his untimely death was intrinsically linked to his occupation.

This chapter will concentrate on the most common types of mining in Britain – coal, tin and copper – although Britain was also mined for lead and iron ore and quarries were mined for slate and lime among other materials. However, the methods for locating records for these various types of mines remain roughly the same whatever type of miner your ancestor was.

Coal Mining

There is evidence that coal has been mined in Britain since Roman times, but the industrial mining of coal that most family historians look to trace their heritage back to has its roots in the eighteenth century. Deep-shaft coal mining, by which miners would be lowered down to an underground pit in a cage, began in the eighteenth century in Britain, but saw its heyday in the nineteenth and early part of the twentieth centuries as industrialization, a shortage of timber

Although the tin mining industry was at its peak in the nineteenth century, the industry was unstable and suffered setbacks in the 1840s following the abolition of the Coinage Laws. Although there was a revival in the demand for tin in the 1870s, tin rushes in several foreign countries in the latter part of the nineteenth century created tough competition within the industry. Tin mining did not provide as much employment as copper mining had done, so migration from tin mining areas was high among workers. By the 1920s there were few tin mines left in Britain, and most of those that survived were mined for by-products like arsenic. The international tin price crash of 1985/6 led to the dwindling of the industry in Britain. The last tin mine closed in Cornwall at South Crofty near Camborne in 1998 (although it is due to re-open and start production again in 2009).

Owing to the importance of metal mining to the Cornish economy, the Camborne School of Mines (CSM) was established in 1888 to teach mining and education in hard rock materials. It now forms part of the University of Exeter. Records of the Camborne School of Mines are held at Cornwall Record Office, while some registers of attendance and admission are kept at Torrington Museum. The existing CSM website has a virtual museum where you can explore the history of mining in Cornwall and the school's heritage at www.ex.ac.uk/cornwall/academic_departments/csm.

for fuel, and the two world wars increased pressure on the demand for fossil fuel.

Coal miners were thought of as a different breed of people. They worked long hours under harsh conditions, and were often portrayed to the general public as being feckless drunks. The first large-scale mines opened in the eighteenth century were all privately owned and the miners who worked for them earned hardly enough money to survive on, and were not guaranteed to earn the same each week. But they were usually provided with subsidized housing built by the mine owner. This benefit could, however, be a disadvantage at times of unrest within the workforce, as employers were known to evict miners during disputes even into the twentieth century. There was rarely enough housing to accommodate all the workers at a colliery so you may find an ancestor lodging with several colleagues at a fellow-miner's cottage when looking at a census return.

In the 1880s coal-cutting machines became available, prior to which time miners had to cut the coal by hand. There were no regulations restricting the number of hours miners were expected to work underground until 1908, when the Coal Mines Regulation Act, commonly called the Eight Hours Act, was passed. Before this time your ancestor may have been expected to work over 12-hour shifts.

During the First World War the State tightened its control over the mining industry and the resources it produced. On 1 March 1917 the government took control of coalfields for the war effort through the

▶ 1950: Three Welsh coal miners emerge after a long shift underground, and head for their homes nearby.

Coal Mines Department. Fewer men were conscripted into the Army from the coal mining industry than most other industries because their work at home was considered just as vital to the war effort. Mines were returned to their private owners after the war, until in 1947 the National Coal Board (NCB) was established following the Coal Industry Nationalization Act of 1946.

The NCB took possession of the assets of over 800 private collieries and their employees, and integrated them into a newly tiered system. The coalfields each had a colliery run by a manager, which were then grouped into 48 geographical areas under the control of an area manager. The areas were in turn grouped into eight geographical divisions, each with its own divisional board that would report to the NCB. This system survived until the 1980s when, amongst much controversy, Margaret Thatcher's government privatized the coal industry, which at that time was considered to be unprofitable. In 1987 the NCB was replaced with the British Coal Corporation, which in turn was superseded by the Coal Authority in 1994.

Mining and Miners' Records

As already stated, employment records for miners can be very difficult to track down because the majority of the industry remained in private ownership through most of its history. Very few records were kept at smaller mines and those that did were rarely retained after the mine was closed. Workers tended to move from pit to pit as labour was often casual and was paid cash-in-hand; therefore lists of employees are hard to come by. If you do find any surviving records of the mine where your ancestor worked it is worth sifting through what is there as you may find a familiar name mentioned among some of the more general administrative records. If your ancestor was affected by a mining disaster then you should have more luck locating some evidence of the event at least, and if they were killed this, ironically, should make finding proof of their employment an easier task.

Mining Records Held in Regional Record Offices

The vast majority of records concerning individual mines and employment papers for miners are located in local and county record offices, and sometimes in family or company archives. Many employment records for miners will not survive because they were not required to be deposited at a national-level archive, and so their continued existence depends on the relevant mining company's decision to deposit them locally or to hold onto them. The only exception to this rule is the records of coal miners who worked after the industry was nationalized in 1947.

Employment records for coal miners who worked under the National Coal Board from 1947 until the 1990s are held by Iron Mountain Records Management at

> Rumer Hill Industrial Estate
> Rumer Hill Road
> Cannock WS11 8EX
> Tel: 01543 574 666

You can request a copy of a miner's records (for free) if you can provide a full name and date of birth as well as their National Insurance number if known and the last colliery they worked at. The documents usually include training records and information about where the miner worked, the type of work he did and any promotions he was awarded with.

The Mineworkers' Pension Scheme holds records about pensions issued by British Coal from 1961 onwards. These pension records give information about a miner's start date, where they worked, the quarries they worked in and any strikes they were involved with. If you have

HOW TO...

... find where a miner worked

• *The easiest way of discovering what mines were in the vicinity of your ancestor's home is to search maps and local trade directories.*

• *The National Archives has many maps showing the locations of mines throughout the ages, including mining maps dating back to 1895 for several counties, which can be located in series POWE 6/85.*

• *The Tithe Maps of England and Wales: a cartographic analysis and county-by-county catalogue, by Roger J. P. Kain, indicates where pits and shafts are shown on the tithe maps produced in the 1830s and 1840s.*

• *There are many region-specific maps that can be found in local record offices as well, so a trip to the local archive may help you to locate a map for the area you are interested in from around the time your ancestors were living there.*

• *Ordnance Survey (OS) maps were produced regularly from the mid-nineteenth century, and these will have mining works and entrances labelled. Copies of OS maps are kept at county record offices.*

USEFUL INFO

David Tonks' book My Ancestor Was a Coalminer *published by the Society of Genealogists has an excellent directory of the local archives, libraries and museums that cover the coalfields of England, Wales and Scotland.*

a mining relative who retired after 1960 you can request a copy of their pension record by writing to

> The Coal Board's Record Office
> Mineworkers' Pension Scheme
> Sutherland House
> Russell Way
> Crawley RH10 1UH

giving the miner's full name and date of birth or National Insurance number.

For coal mining records prior to nationalization in 1947, and for all other types of private mining records, it is usually necessary to know the name of the company or the colliery your ancestor worked at in order to establish whether any records of employment are likely to have survived for them. If you are unsure of either of these facts, perhaps if you simply found your ancestor on a census return described as a tin miner for example, then it will be necessary to find out what mines were within walking distance, or at least reasonably close to their home for them to have worked at. Migration can make this task even more challenging as your family may have moved regularly to work in different pits.

If you are lucky enough to know the name of the company that owned the mine where your ancestor worked then you can search for the location of that company's archives, if they survive, using the National Register of Archives at www.nationalarchives.gov.uk/nra. Simply navigate to the Corporate Name search engine and enter the name of the company your ancestor worked for. If any company records are located the database will tell you what type of records and for what years survive and where they are deposited. For example, the Northumberland Collections Service in Ashington holds the records of the iron miners Allen & Burnhouse Gill Mining Co. from 1877 to 1892.

The description of the records is usually very broad so you should phone the archive in advance of a visit to ensure they hold records relating to employees for the exact period you are interested in. The NRA search engine can be used to search for records by place name as well if you know where the colliery was but are unsure of the name of the company that owned the mine where your ancestor worked. Equally, if you had an ancestor who worked in a mine belonging to a landed gentleman's estate, such as the Duke of Norfolk or Lord Fitzwilliam, you can search for the location of their estate papers using the Family Name search engine. Estate papers for some of the earlier mines may be fruitful for locating records of pay and subsidized accommodation for miners.

Another method of locating records held in local archives for specific collieries is to consult the Access to Archives database at www.a2a.org.uk. Simply enter the name of the colliery where your ancestor worked and descriptions of the documents containing those words will be found. The document descriptions tell you the date range they cover, where they are held and the document references. There are many documents relating to collieries that have been catalogued using the Access to Archives database, but which may not be found using the National Register of Archives, and vice versa, so it is important to consult both.

Not all records deposited in local archives will be found using either the NRA or A2A databases, but some local record offices and archives have catalogued their collections on their own website, so always check the website of the county record office covering the area where your ancestors mined. Cornwall Record Office has a catalogue that can be searched online and where descriptions of the records of Dolcoath tin and copper mine, affectionately known as 'The Queen of Cornish Mines', can be found dating back to 1588.

Local institutions have also made attempts to preserve the history of the local industry in some areas, such as the University of Wales, Swansea Archives, Library and Information Services, which holds the South Wales Coalfield Collection (SWCC), established in 1969 in an attempt to collect and preserve documentation about the history of the South Wales mining community. The South Wales coalfield covered parts of Carmarthenshire, Swansea, North Port Talbot, Bridgend,

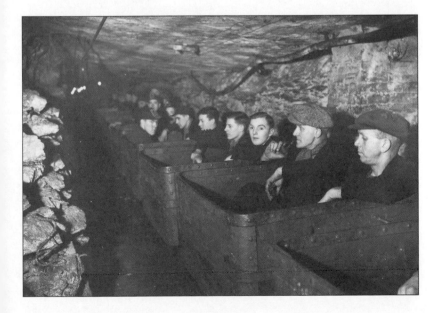

◄ Cramped working conditions down in the pits in 1939. Here coal miners wait in ore cars at the start of their shift.

Rhonnda Cynon Taff, Vale of Glamorgan, Merthyr Tydfil, Cardiff, Caerphilly, Blaenau Gwent, Torfaen and Powys. The archive is open to all researchers and there is an online catalogue to the SWCC records available from www.swan.ac.uk/swcc.

Descriptions on the SWCC catalogue include documents relating to the South Wales Miners' Federation, which became the National Union of Miners (South Wales Area), records from miners' institutes, co-operative societies and documents concerning prominent individuals within the mining community, as well as some colliery records. The SWCC catalogue also contains descriptions of audiotapes, videotapes, photographs, banners and manuscripts deposited between 1983 and 1993. Material deposited before 1983 is catalogued in two paper guides available from the Library & Information Centre. Audiotapes, videotapes and banners are held at the South Wales Miners' Library, also in Swansea. Coalfield Web Materials, a website based on the SWCC, is a fantastic resource that has started to provide digital images of photos and online audio recordings from the collection as well as giving a background history to the South Wales coalfield. The website can be found at www.agor.org.uk/cwm where there is a search engine to find specific items within the collection.

In addition to the South Wales Coalfield Collection, the University of Wales archive also holds documents among its Local Collections relat-

The National Archives

The National Archives in Kew is where the majority of paperwork concerning the administration and government involvement in the mining and quarrying industries is deposited. It does not, however, hold mining personnel records, but some of its records may be helpful if you would like to research the history of a particular mine where your ancestor worked.

● Records of the Home Office contain some papers relating to individual mines and quarries, occasionally found among correspondence in HO 42 between 1782 and 1820, in HO 44 for records dated 1820 to 1861, and in HO 45 for those from 1841 onwards. There is a register in HO 46 for records kept in HO 44 and HO 45.

● Home Office Registered Files for Mines and Quarries between 1887 and 1920 are kept in POWE 6.

● The reports generated by the Mines Inspectorate are held in series HO 87/53 for the years 1851 and 1852 and in POWE 7 for 1850 to 1968.

● Some mining records were formerly deposited in the Land Revenue Record Office, and these can be found in series LRRO. For example a set of accounts for Yorkshire lead mines dating from 1697 to 1831 were deposited in LRRO 3/84, and LRRO 1 contains some maps and plans.

● The Coal Mines Inspection Act of 1850 required colliery owners to make plans of their coal mines and some colliery plans are held in COAL 38.

By the end of the nineteenth century the Board of Trade had

ing to other metallurgical industries in the region, including the Whiterock Copper Works, the Morfa Tinplate Co. and John Player & Sons of Clydach Tinplate Works. To locate records of other mines in Wales, including those that mined the North Wales coalfield, try the Archives Network Wales database run by the Archives and Records Council Wales at www.archivesnetworkwales.info.

Mining Records Held at Major Archives

Occasionally there are references to particular mines among papers deposited at major archives, though these are less likely to be rich in information about individual workers, and if they are you may struggle locating a mention of your own ancestor as the papers are unlikely to be thoroughly indexed. However, if you would like to research the wider history of the mine where your ancestor worked and the general history of the industry, then it is worth visiting some of the following archives.

Mining Records Office

Some records of post-nationalization coal mines under the control of the National Coal Board, British Coal and the Coal Authority are held by the Coal Authority at the Mining Records Office in Mansfield. The

become concerned with the industrial and economic characteristics of mining and quarrying. In 1920 the Board of Trade set up the Mines Department to replace the Mines Inspectorate, whose responsibilities were in turn taken over by a succession of government ministries and departments between 1942 and 1992, including the Ministry of Fuel and Power, the Ministry of Power, the Ministry of Technology, the Department of Energy, and the present Department of Trade and Industry. The Ministry of National Insurance was also responsible for assessing and granting compensation for industrial diseases, while the Ministry of Labour handled issues concerning industrial relations, health, safety and welfare. The records of all of these government departments are held at The National Archives.

In addition to documents found among the records of those departments that worked in conjunction with the mining industry are also some stray records relating to privately owned mines that have been deposited at The National Archives in the form of estate papers, such as those of Lord Derwentwater's forfeited Northern Estate covering Cumberland, Northumberland and County Durham, whose assets went to Greenwich Hospital after he took part in the 1715 Jacobite rebellion. Along with other Greenwich Hospital papers, the records of this estate are kept with Admiralty records in series ADM 75, ADM 79 and ADM 169, and include deeds, plans, reports, rent rolls and other estate records. These types of records may be found by a keyword search using the online catalogue.

public records kept at the Mining Records Office are unlikely to be of much use to family historians, but if you are interested in researching the history of the colliery where your ancestor worked it may be worth a visit. Their collections are principally made up of Coal Abandonment Plans dating back to 1872 depicting areas of coal extraction, the locations of mine entrances and the extent of coaling operations; the Coal Holdings Register, which dealt with the transfer of coal ownership prior to nationalization and contains associated records such as former Coal Commission claim files with plans and mining leases; and a Licence Register with information about licence applications and granted licences with plans showing the area to be mined.

The Mining Records Office is open daily between Monday and Friday. There is a charge of £50 plus VAT to view the Coal Holdings Register and the Licence Register. To make an appointment at the records office contact them on 01623 637233 or write to them at

> The Mining Records Office
> 200 Lichfield Lane
> Berry Hill
> Mansfield
> Nottinghamshire NG18 4RG

National Library of Scotland

The National Library of Scotland is home to some Scottish coal-mining records, including papers relating to Scottish mine-owning families. There are documents and books about the history of mining in Scotland, such as an original draft of R. Page Arnot's book *The History of Scottish Miners* written in 1955. To locate documents for specific mines in Scotland and descriptions of papers held in the Library, the National Archives of Scotland and regional Scottish record offices, visit the Scottish Archive Network (SCAN) at www.scan.org.uk and try their online catalogue.

Mining Museums

Specialist mining museums are the best places to go to find out more about the colliery your ancestor worked at, what conditions would have been like for them there, and to locate photographic and additional archive material. If you are very lucky the pit your ancestor worked at may have been turned into a museum. Greevor tin mine in the west of

CASE STUDY

Lesley Garrett pt 1

Lesley Garrett was born in Thorne, near Doncaster, in the traditional heart of Yorkshire's mining belt. Her love of music came from growing up in a musical family and in a community where music went hand in hand with the industry of the area. Lesley knew that some branches of her family had worked in the mining industry. Yet when she dug deeper, she discovered that a talent for music appears to have run in her family for several generations.

Lesley was intrigued by her maternal great-grandfather, William Wall. Born in 1874, he had worked down the pit most of his life, as had his father before him. However, his son Colin – Lesley's grandfather – was born with a weak chest. Realizing that Colin could never work in a mine, William took the radical step of introducing him to music in the hope he would find alternative employment.

William borrowed a book from the library and taught himself to play the piano. By keeping a few lessons ahead of Colin, he was able to teach his son to play. Colin would become a fine classical pianist and father and son even composed their own music.

Colin won a silver medal in the London School of Music exams in December 1915. He went on to make a good living playing the piano with a small orchestra that gave concerts for the miners and accompanied silent movies. Later, during and after the depression, he played at the White Hart Hotel in Thorne. Though this was a marked change, Colin insisted on playing an hour of classical music before agreeing to accompany the popular performers of the day (the 'turns'). Soon classical music lovers from across the area came to hear this talented musician play.

Although employment records no longer survived for mining members of her family such as great-grandfather Frank Appleton of Beeston, Nottinghamshire, Lesley was able to gain a greater understanding of their lives, and the daily conditions they faced, by visiting the National Coalmining Museum in Wakefield, Yorkshire, and travelling 450 feet underground to the coalface. By seeing first hand the cramped conditions, the equipment they used, the dust and danger they faced, she finally began to understand quite why they made the most of their time above ground – and used music to celebrate their close community spirit.

◀ Lesley's maternal grandfather, Colin Wall, who won a silver medal in the London School of Music exams.

> *Sadly, the easiest way to find records about a mining ancestor is if he died in a pit accident.*

Cornwall is the largest preserved tin-mining site in Europe, turned into a museum in 1993 after its closure following 300 years of industry. The mine has a website at www.greevor.com where there is a timeline following the history of tin mining, but as with all of the museums mentioned here, the best way to get a taste of what life would have been like for your ancestor working as a miner is to visit the museum and see their working environment with your own eyes.

Tower Colliery in the South Wales Valley was one of the world's oldest continuously worked mines. Opened in 1805, it finally closed in 2008. The colliery has a website at www.towercolliery.co.uk where there are pages on its history and a photo gallery as well as information about what it is like mining the colliery in the twenty-first century.

Even if you do locate a local museum that covers the particular mine you are researching, it is still worth visiting the relevant national mining museum as well. The National Coal Mining (NCM) Museum for England in Wakefield, West Yorkshire, has put many of its collections online, accessed via www.ncmcollections.org.uk. From here you can view images of equipment used by miners throughout the ages, badges awarded to miners from various collieries and photographs of men who worked in English pits. The museum has a library stocked full of specialist books about the history of mining and individual mines across England as well as some primary material like accident reports, annual reports, reports of the Inspector of Mines, and mining newspapers. The library catalogue can also be searched from the NCM Collections Online website.

National Museum Wales (NMW) has two branches dedicated to the nation's mining and quarrying industries – the National Slate Museum at Llanberis in Gwynedd and the Big Pit: National Coal Museum at Blaenafon in Torfaen. Links to the two museums can be found on the NMW website at www.museumwales.ac.uk. The National Slate Museum is on the site of the Dinorwig quarry, a picturesque location flanked by Snowdon, where you can visit four quarrymen's cottages rescued from demolition. The Big Pit is a real coal mine that can be explored on an underground tour. You can visit the pithead baths and see the mining galleries and museum exhibitions. For those with mining ancestors from southern Wales there is also the South Wales Miners Museum at Afan Argoed Country Park in Port Talbot where there is a large collection of photographs. Visit www.southwalesminers museum.com to find out more.

The Scottish Mining Museum has a website at www.scottishmining museum.com where a selection of their photographic collection can be viewed and fact files about the history of mining can be downloaded. There is a library and archive at the museum that is open to researchers, but they do not hold any employment records. Scotland also has a museum of lead mining in Wanlockhead, Lanarkshire. The Museum of Lead Mining has a large archive with records concerning many of the families who lived and worked in Wanlockhead, Leadhills and the surrounding villages since Victorian times. You can contact them to find out if they hold records concerning your ancestors by writing to:

> The Museum of Lead Mining
> Wanlockhead
> Lanarkshire ML12 6UT

They also have a website – www.leadminingmuseum.co.uk.

Mining Disasters

Sadly, the easiest way to find records about a mining ancestor is usually if they died following an explosion or accident in a mine. There are many sources you can consult to find the names of those killed and eye-witness accounts or reports about the events that unfolded that day if you know a rough date for the incident, or at least know the name of the mine where the disaster happened.

Most major mining disasters generated the following types of testimony that a family historian can try to locate – an inquest, an official report of the disaster, a mention in the annual report of the Mines Inspectorate, local newspaper reports and possibly a memorial in the parish church. A local Relief Fund may have been set up, and oral and written memoirs of the event may have been recorded.

- If an inquest took place and a record of it survives, the papers should be held at the local county record office. Alternatively, you may have to rely upon newspaper accounts of the inquest.
- Reports about many major mining accidents were published in the Parliamentary Papers, which are now available online via The National Archives website.
- An Explosives Inspectorate, created by the Home Office, compiled a list of mining accidents between 1896 and 1952 with relevant correspondence and papers kept in The National Archives series EF 2/7.
- There are also reports concerning specific incidents filed in various series at The National Archives in Kew, like the records of the

USEFUL INFO

Charles Dickens, as reporter for the magazine Household Words *printed between 1850 and 1859, conducted and published interviews with coal miners who survived accidents and explosions in the North East of England in the late 1840s and early 1850s. Some of these accounts mention names of men, but they are most valuable for their frankness when describing the dangers of working down the pit and the most common causes of many tragic disasters. Carelessness among workers was deemed to be one of the greatest risks to miners' lives, and the employment of very young children in responsible positions made this matter worse, though in reality mine owners rarely invested sufficient funds into maintaining safety in mines. A website set up by a former coal miner at www.pitwork.net has transcriptions of some of these interviews, or copies of the publication from December 1850 can be found at the British Library in London.*

Labour Movements and Trade Unions

Horrendous working conditions, appalling levels of pay in relation to the work undertaken and the constant threat of unemployment gave rise to much tension between miners and their employers, producing trade unions and organized labour movements dating back to the 1700s, which have been consistently active right up to the present day. Early labour movements in the industry formed in response to systems of economic bondage and serfdom, notably in Scotland where coal miners were bonded to their masters in a form of lawful slavery and could face fines or corporal punishment if they attempted to leave their employment. Serfdom and bondage were outlawed in 1799, although they continued in some areas of Britain until much later.

In the coal-mining industry, District Unions were set up to fight for the protection of miners' rights in a business where children as young as four and five were expected to work 16-hour shifts six days a week. The industry was marked by strikes throughout the nineteenth and twentieth centuries, one of the earliest organized strikes taking place in 1844 under the lead of the Miners' Association of Great Britain and Ireland in a call for better wages. In 1889 the District Unions were replaced by the Miners' Federation of Great Britain, which was a forerunner to the National Union of Mineworkers (NUM), established just before nationalization in 1944. If you have an ancestor who was prominent in a trade union movement there should be evidence of this among the papers of the relevant union. Around

Valleyfield Colliery explosion of 1939 in Fife, Scotland, filed in BT 103/105–108, and the report of the Tribunal of Inquiry into the Aberfan Disaster of 1966 in Wales stored in series BD 52/154. These types of records can usually be found via The National Archives online catalogue by doing a keyword search for the name of the mine and restricting the date range to the year of the accident.

- General records concerning mining accidents can be located in the papers of the Safety and Health Division and Inspectorate of Mines and Quarries for 1876–1984 in POWE 8, and registered files on accidents between 1887 and 1920 are kept in POWE 6.

The Coal Mining History Resource Centre website at www.cmhrc.pwp. blueyonder.co.uk is home to the National Database of Mining Deaths in the United Kingdom, 1850–1972. The author of the book *Mining Deaths in Great Britain*, Ian Winstanley, compiled the list of 166,000 names currently contained on the database. His published book contains the name, date, colliery, job and age of over 55,000 miners who were killed or injured. The Durham Mining Museum is a primarily online museum, which has a very informative website at www.dmm.org.uk providing information about mining in the north of England, covering Durham, Northumberland, Cumberland, Westmorland and the Ironstone mines

5,000 trade unions are known to have existed in Britain over the last couple of centuries, with millions of members supporting them. If you can find out which trade union your ancestor was a member of this can give you an insight into their political motivations and daily concerns. Trade Union Ancestors is a particularly informative website run by an enthusiast at www.unionancestors.co.uk, where you can scour a comprehensive trade union directory and pick up hints about how to trace their records. The papers of most trade unions will have been deposited at local record offices. An exception is a collection of records covering 1936 to 1947 for the Mining Association of Great Britain, which had its own trade newspaper and mining college records, which are kept at The National Archives in series COAL 11.

If you have difficulty establishing whether your mining ancestor was a member of a trade union but you believe he may have supported some of the labour movement causes, you can find out more about the miners' concerns, their actions and the consequences by reading local newspaper reports to find out how strikes affected their area. The National Union of Mineworkers is still very much alive and has a website at www.num.org.uk. The website is a great source for the history of trade union movements within the coal industry from the earliest times right up to the present day. You can look at the dates given for major disputes on the NUM website to work out which ones may have affected your relatives. There are records about mining disputes and strikes at The National Archives, including daily bulletins about the coal strike of 1920–21 kept in CAB 27/79 and PRO 30/26.

of North Yorkshire. The website has statistics and reports about accidents, listing the names and biographical details of those killed. There is also a list of mining disasters across the rest of the United Kingdom, some with lists of those who died.

County record offices and local libraries will be useful for finding newspaper reports about mining accidents and deaths if you know a rough date for the disaster. You can check the location of a repository for many local newspapers, most of which have since been discontinued or renamed, from the regional Newsplan databases supported by local libraries in conjunction with the British Newspaper Library at www.bl.uk/collections/nplan.html. The National Library of Wales, the National Library of Scotland and the National Library of Ireland support identical online Newsplan projects for those regions, details of which can be found on the British Library site.

The deaths of hundreds of thousands of miners over the centuries left a staggering number of widows and orphaned children, many of whom would have required financial support. It is worth finding out if a Relief Fund or society was set up for the colliery where the accident happened to see if there is a record among their papers and meeting minutes of financial support given to your ancestor's family. The archive where the mine's paperwork is filed should be able to assist you with this.

Despite the many deaths that occurred in British mines throughout the eighteenth and nineteenth centuries, it wasn't until the Coal Mines Act of 1911 that Parliament decreed all colliery owners should have in place a fully trained Mines Rescue Team rather than a voluntary group of mineworkers on call to assist at times when disaster struck. Rescue stations were set up where men could be trained and specialist equipment stored. When disaster struck there were often a large number of men among the dead who had entered the mine after an explosion to help those inside. The Mines Rescue website at www.heroes-of-mine.co.uk contains a wealth of information about these brave men who risked their lives in an attempt to save their colleagues and friends.

Finding Out More

- Local mining history societies may be able to help you find out more about your mining ancestor's life. The National Association of Mining History Organizations was formed in 1979 to act as a national body for mining history in the United Kingdom and Ireland, and it has a website at www.namho.org. This website is home to a brilliant list of links to associated websites and member organizations, including many societies, clubs and research groups, some of which you may want to consider joining if you can find ones relevant to your area of research.

- While women were banned from working underground in the pits from the 1840s, they still worked above ground at collieries. You can find out more about your maternal ancestors' roles in the industry from www.balmaiden.co.uk, where there is also a database containing the names of 23,000 women who worked in mines in Devon and Cornwall. There is further information about these Bal Maidens (as these women were known) on the BBC website, at www.bbc.co.uk/nationonfilm.

- The BBC's Nation on Film website contains archived film footage that can be watched for free. They have two pages full of clips celebrating the heyday of both the coal and tin mining industries, with interviews of miners recalling their experiences and memories, and images of men and women at work in the mines. The clips can be downloaded from www.bbc.co.uk/nationonfilm.

- The Mining History Network website, supported by the University of Exeter, at www.projects.ex.ac.uk/mhn/welcome is a fantastic online resource with links to pages about British, Irish and Scottish mining

Membership of a trade union can give an insight into your ancestor's daily concerns.

as well as mining in other parts of the world. There are detailed bibliographies directing you to reliable sources for mining history, a directory of mining historians and their contact details should you wish to seek out professional help, and the site supports a mining discussion forum where you can communicate with other researchers and experts about areas of mining research you are struggling with.

- If you have a burning desire to get a real taste of working life underground, the Arigna Mining Experience Centre in County Roscommon, Ireland, takes its visitors on a 45-minute underground tour to demonstrate what it would have been like working in some of the narrowest coal seams in the western world. The Arigna area was mined for iron from the seventeenth century through to the nineteenth, when coal mining became the region's principal form of employment. You can find out more about the Arigna coal mine from the www.arignaminingexperience.ie website.

There is a distinct lack of centralized information and documentation regarding the mining and quarrying industries and, most importantly, the people who worked in them. Some ex-miners and descendants of pit workers have built websites dedicated to the history and memory of the mines where they worked, often rich with photographic archive material. These can be difficult to locate and a general web-search or browse through the list of links on some of the other websites already mentioned is the only real way of finding out if somebody has already researched the history of the mine where your ancestor worked. Some examples include the history of Blaenserchan Colliery at www.blaen serchan.co.uk and the Welsh Coal Mines website at www.welsh coalmines.co.uk, which is packed with information about collieries once found around Wales. A particularly informative site about the history of working in a coal pit, with pages about disasters, mining heroes, pictures and memories compiled by a former miner can be found at www.pitwork.net.

USEFUL INFO

Suggestions for further reading:

- My Ancestor Was a Coalminer *by David Tonks (Society of Genealogists, 2003)*

- The Hardest Work under Heaven: The Life and Death of the British Miner *by M. Pollard (Hutchinson, 1984)*

- Life and Work of the Northern Lead Miner *by A. Raistrick (Alan Sutton, 1990)*

- The History of the British Coal Industry *(Oxford, 1993)*

- The History of Tin Mining and Smelting in Cornwall *by D. B. Barton (D. Bradford Barton, 1967)*

- *The National Archives Research Guide number 35, 'Coal Mining Records in The National Archives', and number 64, 'Sources for the History of Mines and Quarries'*

CHAPTER 16

Occupations: Factories, Foundries and Mills

The greatest embodiment of the Industrial Revolution were the mills and factories that spread over large swathes of the country, and the associated movement from the countryside that created the great towns and cities of the North West. This chapter examines the sources that enable researchers to step back in time and revisit the dry, dusty and noisy factory floor of a typical cotton mill, match factory or similar place of employment.

Historical Context

> *The Industrial Revolution changed the British landscape forever.*

The Industrial Revolution of the eighteenth and nineteenth centuries, coupled with a population boom, changed the British landscape forever and germinated modern society as we know it today. While Britain's rural economy declined, 'those dark satanic mills' springing up all over the north fuelled a booming export industry. Formerly small towns like Manchester and Sheffield exploded in size, and their communication networks spread rapidly. In 1851 Britain invited the world to marvel at industrial masterpieces housed at the Great Exhibition in Hyde Park. Meanwhile the social and economic revolution taking place attracted international attention for both its merits and its woes. Those people who worked long shifts under intolerable conditions and lived in squalor to keep the workshop of the world running were equated to

white slaves by horrified onlookers, like the MP Michael Sadler. This was the age and environment in which Karl Marx wrote the *Communist Manifesto*, urging the proletariat to rise up against their capitalist oppressors, while Friedrich Engels despaired at *The Condition of the Working Class in England*. Whether you trace your family tree back to a humble cotton spinner or a wealthy mill owner, getting to grips with the rich heritage hiding behind our industrial ancestors is a fascinating journey through modern British social history.

Early factories were commonly called mills because they were powered by watermills before steam power was deemed more efficient. The development of the steam engine and subsequent use of steam power at the very beginning of the 1800s marked a dramatic change for manufacturing as mills and factories no longer needed to be located next to rivers. Urban areas that built up around a newly opened mill or collection of factories were known as mill towns, and our ancestors who inhabited them were primarily unskilled machine minders and people trained to use the latest technology being developed. They formed a new breed of working-class people who lived a highly regulated life dominated by clocking on and off, pay rates, supervised labour and monotonous production lines.

▲ The factory chimneys of the area around Wolverhampton, 1866, darkened the skies – giving rise to the name 'Black Country' for the region.

Most factories opened up in urban centres where there was a large workforce to draw upon, but industrial slums developed as local trades and the old cottage industries became obsolete, forcing people from rural areas into the already overcrowded towns. If you find an ancestor employed in a manufacturing industry you may discover that their generation or the generation before them had migrated from a more rural setting in search of work. By the end of the Victorian period, far more people were employed in manufacturing than in agriculture. (In contrast rural Ireland did not experience an industrial revolution in the way the rest of Britain did and as such was sorely hit by the Potato Famine of the late 1840s that killed around one million agricultural workers and led to the emigration of millions more.)

Manchester, Birmingham, Bradford and Leeds were at the heart of industrialization in the early 1800s, dominated by a few key industries.

- *Cotton:* The manufacture of cotton, based principally in Lancashire but also found in parts of Yorkshire, Derbyshire and Cheshire, was Britain's most valuable manufacturing industry from the mid-nineteenth century until the late 1920s, at which time the effects of the Great Depression and tension with its biggest raw cotton supplier and cotton goods consumer, India, led to a decline. The manufacture of cotton goods was a two-phase process. Spinning, which was mechanized by the eighteenth century, was a process that transformed raw cotton into usable threads. The second phase, weaving, manufactured the cloth into clothing or other cotton goods. Weaving was not mechanized until the 1840s, after which time hundreds of thousands of handloom weavers became redundant at an astounding rate.
- *Wool and worsted:* In Yorkshire wool and worsted provided employment for the majority of factory workers in towns like Bradford and Halifax, and this continued to be the case until the mid-twentieth century.
- *Flour milling:* Corn milling was an ancient rural industry that changed drastically in the 1880s when giant flourmills were built in London, Liverpool, Hull and other major ports to take advantage of increasing wheat imports and the use of new roller grinding technology. These gradually replaced the wind and water mills once found all over rural England.
- *Foundries:* Foundries built up around mining areas to manufacture the tin, copper, iron and zinc excavated there and to use the coal from the mines. Sheffield was once world renowned for its steel, a major source of employment in the area until international competition led to most of the foundries closing down in the 1970s and 1980s.

Despite generally rapid industrial growth during the nineteenth century there were downturns in the economy. There was widespread unrest in 1841 and 1842, and in 1846 the whole of Great Britain experienced a temporary industrial depression that lasted until 1848. The textile industry suffered during the American Civil War when imports of raw cotton from the United States stopped, bringing about widespread unemployment and the onset of the Lancashire Cotton Famine in 1861 lasting four years. The end of the 1880s and early 1890s saw another general economic downturn, known at the time as the 'great depression', a term which has subsequently been used for the worldwide depression following the Wall Street Crash in 1929.

Child labour was a problem posed by these new industries, as it was in mining (see Chapter 15). A major piece of legislation, the Factory Act of 1833, was passed after a Royal Commission led by the Tory MP Michael Sadler and the Evangelical social reformer Lord Ashley found evidence of the exploitation of children in British factories. The Factory Act forbade the employment of children under the age of 9, and all children aged between 9 and 13 were limited to a 48-hour week, while children under 18 years old could not work more than 12 hours a day. A Factory Inspectorate was created by this legislation to monitor the working conditions of women and children in factories. There were regular modifications to the Factory Act, including an 1891 Act that raised the minimum age of children employed in factories to 11 years old.

The new industrial occupations of the nineteenth century quickly became gender specific. By far the majority of workers in the cotton, wool, worsted and other textile industries were female, although some men were employed in more responsible and better-paid positions. Many Irish Catholic women, escaping bleak lives in rural Ireland by immigrating to the north-westerly ports of England, were employed in the lower-paying spinning jobs by the turn of the century.

During the First and Second World Wars women were temporarily expected to replace men who had left their factory jobs to fight in the Army. The steel factories in Sheffield were seconded for use as ammunition factories, making the city a prime target for air-raid attacks and putting the women working there at great risk. Royal Ordnance Factories (ROFs) were munitions factories set up by the government during the Second World War and were built in areas considered safe at the time from German bombing, such as south-west England, Wales and Scotland. Most records of the ROFs are kept at The National Archives among records of the Ministry of Aviation in AVIA series.

USEFUL INFO

Manufacturing was chiefly based in the north of England and in Scotland, while the Welsh economy was supported by wool, mining and the manufacture of the materials this produced. The north of Ireland had a more industrial-based economy than the south, relying upon shipbuilding, textiles and rope manufacture.

Tracing Your Industrial Ancestors

CASE EXAMPLE

Tracing your industrial ancestors

Actor Robert Lindsay discovered that his paternal grandfather, Jesse Stevenson, worked in the local Stanton Ironworks as an iron fettler on his return from serving with the Sherwood Foresters during the First World War. Jesse lived in a cramped two-up two-down cottage owned by the company. His situation was typical of many men returning from the war.

In order to research the history of your industrial ancestors in any depth, it is necessary to know the name of the company that they worked for.

You may be fortunate enough to have an elderly relative who can give you some guidance, or you might find a letter from the firm among some of the paperwork passed down through the family. Army discharge papers (if your ancestor served in the forces during either World War) sometimes state the name of the company the soldier intended to work for, or at least their occupation and address (see Chapter 9 for more information about Army ancestors).

If this is not the case and all you have to go on is a census return stating 'cotton spinner' then your research will start by exploring who the main employers were for the area your ancestor lived in. This can be done using detailed historical maps, particularly those updated regularly by the Ordnance Survey, and trade directories, and by reading books about the history of the local area, all of which can be found in the county record office for your ancestral hometown.

Unless your family lived in one of the few industrial villages that adapted to urban changes taking place in neighbouring towns, you are likely to find more than one company who could have employed your ancestors and will probably never be able to pinpoint which is the right one. Even if you are sure of the company they worked for, the likelihood of locating concrete evidence of your ancestor's employment there is very slim. Do not let this deter you because you can still research what working life was like for people living in that area. Your task will be one of local history research, finding out what the main employers of the area were like, how well business did there and whether there were any localized depressions or booms that might explain a sudden change in circumstances or location for your family.

When looking at maps of the area take an interest in the way the streets were formed and the houses were built. You can trace how a village or town expanded over the nineteenth century as industrialization took hold by comparing maps from different decades. Ordnance Survey maps in particular will give you a bird's-eye view of each building and an idea of the house sizes so you can see whether your ancestors lived in one of the two-up two-down terraces that characterized industrial urban centres. Compare this with the number of people you found living there on the census return.

Try to find old photographs among the county record office's collections of the street or urban district where your ancestors lived and analyse the impact industrialization had on their landscape. Look out for smoggy factory chimneys and the names of local companies emblazoned on the outside of mills on the horizon, and be aware of the clothes people in those photographs are wearing. Most working-class women and children, and even some men, wore wooden clogs as late as the twentieth century because they were cheap, hardwearing and protected against the damp factory floors. Draw on as many of these sources as you can find to conjure up a picture of your industrial ancestors' world.

Investigating the exact meaning behind your ancestor's job title will give you a deeper understanding of what their daily routine would have involved. What part did they play in the overall production process? There may have been safety risks involved in the tasks they had to do and the machinery they used. Try to find out how their job was affected by technological advances throughout the industrial age – were more job opportunities created in their field of work or were there widespread redundancies as employers invested in machinery that did their job more quickly and efficiently and was more cost-effective?

Locating Surviving Company Records

The Royal Commission on the Historical Monuments of England (RCHM), which was amalgamated with the Public Record Office to form The National Archives, produced a catalogue to the records of around 1,200 textile and leather businesses of between 1760 and 1914, describing the type of records that survive for each, their covering dates and location. These descriptions can now be searched using the NRA online database at www.nationalarchives.gov.uk/nra along with those of many thousands more manufacturing businesses. If you know the name of the company your ancestor worked for then a quick search of the Corporate Name database may be able to find some records that are worth you looking through. For example a search for the records of Stanton Ironworks Co. Ltd using the Corporate Name search located records dating back to 1878 held at Derbyshire County Record Office, including the diaries of Fred Alvey, a draughtsman.

As with locating the records of mining companies explained in Chapter 15, it is also worth searching the Access to Archives database for company records, at www.a2a.org.uk. For Welsh companies look at

Industrialist Ancestors

Most factories were set up and owned by local entrepreneurial businessmen who were not connected to the traditional aristocratic families associated with old wealth. Factories tended to be self-financing, with investments sought from relatives or friends and profits reinvested into the industry. Banks invested very little money in industrialization. This system meant that British manufacturing companies were relatively small scale, unlike in America, with the exception of a few very wealthy companies such as the nineteenth-century cotton-spinning corporation Arkwright & Peel, the tobacco firm Wills, and Coats who manufactured sewing thread.

Manufacturers and industrialists often dominated local councils right into the twentieth century, and as a major employer they would have had a strong influence on local society. If you believe your ancestor was part of the industrial elite that emerged during the nineteenth century then surviving family and business papers may be of interest to you. These will often consist mainly of accounts, which can give you a perception of how well the business did and how wealthy your ancestor was, but may also contain personal letters.

● The National Register of Archives has a Personal Name search engine to help you locate the repositories of private papers, or a search of the Corporate Name database may be able to find surviving business records.

● Your ancestor may have been a member of an organization for the protection of employers' rights, such as the Oldham Textile Employers' Association that has an archive located in the John Rylands Library at the University of Manchester. Ask at the local county record office and consult the libraries mentioned later in this chapter to find out if there were any such organizations operating locally in your ancestor's area of work. Such records may have details about members and will give you an insight into your industrialist ancestor's economic priorities and beliefs.

● The records of local councils will be held at the County Record Office

the Archives Network Wales website at www.archivesnetworkwales. info, and for records held in Scottish archives try searching the Scottish Archive Network (SCAN) at www.scan.org.uk. If you cannot locate the company records you are looking for using any of these databases then speak to the county record office that covers the area the company was based in to find out if the business is listed in their catalogues and indexes.

While some records that shed light on the employment of individuals do survive, the majority of material you are likely to come across in your investigation will concern the history of the trade your ancestor was employed in, general working conditions at the time, the products that were produced and the buildings they were manufactured in. Local history sites are sometimes a fruitful source for the impact of industry on the area. Check the local council's website for information about the area's history and look out for links to websites set up by local businesses and historians about the background to buildings in and meeting minutes will contain any input your relative had regarding local issues. Search the name index cards available at most local record offices to see if relevant documents about them can be found quickly.

Prior to the Great Reform Act of 1832, which gave all men with property the right to vote, very few manufacturers entered Parliament as MPs. However, after this time their numbers grew steadily. If your industrialist ancestor was a Member of Parliament then a trip to the Parliamentary Archives is in order, where you will find biographies and some photographs of MPs and can use the sources there to find out whether your ancestor tried to oppose any of the factory legislation discussed in Parliament. Hansard's transcriptions of the Parliamentary Debates, covering select debates from 1803 and reporting all in full from 1909, may be useful for this, as will the Parliamentary Papers that are now available online from The National Archives' computer terminals.

The *Who's Who of British Members of Parliament*, published in the 1970s and 80s in several volumes covering MPs since 1832, may contain biographical information about your ancestor. Maurice F. Bond wrote a *Guide to the Records of Parliament* (HMSO, 1971) that is worth consulting if you intend to visit the Parliamentary Archives.

The Great Exhibition of 1851 attracted 6 million visitors to London in the six months it was open between May and October. Over 100,000 exhibits were displayed under the four categories of Manufactures, Machinery & Mechanical Invention, Raw Material, and Sculpture & Plastic Art. You can find out whether your ancestor was among the many manufacturers proudly exhibiting their products by visiting the National Art Library at the Victorian & Albert Museum where plenty of catalogues and descriptive publications from the time are held. The National Archives holds some original documents for the event, which can be searched in the online catalogue, though these mainly consist of administrative records, and the London Metropolitan Archives has a large collection of colourful prints depicting the spectacle.

the region, like www.oldmerthyrtydfil.com, which tells the history of iron foundries and factories in Merthyr Tydfil, Wales. The GENUKI website at www.genuki.org.uk also has advice about locating subject-related records, including occupations, specific to each area of the United Kingdom and Ireland. Have a look at the relevant county page for your ancestor's place of work to find out whether any local sites have been set up specializing in a local industry or company you are trying to research. There is a website directory, with links to local history sites too, at www.local-history.co.uk.

Regional Industry Archives

The majority of records concerning private companies that owned factories, foundries and mills will be found in the county record office for the area in which your ancestor worked, but there are some regional institutions that have archives preserving the history of industries native to the local area, which can be found by searching the ARCHON directory via the National Archives' website at www.nationalarchives. gov.uk/archon.

- Sheffield Industrial Museums Trust is responsible for Kelham Island Industrial Museum, Abbeydale Industrial Hamlet, and Shepherd Wheel. The Trust cares for a unique collection of archive material recording the heritage of the Industrial Revolution in and around Sheffield. Material relating to the people, products and manufacturing processes of steel, iron, cutlery, tool making, silver and holloware (galvanized steel 'hollow' items such as dustbins and buckets) manufacture are stored primarily at Kelham Island Industrial Museum. The documents, plans, technical drawings, books, photographs, equipment, clothing, advertising, machinery

▶ An illustration of cotton spinning at Walter Evans & Company, Derby, in 1863.

National Archives

There is very little concerning industrial trades at national-level archives because they were private enterprises that were not required to deposit their papers with any governmental department. The National Archives in Kew, however, does hold the records of the Factory Inspectorate established as part of the 1833 Factory Act to protect children employed by manufacturers, and later to monitor the conditions of all workers.

The Factory Inspectorate comprised of four inspectors who reported to the Home Secretary on the state of factories within their district and the conditions of young people employed in them. In 1844 sub-inspectors were also appointed to ensure factory regulations were being met. A central office, later known as the Factory Department or Factory Office, was established at this time. In 1878 a chief inspector was appointed who was responsible for the central office and directly answerable to the Home Secretary. In 1946 the Factory Inspectorate was transferred to the Ministry of Labour and National Service.

● Registered files of the Factory Department and Inspectorate, including minutes of meetings and conferences of factory inspectors, files concerning the arrangement of their districts and other records relating to factory inspection, are kept in The National Archives series LAB 15 for the period 1836 to 1975.

● If you find a death certificate of an ancestor who died as a result of lead or anthrax poisoning, then consult the Registers of Lead Poisoning and Anthrax Cases arranged chronologically between 1900 and 1951 in LAB 56.

● There is a large amount of paperwork relating to factory inspectors in HO 45, and Factory Entry Books of out-letters to inspectors of factories and orders relating to the administration of Factories Acts covering 1836 to 1921 are kept in HO 87.

It is also worth browsing through results for Factory Inspectorate in the online catalogue to see if there is anything of interest to your area of research.

and original art that form the Trust's archive can be accessed at the Collections Management Centre if an appointment is made by telephoning 0114 272 2106.

- John Rylands University of Manchester Library in Manchester city centre houses the Greater Manchester Textile Mills Survey Archive, formed between 1985 and 2000. The survey consisted of maps identifying around 2,400 mill sites, field surveys of around 1,000 sites, and research using archives and library material. The findings of the survey were published in *Cotton Mills in Greater Manchester* in 1992. The library also contains records of several industrial organizations based in the area as well as some company papers. Detailed catalogues of the library's archive collections can be searched from the Electronic Gateway to Archives at the Rylands (ELGAR) at http://archives.li.man.ac.uk.

- Spinning the Web is a website developed by Manchester Library and Information Service to give access to around 20,000 items gathered from libraries, museums and archives that tell the story of the

Lancashire cotton industry. The website, at www.spinningtheweb. org.uk, has plenty of digital documents, photos, newspaper articles and original reports that can be searched by keyword as well as in-depth historical accounts of what life was like in the industry at different points in time and from the perspective of mill owners, managers and workers. Use the Full Search facility to locate specific items from the 17 institutions that have been involved with the project, such as the Lancashire textile museums. There is information about both living and working conditions, leisure time and major reforms within the cotton industry. You can look at the effect the Industrial Revolution had on individual towns in Lancashire and how the manufacture of cotton developed during this time, and find examples of the goods that your ancestors made. This is a brilliant resource for anyone researching ancestors involved in the Lancashire cotton industry at any level and can give you guidance about the archives you should visit to find out more.

- Glasgow City Archive, the University of Glasgow and Glasgow Museums have combined their resources into the Glasgow Story website using images from their collections and the expertise of some of Scotland's best writers to tell the history of the city. The website is split into six historical periods, one of those being the Industrial Revolution where you can learn all about the impact of industry on the Glaswegian way of life between the 1770s and 1830s. Iron and steel manufacture, shipbuilding and the textile and leather industries helped Glasgow to rival Edinburgh in industrial and economic growth during the nineteenth century. The project can be found at www.theglasgowstory.com.

Specialist Museums

Whether or not company archives survive for your ancestor's employer, there are plenty of museums dedicated to preserving the heritage of Britain's industrial age found dotted around the United Kingdom, and these are the best places to find out about your ancestor's life and times. Find one relevant to your ancestor's occupation so you can see what their working and living environment would have looked like. Local councils and archives have information about the museums in their area. If you are very lucky there may be a museum dedicated to the history of the local industry in your ancestor's hometown. Here is a selection of some of them.

Derwent Valley Mills

In 1771 Richard Arkwright constructed the first successful water-powered cotton spinning mill in Cromford, Derbyshire. The mills at Cromford had a large workforce with a factory village to house them and became a model for factories across Britain and Europe. To celebrate Arkwright's status as the 'father of the factory system' and to maintain the World Heritage Site at Derwent Valley where the communities who supported Arkwright's mills lived, the Arkwright Society has been working hard to conserve the industrial buildings that have been standing now for over 200 years. You can find out more about visiting the Derwent Valley Mills at www.arkwrightsociety.org.uk.

> ' Specialist industrial museums are the best places to find out about your ancestor's life and times. '

The Weaver's Triangle

The Weavers' Triangle is a modern name given to an area of the Leeds and Liverpool canal, built between 1770 and 1816, that was once the heart of Burnley's textile industry and where a largely untouched Victorian industrial landscape survives. The Weavers' Triangle Trust, formerly the Burnley Industrial Museum Committee, has several museums within the area of the triangle, including a Visitors' Centre where you can find out how cotton was made and have a go at weaving to get a real experience of your ancestors' working life. There is a weaver's cellar dwelling where you can see what living conditions were like for weavers a hundred years ago. A short walk from the Visitors' Centre is Oak Mount Mill Engine House, which produced cotton from 1830 until 1979. The engine house and chimney were built in 1887. As well as the museums there are foundries, warehouses, domestic buildings, workers' houses and a school that have been preserved as part of Burnley's industrial heritage. You can go on guided tours along the canal towpath that take you through all the areas of the Weavers' Triangle. Visit www.weaverstriangle.co.uk to find out about opening times and view the online picture gallery.

Leeds Industrial Museum

While Lancashire was the heart of England's cotton industry, other counties are equally as proud of their textile heritage. There are plenty of museums where you can learn all about your ancestors who were employed on the other side of the border in Yorkshire. The Leeds Industrial Museum at Armley Mills is a living piece of history for the Yorkshire textile industry where the working environment of mill

employees has been reconstructed. The website at www.leeds.gov.uk/
armleymills is great for finding out more about the history of the mills
and the textile industry in Leeds. There is also an interesting web
gallery with an online exhibition looking at the lives of working children in Britain.

Bradford Industrial Museum

Bradford Industrial Museum set in Moorside Mills gives you the opportunity to witness the differences between the lifestyles of mill workers
and their superiors, with a fully equipped Mill Manager's House and a
humble Victorian terrace known as Gaythorne Row decorated in the
style of mill workers' cottages in the nineteenth century, the 1940s,
1950s and 1960s.

Ironbridge Gorge

The Ironbridge Gorge in Telford has no less than ten museums specializing in the industrial history of this part of the world. The preserved
Victorian town of Blists Hill is home to an iron foundry, blast furnaces
and a brick and tile works where you can get an authentic taste of what
it would have been like living in the midst of many industrial businesses. The former homes of members of the Darby family who owned
Coalbrookdale Ironworks are open to visitors for you to see how this
industrialist Quaker family lived. The old Coalport China Works has
been turned into a museum where you can find out how fine china and
glass products were made at the factory and take a tour of the social
history gallery. In the 1880s Broseley Clay Tobacco Pipes Works was
opened nearby and manufactured pipes in the village until the 1950s,
when the factory was abandoned and left untouched until it became a
museum in 1996. The Ironbridge website at www.ironbridge.org.uk
has some brilliant pages about the history of the area and the companies who brought the Industrial Revolution to Telford.

Manchester Museums

Manchester is home to both the People's History Museum and the
Museum of Science and Industry. The People's History Museum
collects and conserves material relating to the history of working
people in Britain and is where the Textile Conservation Studio and
Labour History and Archive Study Centre are based. It has close links

CASE STUDY

Jane Horrocks

Jane Horrocks grew up in Rawtenstall, which today is a small town in Lancashire but throughout the nineteenth century played an important part in the industrial revolution, specializing in the production of textiles from cotton. Indeed, the dozens of cotton mills in the town that operated from the mid-nineteenth century onwards provided the principal source of employment in the town.

Jane was able to construct a basic family tree using standard sources – birth, marriage and death certificates and census returns – and by looking at the occupations listed on each document established that several branches of her family worked in the cotton mills for a living. She was particularly interested in the women in her family tree, including Sarah Alice Cunliffe, her maternal great-grandmother.

Sarah Cunliffe was born in Rawtenstall in 1858, and according to her birth certificate her father John worked as a mechanic in a local cotton mill. This was confirmed by the census returns from 1861 onwards, where Jane was able to trace, decade by decade, the progress of the family. By 1881, Sarah Cunliffe, now aged 23, was working as a cotton weaver, joined by her 15-year-old sister Mary and 13-year-old brother George. Her parents died within two years of each other, in 1888 and 1890, leaving her to look after her younger siblings.

Jane was able to establish what conditions in the factory were like by reading contemporary accounts and examining photographs of Rawtenstall and its associated mills from the nineteenth century, gaining a sense of how harsh conditions would have been by looking at the daily rules and routines for each type of worker, and researching some of the likely injuries that would have come from the back-breaking tasks – risk of amputation of fingers or limbs, a dry cough caused by the fibres in the air, and constant ringing in the ears through the endless pounding of the looms hour after hour in an enclosed space. By visiting a working mill, some of the ferocity of the working environment became clear.

Yet life for the mill owners was in complete contrast to their employees, and the proprietors could afford large houses and a comfortable lifestyle. Jane discovered that one branch of cotton labourers, the Ashworth family, shared the same surname as one of the larger mill owners in Rawtenstall, James Henry Ashworth & Company, who were listed in a trade directory in the mid-nineteenth century. By following both family trees back in time, Jane was able to establish that they had actually shared a common ancestor in the seventeenth century; and that a quirk of fate had led her branch of the family into the mills as poor labourers, whilst the other side ended up owning them.

◀ Sarah Cunliffe, Jane's maternal great-grandmother in the cotton mill where she worked.

USEFUL INFO

If you would like to find a museum for a particular industry in the area where your ancestors lived then search the 24 Hour Museum's database of more than 3,800 museums, galleries and heritage sites at www.24hourmuseum.org.uk using the Advanced Search option. The website's City Heritage Guides give information about which attractions to visit in some of the biggest industrial towns in Britain, including Birmingham, Leeds, Liverpool, Manchester, London and Newcastle.

with the Museum of Science and Industry where there is a special Local History section tracking how Manchester changed from a Roman outpost to the world's first industrial city. The museum is a must for anybody with industrial roots in Manchester. Its collections encompass the textile industry, paper manufacturers, tool production, housing and sanitation and the website has snippets of oral history for you to download. Visit www.msim.org.uk to get a taste of what's on offer.

The Black Country Living Museum

The Black Country Living Museum is an open-air urban heritage park where historic buildings from all over the Midlands have been rebuilt in tribute to the people who once lived at the heart of this industrial area to the west of Birmingham. The Black Country gained its name from the thick smoke pumped out by thousands of iron foundries and forges and the black spoil left over from coal mining in the countryside. Situated a mile from Dudley town centre, the museum is great for understanding what everyday life was like for ordinary workers in small industrial towns, from going to school to watching a silent movie in the 1920s cinema and visiting the many workshops and small factories to watch how people once worked.

National Wool Museum, Carmarthenshire

The National Museum Wales explores one of the country's most important industries at the National Wool Museum found in the former Cambrian Mills in the Teifi Valley, Carmarthenshire. Home to the National Flat Textile Collection, you can visit the restored mill buildings, see the machinery your ancestors would have worked with and have a go yourself at carding, spinning and sewing.

Ulster Folk and Transport Museum

Northern Ireland celebrates its industrial heritage through the projects of the National Museums Northern Ireland. The Ulster Folk and Transport Museum has a special exhibition about the building of *Titanic*, and is home to the Living Linen Archive, set up to record oral history from workers in the industry. Their website contains an index of linen firms about which material has been gathered.

Industrial Unrest and Labour Movements

Some of our working-class ancestors found a voice and an outlet for their frustrations about their appalling working conditions by joining organized movements that could communicate their grievances to employers and local sympathizers. It is unsurprising that labour movements gained popularity in British industries renowned for their monopoly over the world market given the nature of their success. Wealthy industrialists praised globally for their business skills and innovation relied heavily on the cheap labour of unskilled, uneducated people who saw few of the financial rewards reaped by their company's success. In an era of political change when the importance of freedom of speech and universal suffrage were widely debated, workers found the confidence to unite and confront these issues. Those who went on strike did so at the risk of their job and often relied on charity for subsistence while they were on strike. The records left behind from these volatile times include strike fund registers, reports, newspaper articles and specially published journals.

Political Movements

Chartism was a movement that gathered momentum among the industrial working class when economic depression was just around the corner in the 1840s, with its grassroots based in Manchester, Leeds, Sheffield, the Black Country, Middlesbrough and Scotland. The movement was originally started around 1837 by a group of skilled London artisans in response to the Great Reform Act of 1832, which had promised universal male suffrage but in practice gave it only to the propertied classes. Consequently, and after many public meetings and riots, the Chartists drew up their own bill requesting democracy for all men and in 1848 presented it to Parliament with a petition. The propertied classes were fearful at the possibility of revolution spreading to Britain from the Continent, and the government used military force and the powers of the press to ridicule the Chartists into eventual submission. Some Chartist activists were caught, punished and even executed. If you believe your ancestor was one of them then read Chapter 27 about how to locate evidence of a criminal ancestor.

The Independent Labour Party (ILP), a forerunner to the Labour Party, was formed in 1893 as a result of the Manningham Mills strike in

USEFUL INFO

Records concerning the Chartists and other political labour movements are held at the Labour History and Archive Study Centre found in the People's History Museum in Manchester and you can find out more about these men online from the Chartist Ancestors website at www.chartists.net. Chartist Ancestors contains the names of many Chartists drawn from newspaper reports, court records and contemporary books. The Chartist Ancestors website has a sister site for Trade Union Ancestors where information about other industrial labour movements can be found.

Bradford. Manningham Mills (later known as Lister's Mills), built in 1873, was the largest textile mill in northern England. In December 1890 the mill workers there went on strike in response to wage reductions. The strike was to last until April the following year, during which time clashes with the Bradford authorities gave the movement national attention. The strike has been recognized as historically significant because it marked a crisis in the relations and opposing interests of textile employees and employers, and highlighted the imbalance of power. A growing awareness of class hostility and the need for political representation emerged in the West Riding of Yorkshire and key activists in the Manningham movement set up the Bradford Labour Union, which formed the ILP two years later. If you think your ancestor was part of this momentous social and political movement then you may be interested in reading Cyril Pearce's analysis in *The Manningham Mills Strike, Bradford, December 1890–April 1891*. Records concerning the strike are held at West Yorkshire Archive Service at Bradford Central Library.

Trade Union Records

The Trade Union Archives at the Modern Records Centre, University of Warwick, holds the records of some trade unions that protected factory, foundry and mill workers, such as the Amalgamated Association of Operative Cotton Spinners etc. of Lancashire and Adjoining Counties. They also have monthly, quarterly, half-yearly and annual reports of the Friendly Society of Ironfounders, 1850–1920, that give names of new entrants organized chronologically by branch. Most of the trade union records deposited at the Modern Records Centre do not include membership records, but their papers may be of interest if you know your ancestor was involved in a particular movement.

The Working Class Movement Library in Salford houses a collection of books, pamphlets, periodicals, archives and artefacts expressing the concerns and activities of labour movements since the 1700s. The Library Catalogue can be searched by keyword from the WCML website.

The Trades Union Congress (TUC) has been operating since 1868 and has created The Union Makes Us Strong website, where a digitized register of around 700 matchworkers from the East End of London, most of them young women, who went on strike against the Bryant & May Match Factory in 1888, can be searched. The matchworkers went on strike in revolt against the sacking of three of their colleagues as a result of speaking to the radical journalist Annie Besant about their working conditions. There is a great deal of history about the activity of

> *People joined labour movements in an attempt to improve their living and working conditions.*

matchmakers and the strike as well as an abundance of digitized documents, photographs and a search engine to search the Match Workers Strike Fund Register set up by Besant at www.unionhistory.info/match-workers/matchworkers.php. The TUC website also has a timeline of its history where other major strikes can be researched, including the General Strike of 1926. If you would like to find other records of the TUC the London Metropolitan University holds the TUC Library Collections.

Very often working-class people joined labour movements out of frustration in an attempt to improve their living and working conditions. These frustrations were magnified at times of economic depression and hardship. If you think your relatives were affected by economic disasters such as the one that led to the Lancashire Cotton Famine, then poor relief records are worth looking into. Records of Poor Law Guardians are arranged by parish and should be found in the relevant county record office. For more information on locating records for poverty-stricken ancestors see Chapter 24.

Finding Out More

- Organizations and societies that still have links with the industry you are researching and have a thorough knowledge of its history may be able to give you some guidance concerning the location of company archives and research sources.
- Bradford Textile Society, founded in 1893, is still going strong and has a website at www.bradfordtextilesociety.org.uk.
- Sheffield Company of Cutlers is an organization that has been representing cutlery makers in the Hallamshire region for 400 years. If your ancestor was a prominent industrialist in the Sheffield cutlery industry he may have been a member of the Company, perhaps even a Master Cutler. You can find out more by writing to

  The Company of Cutlers in Hallamshire
  The Cutlers' Hall
  Church Street
  Sheffield S1 1HG
- The North West Film Archive, part of Manchester Metropolitan University's Library Special Collections, has over 5,000 titles capturing the industrial, working and home life of communities from the North West of England. You can browse descriptions of the archive's collections at www.nwfa.mmu.ac.uk and contact a member of staff to enquire about viewing any moving image files.

CHAPTER 17

Occupations: Travel and Communications

The Industrial Revolution spawned a range of occupations closely linked to the expansion of the transport network, particularly railways and canals but also improvements to the roads and highways. The records described in this chapter allow you to piece together the lives and careers of workers in these industries, as well as establish more about the impact the new communication networks had on daily life.

Historical Context

' Britain's communications networks changed our ancestors' environment beyond recognition. '

From the eighteenth century a lot of money was invested in improving British transport systems to aid industrial growth. Inland waterways were dug to carry products for export from isolated provincial towns to the docks where they would be shipped by cargo boat. The network of canals constructed from the mid-1600s provided work for travelling labourers known as navigational engineers, or 'navvies', employed by construction companies. The waterways network, which covered around 4,000 miles by the early nineteenth century, was more cost effective and efficient for transporting goods than the alternative of horse-drawn wagons on roads that were poorly maintained and plagued by highwaymen.

Some of the problems of road transport were addressed at the beginning of the eighteenth century when a number of Turnpike Acts were

▲ A station master gives directions to a passenger, 1890.

passed, setting up turnpike trusts to ensure the maintenance and safety of major routes financed by a toll. As more turnpikes were constructed around the country journey times by coach were reduced by up to a half and more people were encouraged to travel further distances as instances of highway robbery fell. Travel was made even more affordable and accessible to a wider section of society with the advent of the railways from the 1830s, allowing ordinary workers the possibility of travelling further afield in search of work and a better standard of life.

Railways were first designed for commercial purposes, transporting finished products at a much faster speed than canal boats so that manufacturers could meet growing demands from consumers. A Cornish engineer, Richard Trevithick, designed the first locomotive-drawn train in 1804, though design failures in the tracks hindered the success of his experiments and it was not until 1811 that John Blenkinsop developed a railway worked by a steam locomotive to transport coal on the Middleton Railway. In the same year William Jessop engineered the Kilmarnock and Troon Railway in Scotland. In 1829 the world's first inter-city line, the Liverpool and Manchester Railway, was

USEFUL INFO

In 1801 London covered a relatively small area, but a hundred years later villages like Leyton and Hampstead had been swallowed up into the mass of suburbs now surrounding the city, a commuter belt feeding the capital with a new class of white-collar workers.

> *Railways allowed people to migrate more easily and industries to grow quickly.*

built and used to test various locomotive systems in the Rainhill Trials, which proved Stephenson's *Rocket* to be the fastest and most advanced design. This marked a turning point in the history of British transport and established rail as the dominant form of land transport for the next hundred years.

Britain's communication networks changed our ancestors' environment beyond recognition during the nineteenth century. The statistics extracted from nineteenth-century census returns demonstrate an unprecedented population boom and, coupled with evidence from maps of the countryside and towns, we can see where this burgeoning population was concentrated, giving us an insight into migration patterns. The expansion of Britain's towns and villages, at times merging to form cities, was principally the work of the railways. These allowed populations to migrate more easily and industries to grow quickly and accommodate more workers.

The British Library has a fascinating collection of London maps that help you to see at a glance how urbanization has spread since Roman times and how communication networks and transport links have been at the centre of these changes. The library's Collect Britain website at www.collectbritain.co.uk has digital copies of the Crace Collection, over 800 maps and plans of the capital from the sixteenth to the nineteenth centuries collected by the Victorian designer Frederick Crace, as well as a link to the British Library's virtual exhibition, London: A Life in Maps.

Improved communication networks had far-reaching effects beyond the obvious impact on population mobility. With transportation carrying goods faster than ever across a larger geographical area, publishers began printing national newspapers to distribute across the country, raising public awareness of central politics and national issues. With the Education Acts of the nineteenth century improving levels of literacy, the popularity of the 'penny post' really took off with the help of the railways. In the Post Office's heyday it was possible to send a letter by post and get a reply the very same day.

The development of turnpikes, inland waterways and railways changed the nature of rural life, altering the British landscape and fuelling urbanization. Improvements in transport had an immense effect on the demography of the population, speeding up industrialization and creating jobs for thousands of people, not just in the transport industry but also in the areas to which people could travel more easily, quickly and affordably. The opening of new transport links was often shortly followed by a surge in migration to that area, so if you've been wondering why your family suddenly upped sticks and moved 20 miles away the answer might lie in the date the local station opened.

Roads and Turnpikes

Prior to the eighteenth century, parishes were responsible for the maintenance of roads, but their failure to keep them safe and in good condition led to turnpike trusts being formed from 1663 with the permission of a local Act of Parliament. Some turnpike trusts improved the roads that were already there while brand-new turnpikes were also constructed from 1706, which totalled 22,000 miles of turnpike road covering the country by the 1830s. The mid-eighteenth century has been described as a period of 'turnpike mania' with over 400 Turnpike Acts being passed between the 1730s and 1760s. The new turnpike trusts financed the maintenance and construction of

Turnpike Records

• Geoffrey N. Wright wrote a guide to *Turnpike Roads* in 1992, which includes a map showing all the turnpikes in 1750.

• Records of turnpike trusts and Highway Boards are kept at local county record offices, where you may find maps, plans, agreements between the parish and the owners of any private land a turnpike adjoined, as well as accounts and records of payments to labourers working on the road and merchants supplying construction materials.

• If your ancestor was a member of a turnpike trust or on the Highway Board their name will probably appear on the committee's minutes. The National Register of Archives at

www.nationalarchives.gov.uk/nra has details of repositories for over 600 turnpike trusts across the United Kingdom and Ireland. The advanced search option under the Corporate Name search allows you to narrow down the search criteria by selecting 'Turnpike Trusts' from the Category list so that you can find records such as those of the Great Western Road Turnpike Trust, whose minutes are held at Glasgow City Archives.

• The National Archives in Kew holds some stray records concerning turnpikes, kept principally among documents inherited by the Transport Departments in series MT. Use the online catalogue to search for these, which include accounts and reports concerning the Holyhead and Shrewsbury Roads in MT 27.

• The National Archives of Ireland has a small number of documents concerning turnpikes kept with records of the Office of Public Works, details of which can be found on the National Archives of Ireland online database.

• The most comprehensive collections of turnpike trust records are held among local collections, however, so use resources such as the Archives Hub found at www.archiveshub.ac.uk to locate records in university and college collections and the Scottish Archive Network (SCAN) at www.scan.org.uk and National Register of Archives for Scotland at www.nas.gov.uk/nras for documents held in Scottish archives and local studies centres.

USEFUL INFO

More affordable coach journeys could be made if a passenger was prepared to travel on the open-air top of the carriage, while those who were better off would pay for an inside seat and the very rich would only travel in a private carriage.

roads by introducing a toll on road users. Tollhouses were built along sections of the road to collect the toll, which began to deter highwaymen who could no longer escape the scene of a robbery unnoticed.

Generally speaking, it was propertied men who managed the turnpike trusts, and their success in constructing a national road network lay in the quality of the materials used and the regularity of maintenance. A coach journey from London to Bath would have taken around three days before smooth turnpikes transformed the boggy paths and made the journey easier for the horses and more comfortable for passengers, so that Bath could be reached within twelve hours in 1779. By the 1780s every region of England and some parts of Scotland and Wales had a local turnpike network that linked into a national system connecting London with modern industrializing centres like Manchester and Glasgow.

Highway Boards made up of groups of parish authorities gradually replaced the turnpike trusts from 1835.

▼ A horse-drawn cab passes a motorized bus in Piccadilly Circus, London, in 1907.

Post Office Records

Descriptions of the administrative and staff records of the Post Office from 1636 to 2000 can be found in The National Archives online catalogue under the POST series, although none of the records are actually held at Kew. Records of employees of the Post Office can be found at the Royal Mail Archive held at

The British Postal Museum and Archive
Freeling House
Phoenix Place
London WC1X 0DL

The archives date back to the seventeenth century, with establishment books listing senior members of staff from 1742, and appointment books from 1831 to 1952 recording all staff appointments. Files about staff pensions survive for all employees from the nineteenth century, with name indexes from 1921 onwards, and the collection includes a large photographic archive. The British Postal Museum website has an online catalogue with descriptions of 50,000 records from its collections at www.postalheritage.org.uk.

The Post Office was also responsible for the national telegraph service, with the entire telephone service being taken over by the GPO in 1912. In 1947 Cable & Wireless Ltd was nationalized and absorbed into the Post Office. The GPO became responsible for more fees and licences, particularly with the growth of the television and media industry, which made it too large and complex to manage effectively and so its powers were devolved and in 1969 the Post Office Act made the GPO a public corporation rather than a government branch, splitting it into Post and Telecommunications. In 1981 the telecommunications side of the Post Office became a separate corporation under the name British Telecom, which was privatized in 1984.

Records of the Telecommunications department of the Post Office and earlier private telephone and telegraph companies have been transferred to the BT Group Archives, whose records date from the nineteenth century. The British Telecom Archives have teamed up with the commercial website www.ancestry.co.uk to digitize historical phone books and make them available to search online.

The Post Office

Charles I established the Post Office in 1635, employing postmasters on the main routes between London and the major cities to collect and distribute mail and to collect revenues on behalf of the Crown. The General Post Office (GPO) was established between 1656 and 1660 in the City of London and comprised the Inland Office for all internal mail, the Foreign Office for overseas mail, and the Penny Post Office for local mail; a Postmaster General oversaw the entire department. The GPO building and the majority of its records were engulfed by the Great Fire of London in 1666, so documentation prior to this date is scarce.

The number of routes and towns covered by the postal network grew during the 1700s as the ever-expanding network of turnpikes made it easier for mail to be delivered further. This necessitated regional

▶ The inauguration of the Penny Post, with the first dispatch arriving at Waterloo station.

▼ An early telegram.

administrators to be established, to ensure local post offices were run efficiently. In 1715 regional surveyors were appointed to fulfil this requirement.

The introduction of the world's first adhesive postage stamps, the Penny Black and the Two Pence Blue, in 1840 gave the postal system a huge boost in the nineteenth century. Increased adult literacy also led to a rise in the amount of mail sent and the Post Office became responsible for telecommunications in 1870 and a banking service.

Canals and Inland Waterways

Although the construction of turnpikes from the late seventeenth century onwards increased the speed at which people and small loads of commercial goods could be transported, the roads did not solve the problem of getting large quantities of industrial produce to the ports. Thus, developing inland waterways was seen as a way to overcome the limitations posed by wagon loads on the roads. Canals were the principal mode of transport for commerce during the eighteenth century and only went into decline from the mid-nineteenth century when a vast network of railway track started to cover the British countryside and took business away from canal companies.

Though few records survive proving that our ancestors helped build these waterways and navigations, there are plenty of records helping us to understand the history of each canal – when construction began, how long it took, how many people were employed on it, how it benefited surrounding areas and how long it proved vital to the local economy. It is possible to find out who invested in canal construction and sometimes the names of those people who worked on the canals as boatmen, as well as information about specific barges and canal boats.

- *The Waterways Archive:* The Waterways Trust administers the Waterways Archive, a collection of material held in several county record offices as well as at two of the three National Waterways Museums located in Ellesmere Port and Gloucester Docks. Ellesmere Port is home to the David Owen Waterways Archive and Gloucester Docks to the British Waterways Archive. The Waterways Archive collections include records of boat owners and registered boatmen, toll records, and large carrying companies' archives, plans, drawings, technical records, documents, books, periodicals, oral history recordings and photographs. The collections from 15 archives, including the British Waterways Archive, David Owen Waterways Archive and National Archives of Scotland, can be searched on the Virtual Waterways Archive catalogue at www.virtualwaterways.co.uk. The Waterways Archives at Ellesmere Port and Gloucester Docks provide a research service at a fee for those unable to visit in person. You can find out more about these services from the National Waterways Museum website at www.nwm.org.uk. While the Virtual Waterways catalogue is a great place to begin your search for canal records, it does not have a comprehensive list of all documents held in British repositories, which is why a search of the county record office or local studies centre nearest to the canal is also wise.

- *Scottish Canal Records:* The first canal to be constructed in Scotland was the Forth and Clyde Canal in 1792, with Scotland's canal network being completed in 1822. The National Archives of Scotland website contains company histories for the Caledonian Canal, the Crinan Canal, the Edinburgh and Glasgow Union Canal, the Forth and Cart Junction Canal, the Forth and Clyde Canal, the Glasgow Paisley and Johnstone Canal and the Monkland Canal, although some records of these companies are also held at The National Archives in Kew. The NAS has an online research guide explaining how to find canal records there with a list of document references for each company at www.nas.gov.uk/guides/canal.asp. The NAS warns researchers that

> *Canals were the principal mode of transport for commerce in the eighteenth century.*

no lists of canal employees survive among its records, though names of canal proprietors and commissioners may be found.

- *Irish Canal Records:* Many labourers employed to construct Scottish canals were in fact Irish. To locate records of Irish canal companies it is worth contacting the National Archives of Ireland, the Public Record Office of Northern Ireland and the Ulster Folk & Transport Museum who have their own archive and library collections.

- *Records in The National Archives:* Railway companies started to take over ailing canal companies and in the 1850s the Board of Trade Railway Department assumed responsibility for them all. This link means that much documentation regarding canals at The National Archives is found with that of railways, docks and roads in the RAIL series.

  A Railway and Canal Division of the Board of Trade was established in 1873 and the administration of canals became the duty of the Docks and Canals Division in 1934. The Transport Act 1947 nationalized the canals as well as the railways, and responsibility for operating inland waterways was transferred to the Docks and Inland Waterways Executive of the British Transport Commission, whose archives now form part of the British Transport Historical Records Section (BTHR) kept at The National Archives.

  The BTHR Section has a card index at Kew where records relating to particular companies, photographs, maps, plans, special collections, books, pamphlets and periodicals can be searched. The records of this section, mainly held in RAIL and AN series, date back much earlier than the 1850s because the rail companies inherited records from canal companies that may have been operating since the 1600s. The earliest substantial canal records held at The National Archives are those for the Wey Navigation, built between 1651 and 1653. Papers consisting of claims to the canal's profits can be found in E177.

- *Parliamentary Archives:* After 1792 the construction of a canal required Parliament's permission, and as such Parliamentary Papers contain local and private Acts of Parliament and the Parliamentary Archives at the Houses of Parliament in London hold some plans and prospectuses for proposed canal constructions from 1794. From 1837 duplicates of canal scheme maps also had to be placed with the Clerk of the Peace and those records may be found among the Quarter Sessions' papers at the local record office. From 1795 the Clerk of the Peace also kept Registers of Boats and Barges for inland waterway craft that had to be registered with the name of the proprietor.

- *River Transport Records:* River transportation for passengers was vitally important before large-scale bridge-building programmes made it possible to cross rivers at several points. The Thames in London, for instance, could only be crossed at London Bridge until 1750 when Westminster Bridge was opened. Watermen operated ferry crossings to carry people between the north and south banks of the river Thames. The Company of Watermen was formed in the 1500s to ensure passenger safety and fair charges. Officers were appointed by the Lord Mayor to issue licences to watermen, printed tables of fares were published from the early eighteenth century and in 1700 the Company encompassed lightermen as well as watermen. Lightermen unloaded cargo from ships and carried it to the port by lighter. Members of the Company of Watermen and Lightermen came from all parts of the river Thames between Gravesend and Windsor, and the Company's records can be found in the Guildhall Library Manuscripts Section in London. The Library's records, dating from the 1600s to the 1940s, include apprentices' bindings and affidavit books, quarterage books, records of contracts and ferry services, records of the Court of Complaint, and registers of lighters, barges and passenger boats. These documents hold the names of watermen and lightermen; some apprenticeship records include dates and places of baptism, while others hold addresses, places of mooring, dates of death, and information about earnings. The Guildhall Library has a research guide explaining more about their records for watermen and lightermen at www.history.ac.uk/gh/water.htm.

▲ Horse-drawn barges navigate a stretch of the Rolle Canal, north Devon, built to move corn from Lord Rolle's estate to the port of Bideford.

   If you are unable to visit the Guildhall then Robert J. Cottrell has compiled indexes of Thames Watermen and Lightermen Apprentice Bindings 1692–1949, Apprentices' Affidavits 1759–1949 and Contract Licences 1865–1926, and offers a search service of his indexes for a fee. Enquiries should be sent to 19 Bellevue Road, Bexleyheath, Kent DA6 8ND.

   Another set of records that you can apply to be searched by post is John Roberts's Waterway Index, containing the names of around

10,000 people employed on inland waterways, from boat builders, canal agents, lock-keepers and toll collectors to boatmen, watermen, flatmen and navvies. The names have been extracted from a number of original sources and Mr Roberts will search his indexes if an enquiry is sent to 52 St Andrews Road, Sutton Coldfield, West Midlands B75 6UH.

The London Canal Museum is a great place to learn all about the heritage of canals in the capital and their website has a useful Family History Checklist for anybody researching canal ancestors, at www.canalmuseum.org.uk/collection/family-history.htm. The website also has a comprehensive list of links to other useful sites for researching the history of canals and waterways. If you are researching ancestors who worked on the Thames in London then take a look at the Bargemen website at www.bargemen.co.uk, which has information about boat builders, owners, watermen, lightermen and dock workers around London and contains some family histories. The site is packed full of research guides for those exploring different types of Thames workers as well as indexes to wills of lightermen, the names of barges involved in the evacuation from Dunkirk, directories for lightermen and bargemen, and other useful sources for family historians compiled from various original documents.

While many canals were filled in or fell into disrepair after their demise, there has recently been a push to conserve some in the interests of heritage, to promote tourism and as a means of exploring a greener method of transport in the modern age. There are many canal associations who work hard to preserve the inland waterways that have weaved their way through the British landscape for around 300 years now, and they can often point you in the direction of surviving records documenting the canal's history and the lives of the people to whom they were important.

Railway Records

Private railway companies operated all lines in Britain until the Transport Act 1947 nationalized the railways in 1948 to form British Rail. The 1921 Railways Act had already merged most railway companies into the 'Big Four' – the Great Western Railway, the London & North Eastern Railway, the Southern Railway and the London, Midland & Scottish Railway. In 1948 the British Transport Historical

Commission collected the records of the several hundred rail, canal and dock companies that were absorbed into the State. The records were initially kept at London, York and Edinburgh but have since been relocated: those records kept in Edinburgh were transferred to the National Archives of Scotland and those records held at London and York are now at The National Archives in Kew.

County record offices also hold a lot of material about the impact of the railways locally, and a search of the main archive databases mentioned already, in particular the National Register of Archives, can unearth many of these collections. The local county record office website for the area where your ancestor worked may shed light on their railway collections, like that of Cheshire Record Office, which has an online index to the staff registers they hold for Cambrian Railway, London & North Western Railway, Great Western Railway, and London & North Western and Great Western Joint Railway. The publication *Was Your Grandfather a Railwayman?* by Tom Richards contains a directory of the record offices around the British Isles that hold papers for various railway companies. Staff records for those people who worked after the Second World War are still with the railway industry for pension purposes, but staff records of workers employed on the railways since the 1960s can be obtained by writing to

The Strategic Rail Authority
55 Victoria Street
London SW1H 0EU

Employee Records

Staff records exist for many railway companies but they are difficult to use and are often poorly indexed, with no central index to the names of all railway employees. The records are arranged by company and some company papers will only give names, details of pay and positions, whereas others give a full service history and some companies have very few surviving staff lists at all.

It is necessary to know which of the 1,000 or so railway companies in operation since the nineteenth century your ancestor might have worked for in order to start looking for any records of his service. Simply knowing that

▼ Passengers from the *SS Mauretania* leave Fishguard on the first passenger boat train, in 1909.

CASE STUDY

Griff Rhys Jones

Griff Rhys Jones made the discovery that his maternal great-grandfather, Daniel Price, worked as an engine driver when he ordered the birth certificate for Daniel's daughter, Louisa Price (Griff's grandmother), from 1891. This one document allowed Griff to start a series of searches for more information. Of most importance was the address where the Daniel lived with his family in 1891 – Garston, Liverpool – which was confirmed by finding the family in the 1891 census. By investigating the local history of the street, Griff was able to establish that the house in which his great-grandfather lived had been built by the London and North Western Railway Company to provide accommodation for its workers. In turn, this led to a search at The National Archives for further information about Daniel's career with the company – but, although some staff records were identified in record series RAIL 410, there were none for the period in which Daniel served.

▼ Griff's maternal great-grandfather, Daniel Price, is shown on the 1891 census.

your great-grandfather worked as an engine driver in Liverpool will not be enough information because there were often several companies operating in most towns and each of their records is kept separately. Most companies merged to form larger companies so the company names changed frequently, and in this case it may be necessary to check the records of more than one company. *Railway Ancestors* by David T. Hawkings lists all the railway companies in England and Wales between 1822 and 1947 in alphabetical order with dates of the merger and details of name changes.

Each company's staff records are divided into departmental registers, such as those for the Locomotive Carriage and Wagon Department, Traffic Department, or Electrical Engineer's Department. There was no uniform way of structuring departments so it can sometimes be tricky working out which department your ancestor's job title fell under. Appendix 6 in *Railway Ancestors* contains lists of staff trades and occupations with the type of department they might come under, and The National Archives' research guide 'Railways: Staff Records' has some more common examples. Within the departmental records there are generally three types of records where you may find a mention of your ancestor:

- Salary registers were kept for clerical salaried staff and tend to include names, dates of birth, career changes, salary increases and bonuses.
- Staff registers were kept for financial purposes and record names, occupations, places of work, starting pay, pay rises, bonuses and allowances.
- Personnel records vary widely in their content from company to company and may cover a whole group of workers with registers of new employees, attendance logs and discipline registers. Usually the records are arranged numerically by staff number though sometimes an alphabetical index will accompany them.

Salary and staff registers can be interesting if they mention a fine incurred for misdemeanours or any allowances paid and time off due to sickness.

You need to be prepared to spend a long time searching among the various types of records for different departments because the records for railway companies are still largely arranged in the format they were when used by clerical staff at the time, and no modern form of indexing has been adopted to make them easily accessible for genealogists' purposes today.

HOW TO...

… find which railway company your ancestor worked for

1. *Census returns for railway ancestors sometimes give the initials of the company they worked for next to their job title, and the* British Railways Pre Grouping Atlas and Gazetteer *published by Ian Allen has a list of these abbreviations to help you decipher the company name they refer to.*

2. *The* Gazetteer *has a map of England, Wales and Scotland showing the lines that ran through each station as well as a list of stations with the companies that owned or operated them. You can establish where your ancestor lived through certificates and census returns then draw up a list of all the possible stations in that area and the companies that your ancestor may have worked for. The next step will simply be a case of working your way through all those companies' records to see if you can find anything.*

3. *The Irish Railway Record Society preserves many historical records about Irish railways. Their website at www.irrs.ie has histories for 14 railway companies operating in Ireland since 1946.*

If you know the name of the company your ancestor worked for, or once you have gathered a list of all his possible employers, there are many resources available to point you in the direction of where to find their records.

- The National Archives' research guide entitled 'Railways: Staff Records' has an appendix listing all the companies they hold records for and the exact RAIL series where those records can be found. Not all records for staff are listed in the appendix, so once you have established what RAIL series those records are kept in it is worth looking at the paper catalogue to get a fuller picture of the series' holdings. There is also a list of all the railway companies and their RAIL series at the beginning of the RAIL catalogues in the Research Enquiries Room, and a keyword search of the online catalogue for the company's name restricting the department code to 'RAIL' should also tell you which series it is held in. It is important to also consult lists provided in the guide *Was Your Grandfather A Railwayman?*, which covers documents for railways in England, Wales, Scotland, Ireland and overseas, and *Railway Ancestors* for railway companies in England and Wales. *Railway Records: A Guide to Sources* by Cliff Edwards takes a closer look at railway documents kept at The National Archives in Kew.
- The National Archives of Scotland not only holds railway records formerly kept by the British Transport Historical Records Department in Edinburgh but also some useful private collections deposited at the archives and modern records deposited by the British Railways Board, British Railways (Scottish Region), Scotrail and Railtrack Scotland. Most railway records are found under NAS reference BR, and their collections comprise staff records, accident reports, station traffic books, letter books, civil engineer records, photographs and much more. The principal employers of Scottish railwaymen prior to the 1921 Railways Act were North British, the Caledonian, the Glasgow & South Western, the Great North of Scotland, and the Highland. The Railscot website at www.railscot.co.uk contains histories of Scottish railway companies as well as a map showing where each company operated and a chronology of Scottish rail history. The NAS has a research guide for people tracing Scottish railway ancestors on their website, and the NAS publication *The Scottish Railway Story* is a must-read.
- The Public Records Office of Northern Ireland (PRONI) holds information about railway companies in Northern Ireland, and the Ulster Folk & Transport Museum has a library and archive containing transport material as well as private collections encompassing rail

CASE STUDY

Sue Johnston

Sue Johnston had been told several legends about her family's period of employment on the railways – the most exciting one relating to the fact that her grandfather, Alfred Cowan, had been the driver for the *Flying Scotsman*. However, by checking details of where he was based throughout most of his career, and comparing this with known information about the route of the *Flying Scotsman* at the National Railway Museum, York, it was fairly easy to disprove this; the *Flying Scotsman* operated on the East Coast Main Line, which originally comprised three earlier railway companies, whilst Alfred Cowan was based in Warrington in the north west, nowhere near.

The other myth related to Alfred's father, James Cowan – and the rumour that he had been station manager at Carlisle Citadel station, one of the busiest in the country throughout the mid to late nineteenth century when James was alleged to be working there.

Sue decided to search for employment records at The National Archives, and struck gold when she found a series of appointment books. They confirmed that he had indeed started work at Carlisle station, but as a porter in 1856, having spent the previous 7 years in some other form of work – unspecified in the documents. James gradually rose through the ranks, and by 1861 he was second assistant platform attendant. However, despite 25 years of service at the station, he never made it to stationmaster, leaving just 7 months after the death of the previous incumbent.

timetables, pictures, books and articles about the history of Ireland's railways. You can search the National Archives of Ireland's online database for their railway collections, which include Railways Companies Returns for the twentieth century.

When searching staff records, in addition to the name of the company and your ancestor's job title it is helpful to know their date of birth, rough period of employment, date of death and place of employment or residence. This information will help you to narrow down your search results and can be found from analysing information given on civil registration certificates and census returns. Look out for changes in the description of your ancestor's employment; for example on his marriage certificate he may have been described as a locomotive fire-man but five years later, on his daughter's birth certificate, he is described as an engine driver, so you know he received a promotion in that space of time. You might need to look at records for different departments if his job description changed. (The covering dates given for records held at The National Archives are sometimes misleading, because the description includes the earliest and latest dates mentioned in the records, which might be a date of birth and death, rather than giving the dates for years of service.)

Railway Police Records

Railway Police were employed by the railway companies from the 1830s to regulate the lines, which included signalling and station duties as well as preventing criminal activity. If you have an ancestor who was a railway policeman prior to the 1920s, his records should be found within the collections of the railway company that he worked for. After the merger of all railway companies into the 'Big Four' in 1921, each had its own police force controlled by a Chief of Police who joined up to form a Police Committee during the Second World War. The National Archives holds most records for railway police from 1921 onwards in series AN.

In 1949, following the nationalization of the railways, the system was replaced by the British Transport Commission Police, encompassing all the former railway company police forces as well as those of some canal and dock companies. British Transport Police was formed in 1962 to replace the British Transport Commission Police, and they still hold record cards for staff dating back to the 1860s, though these are not complete. More information about them can be found on the History Archives pages of the British Transport Police website at www.btp. police.uk. Copies of the *Railway Police Journal* for 1949–81 are kept at The National Archives in series ZPER 61. If you have a particular interest in this subject then read a copy of Pauline Appleby's *A Force on the Move: The Story of the British Transport Police 1825–1995*.

Accident Reports

If you find out from a death certificate that your ancestor was killed by a train accident (and there were very many railwaymen – not to mention passengers – who were killed in this way) there are plenty of surviving reports kept among company papers, and also in the records of the Board of Trade Railway Department, Ministry of Transport Reports and Railway Inspectorate at The National Archives, in series RAIL 1053, MT 114 and MT 29.

The records of the Railway Benevolent Institution, which granted financial aid to those former railwaymen and their families who had paid a subscription, are held at The National Archives in RAIL 1166. This is a useful source for those family historians who are unsure which company their ancestor worked for, because the institute drew subscriptions from railwaymen across the country. The Annual Reports in RAIL 1166/1–80 covering 1881 to 1959 contain lists of supporters as well as reports from the railway's orphanage at Derby. The Railway Benevolent Institution's Minutes are found in RAIL 1166/87–149 for 1858 to 1982. The most useful source for family historians can be found in the records of grants in RAIL 1166/81–86 for the years 1888 to 1919, which give details of grants awarded to railwaymen or their widows and orphans after an accident, illness, death or during old age. Each volume is indexed, but if a grant was given to a railwayman's widow or family after his death then the entry does not always name the deceased worker but may give the name of his company and his position instead.

Armagh County Museum and Library has a Railway Collection with a display highlighting one of Ireland's worst railway disasters, which killed 89 people and injured 400 others in 1889 when a train loaded with Sunday school children bound for a seaside trip was involved in a collision. Contact the library to find out what material and photographs they hold and to book an appointment on 028 37 523070.

Other Resources

The records of the British Transport Historical Collection Library are held at The National Archives in series ZPER and contain railway periodicals that may be of interest to those researching the history of the railways.

- *Railway House Journals* are a potential source of personal information for employees of British Railways, Furness Railway, Great Central, Great Eastern, Great Western, London, Midland & Scottish, London & North Eastern, London & North Western, North Eastern Railway and Southern Railway. These are sometimes indexed and contain photos of staff and personal announcements in the Staff News section.
- *The Railway Gazette* may provide obituaries or details of service on retirement and information about special achievements for senior employees.

Copies of these two journals are held at the National Railway Museum Library and Archives.

> *Salary and staff records mention fines incurred, allowances paid and time off for sickness.*

> *The National Railway Museum is a brilliant source for railway history.*

The National Railway Museum Library and Archives in Leeman Road, York, does not hold any staff records but is a brilliant source for photographs, books, maps, periodicals, timetables and archives valuable to anyone studying railway history. The photographic collection in particular dates back as early as the 1850s and consists of over 1.4 million photos, 200 of which have been printed in Ed Bartholomew's book *Railways in Focus – Photographs from the National Railway Museum*, and a further 50,000 are available to view on the Science and Society Picture Gallery website at www.scienceandsociety.co.uk/galleries.asp. The NRM archive collection contains more technical records of those companies whose staff records are kept at The National Archives and the National Archives of Scotland, and also includes records of railway workers' associations and engineers' drawings.

Finding Out More

- If you hit a brick wall in your research or would like to contact other researchers with railway ancestors, then the Railway Ancestors Family History Society is the place to go. The society helps its members to trace their railway ancestry in England, Scotland, Wales and Ireland, and also to find British railwaymen overseas. It can point you in the direction of records, documents, books and special collections that exist in record depositories around the world and they are constantly discovering and investigating previously unknown sources. The society is currently compiling a Railway Workers Index, and the website – www.railwayancestors.fsnet. co.uk – has a surname index where other researchers have posted messages about the people they are looking for and you can ask others for help when you get stuck. There is also a list of British and Irish railway companies with RAIL references. Another useful point of contact is the Railway and Canal Historical Society (www.rchs.org.uk).
- Ingenius is an online project sponsored by the National Railway Museum, Science Museum and National Museum of Photography, Film and Television that provides free access to 30,000 images including a whole section dedicated to transport, from water and road transport to a large amount of railway imagery, found at www.ingenious.org.uk/See/Transport.
- If you have ancestors who worked on the transport system in the Suffolk area then the Ipswich Transport Museum will be of interest to you. Their collections cover many historical forms of transportation

that we have not had time to look at here, such as horse-drawn and electric trams, horse buses, trolleybuses and motor buses, as well as the railways and airports and the histories of some major companies that monopolized the transport industry in that region.

- Cyndi's List is an online directory to thousands of genealogical sources and has a whole section dedicated to Canals, Rivers and Waterways, and one entitled Railroads. Visit www.cyndislist.com/canals.htm where you will find links to plenty of associations, societies and individuals who have researched the history of canals in their local area, including the Basingstoke Canal Authority, Pennine Waterways, Beverley Beck Canal, Lancaster Canal and Driffield Navigation in Yorkshire, and Wilts & Berks Canal.

- Canal Boating in the UK and Europe, at www.canals.com/biwaterway.htm, is another fantastic online source for locating heritage pages and societies for canals across England, Wales, Scotland and Ireland.

- The GENUKI website has some localized indexes to canal and railway workers, such as a surname index for boatpeople of Wolverhampton compiled from baptism and marriage registers. Check the page for the county in which your ancestor worked for similar transcriptions. The Occupations page also has sections dedicated to Canal People, Railwaymen and Postal Workers where you can find useful links for researching these jobs.

In addition to the publications mentioned, there are plenty of websites run by enthusiasts to give you a quick idea of the general history of transport in Britain. These are often very helpful because they draw upon a number of secondary and primary sources, though you should always verify any information you intend to use as part of your own research.

The Industrial Revolution and The Railway System at www.mtholyoke.edu/courses/rschwart/ind_rev/index.html presents a wide variety of information on the nineteenth-century railway system. General histories about canals and inland waterways can be found on the Waterways History website at www.jim-shead.com/waterways/History.htm maintained by Jim Shead. Mike Stevens has put online the Inland Waterways of England and Wales: Their History in Maps from 1750 to 1950, found at www.mike-stevens.co.uk/maps/index.htm. Canal Junction organizes holidays on canal boats but its website also has some great links to heritage sites, at www.canaljunction.com/canal/heritage.htm. These are great starting places to give you a taste of the world in which your ancestors found themselves.

USEFUL INFO

Suggestions for further reading:

- Railway Ancestors: A Guide to the Staff Records of the Railway Companies of England and Wales 1822–1947 *by David T. Hawkings (Alan Sutton, 1995)*

- Was Your Grandfather a Railwayman? A Directory of Railway Archive Sources for Family Historians *by Tom Richards (Federation of Family History Societies, 4th edition, 2002)*

- Railway Records: A Guide to Sources *by Cliff Edwards (Public Records Office Publications, 2001)*

- The Canal Boatman, 1760–1914 *by Harry Hanson (Sutton, 1984)*

- Transport in the Industrial Revolution *edited by Derek Howard Aldcroft and Michael J. Freeman (Manchester University Press, 1983)*

- Transport and Economy: The Turnpike Roads of Eighteenth Century Britain *by Eric Pawson (Academic Press, 1977)*

- *The National Archives Research Guides, numbers 69, 75, 82, 83 and 97.*

CHAPTER 18

Occupations: Farming and Agricultural Labourers

For centuries, agricultural labourers and small tenant farmers made up the vast bulk of the working classes in society, until the Industrial Revolution changed the face of Britain forever. Many of these individuals lived seasonal itinerant lives, moving from estate to estate in the hope of gaining work at key times of the year, especially during the harvest season. Traditionally, they left fewer traces in official records – yet this chapter outlines ways in which you can unearth information about any relatives who lived off the land.

> *When we find 'agricultural labourer' on a census return we assume that's the end of our hunt.*

Family historians, more often than not, write the agricultural labourer off as 'boring' (with a long sigh) whenever he's found on a census return. Yet you will probably be surprised at just how little you understand of his struggles and the impact he had on the developing modern world. Perhaps your great-great-great-grandfather was thrown off the land his family had farmed for decades by the General Inclosure Act of 1801, or maybe his son was one of the 'Swing Rioters' executed in the 1830s. The life of an agricultural labourer was dictated by the harvests and if there was a bad one he and his family were condemned not only to unemployment but also to starvation. The Great Irish Potato Famine of the 1840s, which lasted five consecutive years, had such devastating effects that they were felt well into the next century.

Historical Context

In the eighteenth and nineteenth centuries Britain underwent an agricultural revolution largely overlooked now because the huge technological advances of the Industrial Revolution have overshadowed it. The Agrarian Revolution was a product of several factors: the increased use of technology in farming such as the seed-drill, the development of the four-field crop-rotation system, the enclosure of common land into fields, the selective breeding of livestock, and a population boom that provided more labour and spurred an increase in food production. The increased mechanization of farming forced many agricultural labourers to find work in urban centres, fuelling the Industrial Revolution, yet without mechanization the farms would not have been capable of meeting the demands for food from the increasing numbers of people in the towns. The general consensus among historians is that the Agrarian Revolution had a negative effect on the lives of agricultural labourers, rendering large numbers of them unemployed and forcing them to find work in the polluted towns. Nevertheless, it was necessary for economic advancement.

▲ Farmers relax with some refreshment after ploughing the fields, 1936.

Prior to the Agricultural Revolution the countryside was run on a feudal system whereby landlords and squires owned large areas of countryside, which was worked on an open-field system with villagers cultivating their own strips of land and grazing livestock on areas of common land, such as the village green. The increased use of agricultural machinery made it more profitable for landowners to enclose these open fields, but until 1801 anyone wishing to do so needed to petition Parliament. In 1801 the first General Inclosure Act was passed, making it far easier to enclose open and common land in England and Wales with the consent of all those affected, but a subsequent Act in 1836 required the consent of just two thirds of those affected. Villagers were given minimal compensation for the loss of their livelihood, and with machinery doing their jobs far more productively many were forced to move to nearby industrializing towns to find work in the factories. Bear this in mind if you find that an urban set of ancestors had their roots in a more rural setting earlier in the nineteenth century.

In 1830 the effects of the Agricultural Revolution culminated in the 'Swing Riots'. Agricultural labourers mainly from the south of England

Poverty and Famine

The agricultural community was subject to several bouts of depression aside from the hardships brought about by the Agricultural Revolution. When the price of food was raised it had disastrous consequences for agricultural workers, for example during the period around the Napoleonic Wars, when bad harvests caused labourers in many parts of England to revolt, with bands of women forcing market sellers to sell their goods at what they considered to be reasonable prices. Agricultural labourers were regularly dependent on parish relief, as work was not guaranteed. Their dependence on relief increased after the old village field systems were enclosed because they had less means to support themselves on the land and were more dependent on their wages, which were too low to survive on. We will look at the way government and local authorities dealt with the parish poor in Chapter 24.

Famine was the greatest threat to agricultural labourers from all parts of the British Isles. In the nineteenth century, as systems of agricultural production altered, the poorest classes relied heavily for their own subsistence on potato crops, which were easy to grow in large quantities on small plots and provided substantial meals. When phytophthora, a plant disease, infected potato crops in Western Europe in the mid-1840s it spread quickly to England, Ireland and Scotland where it had its most devastating effects among communities with the least support from the government and wealthy classes. Crofters in the Scottish Highlands had become dependent on potatoes since the responded to unemployment and underemployment, low wages, low levels of relief and a generally low standard of living by rebelling against landowners. Harvests had been poor in the late 1820s, which meant a cut in wages and an increase in food prices. The purchase of new threshing machines by farmers also threatened the promise of hand-threshing work during the winter months, relied on by so many labourers. The riots began in the autumn of 1830, with protestors sending threatening letters to landlords signed with the pen name 'Captain Swing'. Farmers' threshing machines were sabotaged and protestors petitioned employers for a wage increase. The labourers' pleas were met with little sympathy from the ruling classes. During the period from 1830 to 1832, around 600 rioters were imprisoned, 500 were transported and 19 people were executed, many of them being young men and boys. The Swing Rising is considered one of the most important for agricultural labourers since the thirteenth century, though like so many similar causes it inevitably lost the battle.

Scotland experienced an agricultural revolution around the same time as England but the consequences for ordinary labourers were far more severe compared to their southern counterparts. Raising rents to completely unpayable levels was just one method employed by landlords during the 'Highland Clearances' as they became known. Other

Highland Clearances when the area of land they were allowed to cultivate was minimized, but they did not suffer complete starvation as famine relief programmes were organized and oatmeal rations provided. Agricultural communities in Ireland were not so fortunate, with approximately 2.5 million being left to starve, die from hunger-related disease, or emigrate to avoid starvation without any form of practical support from the British government, which was well aware of their plight. In fact landlords continued to export Irish grain and livestock for their own profit, while the export ban put in place during previous potato crop failures was not imposed.

Ireland's experiences of agricultural 'improvement' had been similar to Scotland's as landlords kept the best pasture for rearing cattle so that beef could be exported to England. Labourers were left with the less fertile scraps of land that remained to produce their own food, causing them to rely heavily on potato crops. Two thirds of the Irish population depended on agricultural work and the majority of those workers received a very low wage from their landlords and would work the land in exchange for a small plot where they could grow barely enough to feed their own family. During the famine of the 1840s over half a million people were evicted from their homes because they could not pay the rent, and those who did not meet a grim end aboard the 'famine ships' were left to rot in the workhouse. Barbara Windsor was deeply moved when she visited a mass grave of famine victims after finding out that her great-grandmother had fled from Cork to the East End of London between 1846 and 1851 to escape the same fate.

landowners resorted to burning tenants out of their homes and several thousand Scottish labourers were forced to emigrate to Canada and North America, which has affected the resonance of Scottish culture ever since. The Scottish Highlands remain more or less deserted to this day, with the relics of stone cottages haunting the landscape.

Corn Laws

The number of people employed in agricultural industries fell from the mid-nineteenth century as labourers migrated to the towns where more consistent employment could be found, machinery increasingly replaced labourers in the fields and more produce was imported from abroad. Corn Laws had kept the price of various types of grain artificially high to protect the interests of British producers since the Napoleonic Wars. This protectionist policy came to an end in 1846 when the Corn Laws were repealed, which made it easier to import foreign grain. The Central Agricultural Protection Society had been established in 1844 to campaign in favour of the Corn Laws, which were opposed by the manufacturing classes who wanted duty on wheat to be abolished so they could buy cheaper food in the towns. Despite the controversy surrounding the repeal of the Corn Laws and the staunch opposition it received, the demands of industry were ranked higher than those of agriculture by the mid-nineteenth century. The number of agricultural labourers and farmers working in corn fields fell rapidly after 1846 as the cost of buying British grain exceeded that of importing it from abroad. By the end of the nineteenth century around 60 per cent of wheat was imported into Britain from America, leading to a severe agricultural depression that lasted from the late 1870s to the end of the century. The vivacity of British farming was not restored until the Second World War forced the nation to become self-sufficient as food imports were drastically reduced and rationing was enforced.

Agricultural Life Collections

When most of us find 'agricultural labourer' in the occupations description on our ancestor's census returns, generally we assume that's the end of our hunt for information for that person, but here we will explore the avenues you can follow to discover more about a humble way of life no longer visible in modern Britain.

The experiences of agricultural labourers would have varied according to where they lived and the period in which they lived. The term 'agricultural labourer' covers a wide range of jobs, so it is necessary first of all to discover exactly what type of farming was prevalent in the area they were from before you can start to establish the type of work they would have been involved in. Then you can get digging to discover how and when nationwide changes such as the Agricultural Revolution, Inclosure Acts and shortages and famine affected their region.

Rural Life Museums and Archives

Some rural history museums have their own libraries and archives where you can start your research into the agricultural history of the place where your ancestral roots grew. To find a rural museum local to your ancestor's village try the Rural Museums Network at www.rural-museumsnetwork.org.uk, which has information on national and regional rural museums across England, Wales, Scotland and Ireland.

The Museum of English Rural Life (MERL)
The Museum of English Rural Life (MERL), founded by the University of Reading, is home to designated collections of national importance that record the changing face of rural life. Here you will find such treasures

as a beautifully embroidered smock sewn by a ferreter's wife for her husband who worked on Lord Hambledon's estate in Leigh, Kent, in the early nineteenth century. It is a major research centre and its archived collections of books, objects, documents, photographs, film and sound recordings can be searched by keyword, such as a place name, via the online database at www.ruralhistory.org/the_collections/index.html. The archive also holds the records of the Royal Agricultural Benevolent Institution, established in 1860 to provide financial aid to farming families in England, Wales and Northern Ireland who were forced out of work by misfortune, old age or ill health.

The museum's library has an extensive collection of books, pamphlets, journals, essays and articles where you can read about all aspects of agriculture and learn about specific rural communities. The Rural History website is a fantastic resource with an Internet Farm and Countryside Explorer (INTERFACE) where you can read about the characteristics of a farm labourer and view digital images and photographs relating to the history of agricultural labourers. There is a specially compiled *Bibliography of British and Irish Rural History* and the online database will search this bibliography as well as the library, archives, photographs and objects held at the museum, providing detailed descriptions and images of some of the objects. An appointment can be made to visit the Reading Room at the University of Reading on Redlands Road by telephoning 0118 378 8660.

St Fagan's National History Museum

St Fagan's National History Museum in Cardiff (formerly the Museum of Welsh Life) is home to the Social and Cultural History Department and the National Agricultural Collection of tools, machinery and equipment. St Fagan's is an open-air museum where native breeds of livestock can be seen in the fields and farmyards and traditional Welsh buildings are used to explore the way Welsh people have lived off the land for the past 500 years. In addition to the museum's exhibitions and displays, the Social and Cultural History Department houses a manuscript archive, photographic archive, sound archive, film archive and library where you will find farmers' diaries and oral recordings taken in the 1950s from Welsh people who were born as early as 1858.

In 2004 the National Library of Wales held an exhibition entitled Life on the Land to celebrate the centenary year of the Royal Welsh Show. This exhibition has been digitized for everyone to enjoy online at http://digidol.llgc.org.uk/METS/XAM00001/ardd?locale=en.

The experience of labourers varied according to where and when they lived.

Scottish Life Archive

The National Museum of Scotland is home to the Scottish Life Archive, whose collections include photographs, letters, documents, diaries and oral recordings. Some of the archive's holdings can be searched online via the Scran Trust website at www.scran.ac.uk. Scran provides digital access to educational material found in Scottish institutions and private archives, including the Scottish Life Archive and the National Archives of Scotland that represent Scottish culture and history. To search the Scran database for free simply enter a keyword into the search box on the homepage. Results will only display the object's title and a thumbnail image with details of the repository, unless you pay for a subscription that allows you full access to the catalogue. If you are unable to find anything useful online then you can visit the Scottish Life Archive at the National Museum of Scotland on Chambers Street in Edinburgh.

The NMS also founded the Museum of Scottish Country Life in East Kilbride just outside Glasgow, where you can visit a Georgian farmhouse and learn about farming techniques introduced in the eighteenth century that revolutionized Scottish agriculture, and how land clearances affected the lives of Scottish labourers.

The Ulster Folk and Transport Museum

The Ulster Folk and Transport Museum has a Folk Collection comprising archived material documenting the history of ordinary northern Irish people through images, sound recordings, written documents, artefacts, folkloric and oral history collections. The Folk Collection has a section specifically dedicated to agriculture covering arable and pastoral farming and rural society. The museum has an extensive range of agricultural equipment on display and a working farm with indigenous livestock. The museum is currently putting items from its Folk Life Archive online via the museum website but you should also contact them by telephone on 028 9042 8428 to find out if they hold any additional material in their archive and library relevant to the area your family were from.

The Irish Agricultural Museum

The Irish Agricultural Museum, set in the grounds of Johnstown Castle in County Wexford, is worth a visit if you had Irish agricultural ancestors. Though it does not have its own archive, you can see the museum's extensive displays on farming activities and households and their exhibitions about the effects of the Great Famine.

CASE STUDY

David Tennant

David Tennant decided to investigate the background of Archie McLeod, his grandfather, who had left Scotland to earn a living in Northern Ireland as a professional footballer, only coming back to his native Glasgow after his playing career was ended through injury to work in a shipyard. Yet his family had not always been based in Scotland's industrial heartland. Archie's grandfather, Donald, had been born and raised in Mull, a rural setting where for centuries farming had been the main profession. Using census returns and birth, marriage and death certificates – located on www.scotlandspeople.gov.uk – David was able to trace the old parish registers for the family (also online) and discovered that Donald was born there in 1819, one of ten children. His parents, Charles and Catherine, lived in a small stone cottage on the estate of the local landlord in the parish of Kilninian and Kilmore, Mull. Charles's occupation at the time of his marriage in 1806 was listed as farmer.

David visited Mull, and investigated the social history of the community further in order to gain a greater understanding of the lives of Charles, Catherine and their family – and why they eventually ended up in Glasgow, as shown by Donald's place of residence on later census returns. The turning point for the family came in 1832, when David's research in local archives revealed that the economic conditions of the day resulted in landlords attempting to maximize profits from their land by switching to sheep farming – forcing out small crofters and minor tenants to create large farms. To facilitate this process, rents were raised and tenants who

could not pay were forcibly evicted. This would appear to have been the case for Charles and Catherine, as their children all left Mull and started new lives elsewhere. This would appear to be a general pattern affecting whole communities, and David was shocked to discover in history books describing the period that by the end of the nineteenth century the Scottish Highlands were one of the most sparsely populated parts of Western Europe. Families such as the McLeods were forced to journey south to places such as Glasgow to earn a living. Their stone cottages still litter the empty countryside to this day.

▼ The wedding of David's grandparents, Archie McLeod and Nelly Blair.

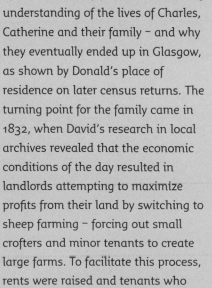

The Mills Archive

Rural mills played an important role in agricultural life for many centuries as the wheat grown in the fields was taken to the local mill to be ground into flour. Traditional wind and water mills slowly disappeared from the British countryside and were replaced by factories as a consequence of industrialization in the nineteenth century. The Mills Archive in Watlington House, Reading, is home to a large collection of historical material relating to traditional mills and ancient milling processes collected by a small group of volunteers who are members of the Mills Archive Trust. To aid family historians, the Trust is compiling a database of mills, millers and millwrights from the records in its archive complete with digital images for all to share. The archive is very small and visits are strictly by appointment only, which can be booked by emailing info@millarchive.com. If you cannot visit the archive in person or find what you need on their online database, the Mills Archive Trust will answer queries that take less than 5 minutes of their time for free but a fee is charged for more time-consuming queries sent to the same email address.

In the first instance, before emailing the Trust or making an appointment at the archives, you should search the Mills Archive online database from www.millarchive.com. Once you have registered for free you can search the database using the People Index, the Mill Index, or by doing a simple keyword search. The online collections contain the names of around 30,000 individuals connected with the milling trade and are a great source for family historians, although there is still a lot of material waiting to be catalogued. The Mills Research Group has also published a series of books containing articles about specific mills and milling processes, an index to which can be found on their website at www.millsresearch.org.uk where you will find instructions about how to order copies.

Researching Agricultural Communities

Once you have a general understanding of the farming industry in your ancestor's region and the type of work they undertook, you can start a more in-depth search for documents specific to your ancestor's village and some of the events that may have changed their world. Manorial records kept by landlords may mention your ancestor's name, while enclosure maps for the village will show how the open fields where they worked and lived were divided and tithe records may even give you a description of their property and the land on which they worked.

Manorial and Estate Records

As always, a trip to the local record office or studies centre is in order, where you should look for estate papers for the land your ancestor worked on. It is not always an easy task establishing which manor – or indeed whether any manor – owned the parish where your ancestors lived. If there is more than one possible landlord it will be necessary to search for manorial records for all those estates. Contemporary maps from the time your ancestor lived in the village may help you establish who the landlords in the area were and where the boundaries of their land lay. Local history books should also be of use, particularly the *Victoria County History*, an encyclopedic record of England's places since the earliest times. Some editions of the *VCH* have been digitized and are available to read for free at www.victoriacountyhistory.ac.uk, where you can also find details of other editions not yet available online.

There are several publications you can consult to find out the names of Scottish landowners for various places at different points in time, including Francis H. Groome's 1883 *Ordnance Survey Gazetteer*, Loretta Timperley's *A Directory of Landownership in Scotland c.1770*, and the *Statistical Account of Scotland* compiled by the ministers of the Church of Scotland from 1791 to 1845, available online from Edinburgh University's EDINA website at http://edina.ac.uk/ stat-acc-scot. The names of landlords in Ireland can be found from Griffiths' Valuation Books (see Chapter 6). The name of the landlord appears in the column headed 'lessor'.

Agricultural labourers usually rented their homes and the land they worked from their employers, and records of estates often include rent rolls showing any money received by the landlord. Manorial records vary widely in their content but can sometimes give valuable information about generations of families who were tenants on the land, including dates of death or remarks about a tenant's character, while information among lease records can contain correspondence that throws light on the grievances of tenants such as pleas to reduce the rent.

▼ Raking and forking the hay *c.*1600 – traditional activities for agricultural labourers around harvest time that survived into modern times.

- The Manorial Documents Register covering England and Wales is a card index held at The National Archives containing details of the location of manorial records and is also available online from www.nationalarchives.gov.uk/mdr.
- Try the National Register of Archives' Family Name search engine to locate English and Welsh manorial records not yet uploaded to the MDR database and for the papers of landed estates in Ireland and Scotland.
- Scottish estate papers can also be searched

under the name of the landowner using the online catalogue of the Scottish Archive Network (SCAN). A guide about locating Scottish estate papers can be found on the National Archives of Scotland website at www.nas.gov.uk/guides/estateRecords.asp.

- The National Archives of Ireland has a guide to the private estate records among its collections at www.nationalarchives.ie/geneal-ogy/private.html, but if they do not hold the estate papers you are looking for the National Library of Ireland (NLI) might have them among its Manuscript collection, details of which can be found on the NLI website.

- The Public Record Office of Northern Ireland holds some private papers including estate records, many of which were deposited following an appeal to prominent families in Northern Ireland in the 1920s. A guide to estate records at PRONI can be found on their website at www.proni.gov.uk/records/landed.htm, which also contains an index to some of the larger estates.

See Chapter 28 for a more detailed look at manorial records.

The Board of Agriculture published agricultural county reports between 1793 and 1817 covering England, Wales and Scotland, with indexes and references to particular parishes. They can be particularly enlightening about agricultural conditions in the places where your ancestor lived and worked. Copies of these out-of-print books can be found in the library of the Museum of English Rural Life, but have also been digitized and are available online from http://books.google.co.uk. Simply type into the Book Search 'General View of the Agriculture of ...', stating whichever county you are researching. These digitized copies are free to view online and the text within them can be searched by keyword.

Tithe Records

Agricultural communities in England and Wales were subject to pay tithes for the upkeep of the local parish church, an ancient tax that continued into Victorian times. One tenth, or a tithe, of the products arising from the ground such as grain, vegetables and wood, and all things nourished by the ground including livestock, dairy products and wool, and the produce of human labour such as the profits from milling and fishing, had to be paid in kind to the local clergy until the dissolution of the monasteries. At this point many tithes were sold off to non-clergy ownership and money payments began to be substituted for

Tithe payments – for the upkeep of the parish church – continued into Victorian times.

Enclosure Maps

Acts of Enclosure and the accompanying maps produced from the eighteenth century (and earlier, though these records are scarce) are held in a variety of repositories. The best way of finding out whether details of an enclosure for your ancestor's parish exist, when the enclosure award was granted and where the records are held is by checking the directories in Tate's *A Domesday of English Enclosure Acts and Awards*, J. Chapman's *Guide to Parliamentary Enclosures in Wales* and Kain, Chapman and Oliver's *The Enclosure Maps of England and Wales: 1595–1918*. The latter index can also be searched online by place name on a website funded by the University of Exeter at http://hds. essex.ac.uk/em/index.html.

Enclosure documents rarely list many names other than those of the people to whom the land was allotted, but they are an insight into how the wider social issues of the time must have affected your ancestor's way of life. Steven Hollowell has written *Enclosure Records for Historians*, containing plenty of example documents to illustrate the type of records that survive, with case studies of opposition among commoners being of particular interest.

payments in kind. The Inclosure Acts of the nineteenth century substituted the payment of tithes by allotting land or introducing a fixed monetary payment or a payment that varied according to the price of corn, known as a corn rent. Those areas that did not arrive at an alternative to tithe payments as a consequence of enclosure underwent tithe commutation in 1836 when the government decided that everywhere in England and Wales should substitute tithe payments for corn rents, known as tithe rentcharges.

A Tithe Commission was established to survey the extent to which tithe commutation had already taken place as a result of enclosure. Enquiries were sent to all the parishes in England and Wales, the results of which can be found among the Tithe Files at The National Archives in series IR 18, though the content of these files varies from parish to parish.

Those parishes that were still liable to pay tithes were then assessed property by property and plot by plot to establish how much rentcharge they should pay. These assessments are known as Tithe Apportionments and can be found in county record offices for the local area, with duplicates kept at The National Archives in series IR 29 and corresponding maps in IR 30. They are of use because they name the owners of land and property as well as the tenants and give a description of the property.

If your ancestor lived in a parish that had been enclosed then Tithe Apportionments and maps will not have been produced. To find out whether tithe records exist at for your ancestor's parish you can search The National Archives online catalogue by entering the name of the parish and restricting the department or series code to IR 29. The reference for the maps will have the same ending as that for the apportionment, but will begin with IR 30 instead of IR 29. If you know the location of your ancestor's house you can find it on the map and use the plot number to locate the apportionment entry, otherwise you will need to scroll through the whole table of apportionments for that parish until you find the name that you are looking for. Duplicate copies can usually be found at the relevant county record office.

Tithes in Ireland

Ireland had a similar system of tithe payments that was reformed between 1823 and 1838. To make the collection of tithes easier the government sought to substitute payments in kind for monetary payments with the Tithe Applotment Act of 1823. All the agricultural land liable to pay tithes therefore needed to be surveyed to establish how much they should pay, a process that took 15 years to complete. The tithe tax was so unpopular, partly because people of all religious denominations were expected to pay tithes to the Established Anglican Church of Ireland, that a 'Tithe War' broke out between 1831 and 1838, when thousands of people refused to pay and held protests, some of them extremely violent.

The Church of Ireland clergy recorded the names of tithe defaulters in 1831 so that they could claim for tithe arrears from the government's Clergy Relief Fund. Many of the 29,027 people recorded in the index of defaulters taken from 232 parishes in Carlow, Cork, Kerry, Kilkenny, Laois, Limerick, Loth, Meath, Offaly, Tipperary, Waterford and Wexford Counties were the same generation of people affected by famine and emigration a decade later. The Tithe Defaulters Schedules list the names, addresses, occupations and some observations and genealogical information about the extended family of those people who did not conform. The Schedules are held at the National Archives of Ireland among the Official Papers Miscellaneous Assorted (OPMA) files and there is an index containing the names and addresses of those listed in the Schedules. The Schedules can be purchased on CD from the Origins Network online store at www.originsnetwork.com, where you can also access the index online by purchasing a subscription.

In 1838 the Tithe Rent Charge Act was introduced in Ireland to absorb tithe payments into ordinary rents payable to the landlord in an attempt to pacify protestors.

The Tithe Applotment Books compiled between 1823 and 1838 for every parish in Ireland, except any land owned by the Church and any land that was of such poor quality a charge could not be levied, are held at the Public Record Office of Northern Ireland and the National Archives of Ireland for their respective areas. The books are arranged by parish and give the names of occupants, the area subject to tithe, a valuation of the property, and details of the quality of the land. They are the last agricultural survey available prior to the Great Famine of the 1840s. The National Library of Ireland produced a Householders Index listing the occurrence of surnames in the Tithe Applotment Books for all the counties in Ireland. Copies of the Householders Index can also be found at PRONI and the National Archives of Ireland.

Farm Surveys

The Second World War increased the importance of agriculture in Britain as food imports were drastically cut and self-sufficiency was vital if the country was to have a strong, healthy population with which to defeat the Nazis. The Minister of Agriculture and Fisheries for England and Wales established County War Agricultural Executive Committees whose duty it was to increase agricultural output. The Committees could direct what should be grown, inspect farms, seize control of land, and mobilize groups of labourers. Their first job was to organize large expanses of grassland to be ploughed up, which insti-gated a farm survey in June 1940 to establish the extent of farms and their productivity as well as the efficiency of the farmer. Between June 1940 and early 1941 around 85 per cent of the agricultural land in England and Wales was surveyed, including all but the smallest farms. Once food production had increased a second more thorough National Farm Survey was planned to record the conditions of tenure and occu-pation, condition of management, the crop acreages and livestock numbers, the fertility of the land, the adequacy of the utilities, and a map of the boundaries for any farm over 5 acres. The findings of this survey are held at The National Archives in Kew.

The National Farm Survey began in the spring of 1941 and was completed in 1943 after 300,000 farms had been visited. Each farm surveyed has its own file in MAF 32 with maps in MAF 73. The records in MAF 32 are arranged by county and then alphabetically by parish. If you know the parish of the farm your ancestor worked on you can find the correct reference on the online catalogue by typing in the name of the parish and restricting the Department or Series Code to MAF 32. The farm record will contain the name of the farm, the farmer and his address. It will not list the names of everybody who worked on the farm

> 6 *The National Farm Survey in 1940 recorded all but the smallest farms.* 9

but should state how many labourers were employed, how efficiently the farm was run and what was produced there. Controversially, the farm needed to be classified from A to C (A meaning the farm was run well, B meaning it was run relatively well and C meaning there was poor management). Only 5 per cent of the farms surveyed were C classified, while 58 per cent were A classified. Farms that were classified lower than A due to 'personal failings' have additional notes as to what these failings were.

- If you want to find a specific farm using the maps in MAF 73 then consult the index map sheets in MAF 73/64 kept on open shelf in the Map and Large Document Reading Room, arranged in alphabetical order by county.
- No individual farm records of the earlier survey taken in 1940 appear to survive but county summaries can be found in MAF 38/213.
- Parish Summaries of Agricultural Returns can be found in MAF 68.
- A statistical analysis of the data was published in 1946, and a proof copy of the report can be found in MAF 38/216 with county-by-county statistical analysis kept in MAF 38/852–863.

The National Farm Survey 1941–1943: State Surveillance and the Countryside in England and Wales in the Second World War by Foot, Watkins, Short and Kinsman reviews the National Farm Survey with illustrations. A similar but more limited farm survey was also carried out in Scotland between 1941 and 1943 and the resulting Farm Boundary Maps and records of the Scottish Agricultural Executive Committees can be found at the National Archives of Scotland. An abridged report to the Farm Survey in Scotland can be found at The National Archives in Kew in MAF 38/217.

Agricultural Disasters and Crises

It can be tricky trying to ascertain whether or not your ancestor's family was directly affected by some of the major political and social events that have made it into the history books. This type of research will require you to study a variety of material, much of it secondary sources, and very often only circumstantial evidence will give you a clue as to the answer. Here we will look at methods of researching the effects of the Swing Riots, the Scottish Highland Clearances and the Great Irish Potato Famine.

> *Court records in county record offices will have information on participants in the Swing Riots.*

The Swing Riots

If your ancestor was an agricultural labourer in the south of England, particularly in Kent and Sussex, they may have joined in the protests during the Swing Riots of 1830 to 1832. The best place to find out is at the county record office where snippets of information about ancestors tried for riotous behaviour can be found in records of Quarter Sessions. Kent, Sussex and the south were not the only areas affected by the riots. The Family and Community Historical Research Society has researched the Swing Riots to determine the true extent of their effect throughout England and have concluded that their influence extended over a wider geographical area than was previously thought. The FCHRS findings have been published in Michael Holland's book *Swing Unmasked*, which can be purchased on the FCHRS website at www.fchrs.com/swing/swing_project. The website also contains a list of all the counties where Swing Riot activity is recorded.

A number of books have been written about the Swing Rioters and their concerns, including E. J. Hobsbawn and George Rude's *Captain Swing*, David Kent's *Popular Radicalism and the Swing Riots in Central Hampshire* and Mike Matthew's *Captain Swing in Sussex and Kent*. It is worth consulting these books before attempting to locate original documents because they will give you an idea of exact dates when uprisings affected the area you are researching and you may even be lucky enough to find your ancestor's name is mentioned.

The Highland Clearances

The Scottish Highland Clearances began around the 1760s and occurred in waves until the 1870s. If your ancestor was evicted during the Highland Clearances then the estate records may give some information about this. The National Archives of Scotland has a Highlands Destitution series in HD and also holds records of Crofters and Clearances, which can be search via the Online Public Access Catalogue (OPAC) on the NAS website, and via the Scran Trust website.

In 1883 a Royal Commission on the Highlands and Islands (also known as the Napier Commission) was set up to hear the evidence of 775 crofters and cottars who were living in a state of serious destitution as a result of the Clearances. Details of the names, addresses and families of the labourers from 61 places were recorded in a manuscript now available at the NAS in series AF50, and the commissioners' report has been printed as a separate document with other parliamentary papers. Some tenant farmers were forcibly put aboard ships heading for Nova

Scotia (or 'New Scotland'), Ontario, the Carolinas and Australasia. Ships' passenger lists can help you to find out about ancestors who emigrated as a result of eviction and you can find out more about locating these in Chapter 23. An index to passenger lists of the Highlands & Islands Emigration Society that assisted around 5,000 people to leave western Scotland for Australia between 1852 and 1857 can be searched in the National Archives of Scotland's search rooms or on the SCAN website at www.scan.org.uk/researchrtools/emigration.htm.

More general research around the topic may be the only way of gathering circumstantial evidence as to the extent of the Clearances in your ancestor's region. You might be able to find out how and whether the Clearances affected your ancestor's parish by comparing the first and second editions of the *Statistical Account of Scotland*, the first edition produced between 1791 and 1799 and the second written in 1845, available online from Edinburgh University's EDINA website at http://edina.ac.uk/stat-acc-scot.

To find out more about the impact of the Highland Clearances visit www.theclearances.org. The Clearances project is designed to tell some of the stories of the millions of labourers directly affected. It has a search engine to find the names of people mentioned on the site, as well as a search facility for place names, parishes, ship names, ports and destinations.

The Great Irish Famine

There is no centralized data for the period of the Great Irish Famine of the late 1840s and with many parishes not having kept burial registers prior to the 1850s, combined with civil registration not beginning until 1864 and those records being lost in 1922, it is impossible to calculate the precise scale of the death toll. Historians have arrived at estimated death and emigration figures for the period by comparing statistics from the few population surveys available just before and just after the famine, such as the Tithe Applotment records of the 1820s and 1830s and Griffith's Valuation of the 1840s to the 1860s.

The Famine Immigrants: Lists of Irish Immigrants Arriving at the Port of New York 1846–1851, published in seven volumes in 1983, can be found on the shelves of the Public Search Room at the Public Record Office for Northern Ireland (PRONI) and at the National Archives of Ireland and the National Library of Ireland, but for information about researching emigration to other destinations you should consult Chapter 23.

There are several Famine Museums established to tell the horrors of this period, such as the Doagh Famine Village and Museum in County Donegal, the Famine Museum at Stroketown Park House in County Roscommon, and the Donaghmore Workhouse and Famine Museum in County Laois. You can find out how to investigate whether your labouring ancestor was reduced to entering the workhouse during this time by reading Chapter 24. James S. Donnelly published *The Great Irish Potato Famine* in 2001, worthy of a read by anyone interested in this dark period of Ireland's history.

Finding Out More

- Anyone with ancestors involved in farming may want to consider joining the British Agricultural History Society. The BAHS was founded in 1952 to promote the study of agricultural history and rural society, and works hard to preserve documents of national importance. Its website can be found at www.bahs.org.uk where digital copies of recent issues of *Rural History Today* can be downloaded giving tips about developing research resources and where interesting articles about agricultural history can be found. The society holds regular conferences and *Rural History Today* publishes details and dates of talks held at institutions around Britain that may cover one of your areas of research.

- In 1911 Barbara and John Lawrence Hammond wrote a pioneering study of the village labourer and the role that industrialization played in crippling the poor labourer and smallholding farmer. At the time their work was viewed as highly controversial, not least because of their unfashionable theory that the Industrial Revolution created social misery for the poorer classes and their critical analysis of the class struggle taking place in the 1830s. The study has been published several times since then, and it can be found under various titles. However, transcriptions from the 1920 edition, *The Village Labourer 1760-1832: A Study of the Government of England before the Reform Bill*, are available online via the University of Melbourne's Archive for the History of Economic Thought (HET) at http://melbecon.unimelb.edu.au/het/hammond/village.html. This book is crammed with extracts from original sources analysing the impact of enclosures and industrialization and takes a detailed look at the Swing Riots, giving the names and fates of some of the main ringleaders.

USEFUL INFO

Suggestions for further reading:

- My Ancestor Was an Agricultural Labourer *by Ian Waller (Society of Genealogists, 2007)*

- The Village Labourer *by Barbara Hammond and John Lawrence Hammond (Nonsuch Publishing Ltd, 2005)*

- Lord and Peasant in Nineteenth Century Britain *by Dennis Richard Mills (Croom Helm, 1980)*

- Farmworkers: A Social and Economic History 1770–1980 *by Alan Armstrong (Batsford, 1988)*

- *The National Archives' Research Guides – Legal Records Information 1: 'Manorial Records in The National Archives'; Domestic Records Information 59: 'Agricultural Statistics'; Domestic Records Information 74: 'Common Lands'; Domestic Records Information 86: 'Enclosure Awards'; Domestic Records Information 106: 'National Farm Surveys of England and Wales, 1940–1943'; Research Note 7: 'Hedgerows'*

CHAPTER 19

Occupations: Professional Classes – Private Sector

Britain's industrial wealth grew dramatically in the eighteenth and nineteenth centuries, and this boom created a multitude of job opportunities in business and the professions for the burgeoning middle classes. Merchants, insurance agents, lawyers, bankers and accountants all flourished in the private sector, and often left a wealth of evidence for family historians to explore. It is these records that we'll examine here; records for civil servants, doctors, schoolteachers, clergymen, etc., will be looked at in the next chapter.

> *Many occupations have their own professional bodies, guilds or associations – which keep records of members.*

When researching the lives of professionals bear in mind that there will often be evidence of their work in more than one field. For example clergymen and people in the legal and in some medical professions usually studied at a college or university. Records you find concerning their career may mention where they were educated, which will help you to locate university records. *Alumni Oxonienses* and *Alumni Cantabrigienses* give biographical notes about Oxford and Cambridge graduates from the thirteenth to the nineteenth centuries, and similar publications have been produced for graduates of the major universities in Scotland and Ireland. People employed in public sector services could supplement their salary by taking on a separate job in the private sector, and men who were highly respected businessmen in the private sector were sometimes rewarded with a position in government. For example lawyers were also likely to be Members of Parliament or to work as civil servants, so records for those occupations may be worth searching too. Some people with highly respected positions are listed in Kelly's *Handbook to the Titled, Landed and Official Classes*, which gives a potted biography of entrants from 1883 to 1977. Very often family businesses were handed down through the generations, which can make locating records over a long period of time a bit easier.

Lawyers, judges, civil servants, schoolteachers, clergymen, freemen of the City of London livery companies, Members of Parliament, some medical practitioners and university members all had to declare an oath of allegiance to the Crown and the Church of England from the 1660s until the nineteenth century, and their names may be found on Oath Rolls kept at The National Archives. Roman Catholics and non-conformists who refused to take Holy Communion in a Church of England ceremony were barred from holding official positions between 1661 and 1828. After this date non-conformists could submit a declaration, found in The National Archives series C 214, and could then take an oath of allegiance.

Most of the occupations described in this section will have their own professional bodies, guilds or associations, which members of that profession could join, so it is worth seeking out the organization that covered their geographical and occupational jurisdiction to establish what archives and records survive. Official trade magazines and newspapers for many professions contain advertisements posted by competitive companies and may contain biographical information

◀ City of London Stock Exchange, 1931.

about influential individuals in the business. Professionals were also likely to leave some property and equity after their death, and so it is always worth searching for a will or letters of administration.

Businesses – Records and Repositories

There are general records that will be useful for whatever type of business you are researching, and these can be found in a range of repositories and libraries.

- Trade and business directories are an invaluable source, listing the names of directors, where the company had its offices, and helping you to trace how long the company was in business.
- Advertisements found in newspapers and directories add a bit of colour to the company's background, giving a vivid description of exactly what type of products and services your ancestors provided. Directories sometimes have an index to their advertisements to save you having to flick through the entire volume.
- The *London Gazette* and major broadsheets such as *The Times* contained a lot of business news, like notices of bankruptcies, the dissolution of partnerships and businesses that had ceased trading.

For a general overview of locating and using business records, John Orbell's 1987 publication *A Guide to Tracing the History of a Business* is still a useful source. If you are researching the profession of an Irish ancestor then consult the chapter on occupations in John Grenham's *Tracing Your Irish Ancestors*. This is a reference guide to sources available in Irish repositories for researching professionals in many occupations, including attorneys, barristers, policemen, apothecaries, surgeons, doctors, teachers, clergymen and more.

Complete business records containing personnel files rarely survive, but if they do they are generally deposited at the local record office or are still in the company's archive. County record offices often hold the records of small, local businesses, the location of which can be found by a search of the company name using both the Access to Archives and the National Register of Archives databases. The NRA has a specific Corporate Name search engine containing the names of over 29,000 businesses from across the UK and Republic of Ireland, with

Company Registration Papers

Company registration records do not contain a great deal of information interesting to family historians, but they may be of use to the descendants of those people who formed the company or were a director or major shareholder, giving financial information about how well the business did. The names and biographical details of many company directors have been published annually since 1879 in *The Directory of Directors.*

The Companies Acts have allowed for companies to be incorporated by registration since 1844, prior to which a company could only be incorporated by Royal Charter or by statute. Businessmen usually chose to incorporate their businesses so that any debts were liable to the company and not to them personally. Registered companies needed to file administrative records with the Registrar of Companies, including details of the company's shareholders and directors, an annual summary of accounts, and a register of mortgages granted by the company. Over one million companies were registered between 1856 and 1976 and the records of those that are still in business in England and Wales are kept at Companies House in Cardiff. Companies House has a London search room at 21 Bloomsbury Street, where microfiche copies of these records can be viewed, and

The National Archives in Kew has indexes to the companies incorporated between 1856 and 1920, and to those on the register in 1930 and 1937.

The Companies House website at www.companieshouse.gov.uk has an online index where you can search for the names of current limited companies and those dissolved in the last 20 years. The Registrar holds on to the files of dissolved companies for 20 years after which time a sample is passed to The National Archives and the remainder are destroyed. Those that make it to The National Archives are stored in series BT 31 and BT 41, and there are brief details about the dissolved companies whose records were destroyed in series BT 95, arranged chronologically by incorporation date. BT 31 has an alphabetical index to company names giving each one's company number so that it can be found in numerical records. The index is on the open shelves at Kew, dating from 1855 to 1976. Companies House in Cardiff has an index to the names of companies dissolved prior to 1963 with the date they dissolved and a note about whether their records went to The National Archives. An online guide can be found on The National Archives' website entitled Domestic Records Information 40: 'Registration of Companies and Businesses'.

Companies registered in Scotland are subject to the same laws as England and Wales, but their records will be found with the Registrar of Companies in Edinburgh. Companies House at 37 Castle Terrace, Edinburgh, EH1 2EB holds records relating to over 50,000 companies with a registered address in Scotland as well as those of some dissolved companies. More records of dissolved companies in Scotland can be found in the National Archives of Scotland.

The Department of Enterprise, Trade and Investment (DETI) in Northern Ireland has a Companies Registry section where Northern Irish companies register their details. To find out about locating registration documents for a Northern Irish company telephone 0845 604 8888 or email them at info.companiesregistry@detini.gov.uk.

The registration of a company in the Republic of Ireland is similar to the process in England, Wales, Scotland and Northern Ireland. Companies need to register their business at the Companies Registration Office in Dublin. To find out about locating registration documents for a company in the Republic of Ireland contact the office by email on info@cro.ie explaining what you know about the company, when it was in business and whether it has been dissolved.

▲ The Great Hall of the Bank of England, from an early nineteenth-century illustration.

descriptions of their archives and where they are deposited. Eric D. Probert has written a guide to *Company and Business Records for Family Historians*, in which he gives examples of the types of business records found in local record offices and reviews the *London Gazette* index to the notices of partnership dissolutions between 1785 and 1811, as well as the indexes available at Companies House.

Major record offices within the UK and Ireland tend to hold some form of records for businesses. The National Archives in Kew is home to a collection of amalgamated business records in the form of documents that were required to be filed with the Board of Trade and its successive government departments. It has produced several research guides covering this field, including Domestic Records Information 122: 'Sources for Business History'. The Public Record Office of Northern Ireland has one of the largest collections of business records in the UK, from small companies' papers to the records of huge firms responsible for making Ulster's fortune. PRONI has produced an information leaflet, leaflet number 18, about its business records dating back to the seventeenth century, found on the website under 'Family history'.

In addition to collections held by local and major record offices, there are a number of specialist business archives and repositories that are worth consulting.

- The Guildhall Library in London is home to a rich collection of business records. Joan Bullock-Anderson has published *A Handlist of Business Archives at Guildhall Library* listing the names of around 800 businesses whose records are stored there, including those of fire insurance companies and City of London merchants. The Guildhall has also put an index online to the trades they hold records for and the names of each company within those trades. Visit www.history.ac.uk/gh/busimnu.htm to find the Subject Index, from accountancy firms to wool manufacturers.
- The Business Archives Council (BAC) works hard to promote the preservation of company archives. It does not hold any business records but may be able to help you locate the records of large and

medium-sized companies. The BAC published a *Directory of Corporate Archives* compiled by Lesley Richmond and Alison Turton in 1997. The BAC Business History Library is now located at its sister site, the Centre for Business History in Scotland, based at the University of Glasgow, which you can make an appointment to visit by telephoning 0141 330 6890.

- If you are researching the history of a Scottish company then the Scottish Business Archive, backed by the Business Archives Council of Scotland (BACS) and run by Glasgow University Archive Services, may well house the company's records. Contact the duty archivist by telephoning 0141 330 5515.

Bankruptcy and Insolvent Debtors

The middling and professional classes were at constant risk of falling into debt, particularly if they had invested a lot of money into setting up a business. These were people who worked hard to attain a good standard of living, yet didn't have the financial security of the upper class and were in constant fear of sliding down the social scale. One option for business people in debt was to declare themselves bankrupt. However, until 1844 bankruptcy could only be claimed by traders who bought and sold goods and owed over £50, after which time companies could also claim bankruptcy. It was not a criminal offence to declare oneself bankrupt, but a trader's creditor could file a legal petition to declare someone bankrupt in an attempt to settle the debt in court. Once the court case was resolved the debtor was no longer a bankrupt, though his reputation may have been damaged.

Insolvent debtors on the other hand were individuals who could not claim bankruptcy. Being an insolvent debtor was a criminal offence, punishable by a prison sentence until the debt was paid off. If your ancestor could never afford to pay off the debt then he risked spending the rest of his life in debtors' prison. The professional classes were in a precarious financial position, at risk of being thrown into debtors' prison because their status enabled them to borrow sums of money and material goods on credit, which they might not be able to pay back if they didn't manage their money carefully.

Insolvent debtors were held in local prisons, but The National Archives holds the records of some major debtors' prisons in London, including Marshalsea, Fleet, King's Bench and Queen's Bench prisons in series PRIS. From 1861 insolvent debtors were allowed to apply for bankruptcy, and as of 1869 debtors were no longer routinely sent to jail.

CASE EXAMPLE

Bankruptcy and insolvent debtors

John Hurt's great-great-grandfather, William Richard Browne, a respectable customs officer in the Port of London and reputed descendant of the Marquess of Sligo, found himself in a complicated financial arrangement with creditors. Unable to pay his debts or come to a reasonable agreement with his creditors, he was imprisoned in the King's Bench Debtors' Prison on two occasions in 1828 and then again in 1836. There are records of his petition to the Court for the Relief of Insolvent Debtors in series B 6 at The National Archives and very long pleadings regarding the case in both the Court of King's Bench and the Court of Chancery where the case was eventually settled.

Bankruptcy Records

- Most case files for bankruptcy hearings do not survive, though The National Archives has some records in series B 3 and B 4, and more records pertaining to bankruptcies can be found in B 1 to B 10.
- Notices of bankruptcies and insolvent debtors' cases can be found in local and national newspapers, and Commissioners of Bankrupts published notices in the *London Gazette*.
- The Court of Bankruptcy was established in 1832, but before this a creditor had to petition the Lord Chancellor to commission a bankruptcy case. Proceedings in the Court of Bankruptcy under the Joint Stock Companies Acts of 1856 and 1857 are in series B 10 for the period 1857 to 1863. After 1869 records of the Board of Trade contain bankruptcy proceedings.
- The records of district bankruptcy courts set up after 1842 are held at local record offices.

The National Archives research guides, Legal Records Information 5: 'Bankrupts and Insolvent Debtors 1710–1869' and Legal Records Information 6: 'Bankruptcy Records after 1869' take a detailed look at all the records for English and Welsh debtors. The Society of Genealogists has two Bankrupts Directories for 1774–86 and 1820–43, microfiche copies of which are at The National Archives.

Similar records for Scotland can be found at the National Archives of Scotland, and information about tracing the records can be found via an online research guide under Court of Session – Sequestrations, as this court handled the sequestration of bankrupts' possessions under various laws passed in the nineteenth century.

Merchants

Merchants were entrepreneurial businessmen whose prosperity grew from the seventeenth century as they sought trade with far-off places and brought riches to British ports. Renowned enterprises like the East India Company grew from humble origins, with a small group of London merchants forming the company in 1600, to being one of the most powerful companies in the world. Prior to the nineteenth century those people who traded in goods abroad also traded in people, with the booming slave trade supporting the wealth of many British merchants. The National Archives has produced research guides for people wanting to trace their ancestors' involvement in the British

Nigella Lawson

Nigella Lawson already knew a fair amount about her family's background; her mother, Vanessa Salmon, was an heiress born into the Lyons Coffee House dynasty and, given the high profile of Lyons cornerhouse tearooms as a quintessentially British institution, Nigella was acquainted with the history of the company – for example, the little-known fact that instead of trading in tea they originally sold tobacco under the name Salmon & Gluckstein. Yet this information enabled her to uncover business archives surrounding the family, and learn more about some of the founders of the company to whom she was related.

The starting point for Nigella's investigation was her mother's family, the Salmons, and their links to Lyons. Nigella already knew that her grandfather, Felix Salmon, was instrumental in running the company and shaping its direction towards the famous Lyons cornerhouse tearooms. Travelling around London, many of today's famous landmarks such as the Trocadero and Hard Rock Café were formerly in the possession of Lyons. Yet despite his success, Felix Salmon came across in family stories as a melancholy man, and one possible cause was his role during the Second World War. Although he was in the catering corps, Nigella discovered from research at the Imperial War Museum that he was likely to have been attached to one of the regiments that liberated the German concentration camp at Belsen. One cannot begin to imagine the trauma of the event, particularly since he was also Jewish, and worked in the catering corps responsible for famine relief for the liberated inmates.

Nigella continued to investigate the history of the company, and turned to Salmon & Gluckstein, the tobacco sellers, who claimed to be the largest in Europe at their launch in 1873. One of the founding fathers was her 2 x great-grandfather, Barnett Salmon. His surname was originally Solomon, and on the 1841 census it transpired that his father Aaron was a clothes dealer in the East End. Barnett started work as a travelling tobacco salesman. He married Helena Gluckstein in 1863, and went into business with his father-in-law Samuel Gluckstein. From consulting trade directories, where the business was advertised, and material in institutions such as the London Metropolitan Archives, Nigella was able to trace the success of the tobacco company, and the decision to branch out into other lines of business – and from these origins, Lyons was born in 1889. To ensure none of his family was ever threatened with poverty, Barnett set up a family fund, but equally insisted that none of the women were allowed to work. According to his will, Barnett was worth £3.5 million in today's money when he died, and – true to his word – he set up a trust fund for his wife worth £¾ million.

◀ One of the famous Lyons tearooms, along Piccadilly, London, in 1953.

Bankers

Most major banks have an archive that they administer privately, such as the Lloyds TSB Group Archives. The records retained by banking archives vary greatly in content from complete staff and customer records to very little early material at all, and gaining access to information about customers' accounts is not always easy, even if the client has died and the information is very old. The Bank of England has an archive in Threadneedle Street, which holds staff records and details of some of its early customers' accounts dating back to 1694. There is more information about the Bank of England Archive on their website at www.bankofengland.co.uk/about/history/archive. The Royal Bank of Scotland has a fantastic archive containing material dating back to the seventeenth century about its employees and customers. More information can be found at www.rbs.com under 'About Us' and 'Our Heritage'. Similarly, the banking firm established by the Rothschild family two centuries ago has extensive archives relating to the family's businesses in London, Paris, Frankfurt, Vienna and Naples. The Rothschild Archive has a website at www.rothschildarchive.org where you can learn more about their collections.

You can establish whether the bank that your ancestor worked for, or had an account with, has an archive or has deposited its records at a public archive by consulting the National Register of Archives. Further guidance can be found in *British Banking: A Guide to Historical Records* by John Orbell and Alison Turton (Ashgate, 2001); for those researching early modern Scottish bankers try *The Scottish Provincial Banking Companies, 1747–1864* by Charles W. Munn.

slave trade, in Overseas Records Information 22: 'British Transatlantic Slave Trade: Introduction', Overseas Records Information 23: 'British Transatlantic Slave Trade: Britain and the Trade' and Overseas Records Information 26: 'British Transatlantic Slave Trade: Abolition'.

Records of the East India Company are kept on open access at the British Library in the archives of the India Office, and are described in more detail online at www.bl.uk/collections/orientaloffice.html and in the publication *India Office Library and Records: A Brief Guide to Biographical Sources* by C. J. Baxter.

An Australian historian, Dan Byrnes, has compiled information about international merchants and bankers from before 1400 to 2004. The site at www.danbyrnes.com.au/merchants provides genealogical and historical information about merchants and traders in a timeline format since the times of the Crusaders, and cites any sources used, providing a detailed bibliography. The Ulster Historical Foundation similarly has a database of the names of merchants and traders in Belfast, Londonderry, Lurgan and Armagh on its website at www.ancestryireland.com. Guildhall Library in London contains the records of City of London merchants and holds many directories of London merchants and bankers dating back to 1677.

Accountants

The world's first professional body of accountants awarded a Royal Charter was the Edinburgh Society of Accountants, established in 1854. Since then there have been many accountancy institutions, societies and associations set up to give their members added status and legitimacy in the profession, many of which were eventually granted a Royal Charter enabling their members to use the title Chartered Accountant.

The principal institutions for accountants in Great Britain are the Institute of Chartered Accountants in England and Wales (formed in 1880 as an amalgamation of the Institute of Accountants, the Society of Accountants in England, the Incorporated Society of Liverpool Accountants, the Manchester Institute of Accountants and the Sheffield Institute of Accountants), the Institute of Chartered Accountants of Scotland (formed in 1951 as an amalgamation of the Edinburgh Society of Accountants, the Glasgow Institute of Accountants and Actuaries and the Aberdeen Society of Accountants), and the Institute of Chartered Accountants in Ireland (formed in 1888 covering both Northern Ireland and the Republic of Ireland). To gain membership an entry exam needs to be passed and a certain amount of work experience is required, with various other restrictions applying as well. Accountants who could not gain access to one of the chartered institutes often joined another body whose membership requirements were not so rigid, such as the Scottish Institute of Accountants, formed in 1880, or the Society of Accountants and Auditors, formed in 1885. The first woman was not admitted into the Institute of Chartered Accountants in England and Wales until the 1919 Sex Disqualification (Removal) Act made it illegal to refuse women membership, although applications from female accountants had been received since the 1880s.

The Institute of Chartered Accountants in England and Wales, which is represented by 22 district societies and 27 branch societies around the country, has retained the main set of membership records and indexes from 1870, and any enquiries should be sent in writing to:

> The Registrar
> Institute of Chartered Accountants in England and Wales
> Gloucester House
> 399 Silbury Boulevard
> Central Milton Keynes
> Buckinghamshire MK9 2HL

You may find a record of your ancestor among fire insurance policies.

The Institute's membership records include when an accountant was articled, when they passed their exams, and the firms that they worked for. Once you know the names of the firms your ancestor worked for you may then go about locating the company records. Details of other records of the Institute of Chartered Accountants in England and Wales kept by the Institute may be found in Wendy Habgood's guide to the company records of around 180 chartered accountants in *Chartered Accountants in England and Wales: A Guide to Historical Records* (Manchester University Press, 1994). The book contains photographs of partners, records of salary books, accounts and partnership agreements. Some of the institute's records have been deposited at the Guildhall Library, comprising limited membership records for 1880–1942, some examination records for 1882–1949, staff registers for 1889–1933 and other material that is detailed in the Guildhall's online leaflet at www.history.ac.uk/gh/ghinfo9.htm. Unfortunately no records for the Society of Accountants in England, which formed in 1872 before it merged with the Institute of Chartered Accountants in England and Wales in 1880, have survived, though records of the Incorporated Society of Liverpool Accountants, the Manchester Institute of Accountants, and the Sheffield Institute of Accountants, set up between 1870 and 1877, are held by the district societies.

The Institute of Chartered Accountants in England and Wales has a section on its website to help genealogists who are trying to find out about their accountancy ancestors. The Accountancy Ancestors page can be found at www.icaew.co.uk/library. There is an obituaries index, a photograph index, a selection of life stories, and reports about accountants who fought in the First World War.

The historical records of the Institute of Chartered Accountants of Scotland (ICAS) and the Edinburgh, Glasgow and Aberdeen Societies that amalgamated to form the Institute, have been deposited at the National Archives of Scotland. Dr. Stephen P. Walker wrote *The Society of Accountants in Edinburgh 1854–1914*, containing genealogical information about some members. The Institute of Chartered Accountants in Ireland has offices and a library in both Dublin and Belfast. You can contact the office that covers the area where your accounting ancestor worked to find out where records of their membership and accountancy firm are held. The Belfast office in The Linenhall can be telephoned on 028 9032 1600 and the Dublin office at CA House can be contacted on +353 (0)1637 7227. The National Library of Ireland has some directories to members of the Institute of Chartered Accountants in Ireland, including one for 1917, among its collection of books and periodicals.

Insurance and Insurance Agents

The association of insurance underwriters, Lloyd's of London, took its name from Edward Lloyd's coffee house and became incorporated as the Corporation of Lloyd's in 1871. Edward Lloyd, a coffee merchant, set up Lloyd's Coffee House in London in the late seventeenth century to serve as a meeting place for merchants to exchange news and for merchants and underwriters to negotiate insurance for ships and cargo. He then went on to publish shipping news in *Lloyd's Lists*, which were taken over by the underwriters after his death, so the insurance business has always had strong links with merchant shipping.

The British middle class were a cautious bunch, always in fear of losing their carefully built-up wealth. Therefore even if your ancestors did not work for an insurance company, you may find a record for them among fire insurance policies from the seventeenth century onwards, if they took one out to protect their home and belongings. Some fire insurance company records are at county record offices, but since many companies were based in London a large collection of fire insurance records can be found at the Guildhall Library. The Guildhall holds records for around 80 London-based fire insurance companies and has produced an online leaflet describing these at www.history.ac.uk/gh/fire.htm. David Hawkings has also written a comprehensive guide in *Fire Insurance Records for Family and Local Historians*.

> ## USEFUL INFO
>
> The British Insurance Business, 1547–1970: A Guide to Its History and Records *by Hugh A. L. Cockerell and Edwin Green is a comprehensive guide to researching the records of companies dealing in marine insurance, fire insurance, life assurance and accident insurance, and contains a directory of the archives of British insurance companies listing the records that have survived and a table listing local insurance agencies, where they were based and where their records are now held.*

The Legal Profession

There are a multitude of directories and indexes to the names of people who worked in the legal profession, so finding evidence of your ancestor's employment and some background information about their education and career should not be too much of a challenge. The information will hopefully lead you to records of the court in which they served and the firm they worked for, but locating detailed information about the cases they worked on will be more complicated. Biographical information and obituaries may give you some clues as to any important court cases they witnessed, but then a search through court records will be necessary to uncover further evidence, and these are not always easy to locate or very descriptive. *Law Reports* published annually summarize each court case and are a good starting place if you know a rough date for a specific case. Consult Chapter 27 for further guidance on researching court records generated by criminal convictions.

Records for Lawyers in England and Wales

In England and Wales lawyers were known as proctors or advocates until the mid-1800s, the equivalent of barristers and solicitors today. In the fifteenth century advocates formed an association known as the College of Advocates. The premises where the college was eventually based, just south of St Paul's Cathedral, were close to many church courts and became known as Doctors' Commons. George Squibb wrote *Doctors' Commons: A History of the College of Advocates and Doctors of Law* in 1977, listing college members with some biographical detail and describing the college records held at Lambeth Palace Library.

- Appointments of proctors can be found in the Act Books of the Archbishop of Canterbury, for which there is an index to the entries between 1663 and 1859 in volumes 55 and 63 of the British Record Society.
- The National Archives has published two research guides to help family historians in this field, Domestic Records Information 36: 'Lawyers: Records of Attorneys and Solicitors' and Legal Records Information 18: 'Sources for the History of Crime and the Law in England'. These list sources kept at The National Archives in Kew as well as records held elsewhere.

Barristers act as advisors on specialist points of law and are admitted to practice through one of the four Inns of Court – Lincoln's Inn, Middle Temple, Inner Temple or Gray's Inn. Most of the admission registers for the Inns of Court have been published and each Inn has its own library and archives based in London; however, access to people not in the legal profession is highly restricted. Middle Temple Library will admit non-members by written appointment only, for example, and may charge a fee.

- In 1896 Lincoln's Inn published their admission registers and the chapel registers of baptisms, marriages and burials that took place at Lincoln's Inn in *The Records of the Honourable Society of Lincoln's Inn; vol. I from 1420 to 1799, vol. II admissions from 1800 to 1893 and chapel registers*. Later admissions to Lincoln's Inn up until 1973 are listed in two volumes held at Lincoln's Inn Library.
- In 1877 Inner Temple published *Students Admitted to the Inner Temple 1547–1660* and, more recently, put an Inner Temple Admissions Database online covering 1660 to 1850, which can be

searched free of charge from www.innertemple.org.uk/archive/itad/index.asp. Admissions books for later dates are at the Inner Temple Library.

- In 1949 H. A. C. Sturgess published three volumes containing the *Register of Admissions to the Honourable Society of the Middle Temple from the 15th Century to 1944*, and a further two volumes published in 1977 by the Hon. Mr Justice Bristow cover the period between 1945 and 1975. The third volume contains an index to volumes I to III, which are arranged chronologically by date of admission, while the appendix in the fifth volume contains an alphabetical index to the chronological records of volumes IV and V.

- In 1889 J. Foster published *The Register of Admissions to Gray's Inn 1521–1889 together with the marriages in Gray's Inn chapel 1695–1754*, and admissions to Gray's Inn since 1927 have been recorded in the journal *Graya*, along with obituaries and other biographical notes on members.

Prior to training as a barrister at one of the Inns of Court, a law student may have studied at one of the Inns of Chancery, such as Clifford's Inn, Clement's Inn, Barnard's Inn, Furnival's Inn, Thavie's Inn, New Inn or Staple Inn. Clifford's Inn was the last of the Inns of Chancery to be closed in 1900.

- The National Archives in Kew holds admission registers for Clement's Inn between 1656 and 1883.

- The Seldon Society published C. Carr's *Pension Book of Clement's Inn* in 1960 and more recently published C. W. Brooks' *The Admissions Registers of Barnard's Inn 1620–1869*.

- In 1906 E. Williams wrote *Staple Inn, Customs House, Wool Court and Inn of Chancery; Its Medieval Surroundings and Associations*, including a list of students admitted to Staple Inn between 1716 and 1884.

- Middle Temple Library contains the admissions books for New Inn from 1743 to 1852.

Men in the legal professions between the sixteenth and nineteenth centuries, including judges and members of the Inns of Court and Doctors' Commons, needed to take an oath of loyalty to the Crown and the Church of England. In addition to the general series containing Oath Rolls, the court where a lawyer wished to practise also has declarations and oaths among its records at The National Archives. Those for Chancery are in series C 214 and C 217; for Common Pleas look in CP 10;

> Men in the legal professions had to take an oath of loyalty to the Crown and the Church of England.

for the Exchequer in E 3, E 169 and E 200; for the King's Bench search KB 24 and KB 113; and for the Palatinates of Chester, Durham and Lancaster look in CHES 36, DURH 3 and PL 23. After 1868 the Barrister Rolls contain oaths in KB 4. There is an index to the names of High Court judges, recorders and magistrates who took an oath from 1910 in KB 24.

There are a wealth of histories and directories that may lead you to information about ancestors who worked in the legal professions, such as Desmond Bland's *A Bibliography of the Inns of Court and Chancery* and the British and Irish Association of Law Librarians' *Sources of Biographical Information on Past Lawyers* by Guy Holborn, reviewing over 500 sources that contain biographical information about lawyers in England, Wales and Ireland. In 1870 Edward Foss published *Biographia Juridica: A Biographical Dictionary of the Judges of England from the Conquest to the Present Time 1066–1870*, which can still be found in major libraries and archives.

For more general guidance on tracing ancestors in the legal profession the Law Society has an online guide about sources in their library that are useful to family historians. Unfortunately the library is only open to solicitors and their staff; however, the 'How to Trace a Past Solicitor' guide (found with other research guides at www.lawsociety. org.uk under 'Products and Services' and 'Library Services') lists the types of sources the library holds, many of which can be found in other repositories. The Law Society recommends using Lists of Attorneys, the Roll of Solicitors to the Court of Chancery, and the *Law List* directory of practising lawyers from 1775 onwards. *My Ancestor Was a Lawyer* by Brian Brooks and Mark D. Herber, published by the Society of Genealogists, gives another general overview of the types of legal sources useful to genealogists.

Records for Lawyers in Scotland

In Scotland there are two types of lawyers – advocates, who may plead in the Court of Session, and solicitors (formerly known as writers and sometimes as law agents). Law agents pleading in the smaller courts such as the sheriff courts are also called procurators. The Scottish Bar is known as the Faculty of Advocates, the Lord Advocate is the principal law-officer and judges usually worked as advocates before their appointment.

There are several publications and directories with genealogical information about members of the legal profession from all ranks that should be your first port of call, particularly the annual *Scottish Law*

List, previously called the *Index Juridicus*, dating from 1848. In 1944 Sir Francis J. Grant wrote *The Faculty of Advocates in Scotland, 1532–1943 with Genealogical Notes*, and Brunton and Haig's *Senators of the College of Justice* lists judges of the Court of Sessions up until 1832. A number of Scottish solicitors can be found in *The Register of the Society of Writers to Her Majesty's Signet*. To locate court records for cases your ancestor worked on consult the comprehensive chapter about lawyers in *Tracing Your Scottish Ancestors* by Cecil Sinclair.

Records for Lawyers in Ireland

Dublin's legal quarter is known as the Four Courts, so named because it is where the courts of Chancery, King's Bench, Exchequer and Common Pleas were housed from the beginning of the nineteenth century. The Irish legal system was revised in the mid-nineteenth century and again by the Irish Free State in 1922 to form the Supreme Court, High Court and Central Criminal Court, all located at the Four Courts. The King's Inns occupied the site prior to the nineteenth century, controlling the entry of barristers to the Irish justice system. In 1982 the *King's Inns Admission Papers 1607–1867* were published, and Colum Kenny has written a couple of guides about the history of King's Inns and their surviving records. The Bar Council Law Library of Ireland is unfortunately restricted to members only; however, their website has an interesting history of the Irish legal system and the librarian may be able to point you in the direction of useful sources. Dublin Directories from the late eighteenth century contain details of Irish attorneys and barristers and the Irish Legal History Society may also be a useful source for advice in tracing Irish ancestors who worked in the legal profession.

In 1922 a separate Bar Council for Northern Ireland was established, and in 1926 the Inn of Court of Northern Ireland was founded so that Northern Irish lawyers could practise from the Bar Library in Belfast. The Public Record Office of Northern Ireland is your best bet for finding records about an ancestor in the Northern Irish legal profession. PRONI has a large Solicitors' Collection, comprising papers deposited by around 140 Northern Irish solicitors' firms. The records contain information about the firms' employees as well as their clients and the cases they worked on. A guide to these records has been published on the PRONI website.

USEFUL INFO

Suggestions for further reading:

• A Guide to Tracing the History of a Business *by John Orbell (Gower, 1987)*

• British Banking: A Guide to Historical Records *by John Orbell and Alison Turton (Ashgate, 2001)*

• Chartered Accountants in England and Wales: A Guide to Historical Records *by Wendy Habgood (Manchester University Press, 1994)*

• The Society of Accountants in Edinburgh 1854–1914 *by Stephen P. Walker (Garland Publishing Inc., 1988)*

• The British Insurance Business, 1547–1970: A Guide to Its History and Records *by Hugh Anthony Lewis Cockerell and Edwin Green (Continuum International Publishing Group, 1994)*

• My Ancestor Was a Lawyer *by Brian Brooks and Mark D. Herber (Society of Genealogists Enterprises Ltd, 2006)*

CHAPTER 20

Occupations: Professional Classes – Public Sector

The employment of individuals for the public benefit has grown as a phenomenon since the Victorian era, even more so since the development of the Welfare State at the end of the Second World War. Here we will explore the records available for researching the careers of people employed directly by the State and people who devoted their working lives to helping the rest of society, including civil servants, government officials, the police, people employed in the medical professions, schoolteachers and clergymen.

> *Being government employees means that public sector workers leave behind a lot of evidence in the archives.*

Civil Servants and Government Officials

Complete personnel records for civil servants are generally destroyed once the employee reaches 72 years of age, though the Ministry of Defence retains records of their employees until they would have been 100 years old. Those that have been kept will have been transferred to the national archive for the country concerned, that is to say The National Archives in Kew for civil servants in England and Wales, the National Archives of Scotland for Scottish civil servants and the National Archives of Ireland and Public Record Office of Northern Ireland for Irish civil servants and government employees.

The National Archives at Kew holds some records for famous or high-ranking civil servants in series CSC 11, as well as some records of ordinary staff who worked for HM Treasury between 1891 and 1976, kept in T 268. The National Archives' research guide, Domestic Records Information 117: 'Civil Servants Personnel Records', suggests more documents where random files on civil servants might be found and where the names of female employees might be located, though the latter are very limited owing to the ban on married women working prior to the Second World War. It is often necessary to have an idea of the department and sometimes the division your ancestor worked in, as well as a rough timescale. Individuals are named in files that have a very general description in the catalogue, so you will need to order general files relevant to the timescale and department you are looking for to search for any names. You will have more luck if your ancestor was a high-ranking civil servant; these individuals are listed in the *British Imperial Kalendar* from 1809 to 1972, when it became the *Civil Service Year Book*. Civil servants who enjoyed high-ranking positions in the overseas service can be found listed in the *Foreign Office List*. The National Archives' research guide contains a list of similar publications for a number of specific government departments.

▲ Civil servants at work in the House of Commons committee office, London 1919.

You may have more luck if your ancestor served in an official position between the sixteenth and nineteenth century, when many individuals were required to swear an oath of allegiance to the Crown, as indeed were many lawyers. The resulting rolls are scattered across a number of different places – oaths sworn before Justices of the Peace will be found in Quarter Session returns, whilst records of lawyers sworn before the various Crown Courts will be at The National Archives. Further information can be found on The National Archives website.

The Society of Genealogists is currently indexing records formerly contained in the Civil Service Commission: Evidences of Age file at The National Archives in series CSC 1. This comprises birth and baptism certificates of civil service applicants. Contact the Society of Genealogists at www.sog.org.uk to find out more. Copies of civil service staff directories can be found at major archives and libraries, containing limited information on the majority of civil servants, such as their appointments, how much they were paid and where they worked.

The post of customs officer is of ancient importance in the UK because of the islands' dependence on international trade. The

CASE STUDY

John Hurt

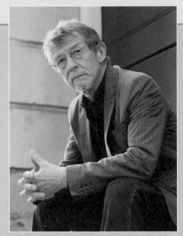

Although John Hurt was fascinated by the family legend that – somehow – he was related to the Marquess of Sligo, proving it was a much harder matter. Along the way, he had come across his great-grandmother Emma Stafford, believing her to be the elusive connection; but by talking to his cousin, John discovered that the link possibly lay through Emma's husband Walter Lord Browne and his family. By investigating details surrounding their marriage, John discovered a notice in the local paper announcing the impending wedding that claimed Walter's father, William Richard Browne, was the head of the Bond Office in London.

This information was fairly specific, and by researching what the Bond Office actually was, John was able to pinpoint William's career as a civil servant in the Customs Office, a major national institution which incorporated the work of checking and issuing bonds for vessels unloading goods cargo in London. Records of employment for customs officers survive in The National Archives, where John was able to trace William Richard Browne's career. He was somewhat surprised to learn that, instead of being listed as the head of the Bond Office – an important, prestigious and well-paid job – he was actually only a clerk, still enjoying an annual salary but by no means as lucrative a position.

Indeed, further investigations at The National Archives revealed that he had run into financial difficulties and, as a result, been declared bankrupt. His customs office pension was used to pay off his creditors and he ended up in court and, eventually, debtors' prison. Having consulted material at the modern Customs House, John gained a greater understanding of the work William Browne would have undertaken before his fall from grace.

◀ Walter Lord Browne, John's great-grandfather (standing rear centre) who was headmaster of Westport House School.

National Archives in Kew and the National Archives of Scotland hold records for customs staff as detailed in The National Archives' research guide Domestic Records Information 38: 'Customs and Excise Officials and Tax Collectors'. The NAS has a research guide to its customs and excise records at www.nas.gov.uk/guides/customs.asp. *Family Histories in Scottish Customs Records* by Frances Wilkins is an interesting read for anyone with Scottish ancestors in the customs and excise profession. These records are covered fully in Chapter 14.

Police

One of the first professional, trained police forces was Sir Robert Peel's Metropolitan Police Force of London, established on 29 September 1829, prior to which time London was policed by the Bow Street Foot and Horse Patrols and by parish constables, responsible for law and order in the provinces. There is a service register covering 1821 to 1829 for the Bow Street Foot and Horse Patrols at The National Archives in MEPO 4/508.

Local police forces were required to be set up in boroughs and counties under a similar system to the Metropolitan Police Force from 1856 in England and Wales, and from 1857 in Scotland, though some areas had been building a police force since the 1830s. There have been hundreds of police forces since the mid-nineteenth century, many of which have since merged, but there is no central police archive for all their records. Some police force records will be found at the local county record office while other forces have kept their own archives, such as that of the Metropolitan Police Force. Police records generally comprise attestation papers and personnel books or registers listing policemen's names, ages, dates and places of birth and notes about their career.

The National Archives holds personnel records for the Metropolitan Police from 1829 up until around 1933 in series MEPO, described in more detail in the research guide Domestic Records Information 52: 'Metropolitan Police (London): Records of Service'. Any research enquiries for records of the Metropolitan Police Force not held at The National Archives should be sent to:

> The Metropolitan Police Archive Service,
> Wellington House,
> 67/73 Buckingham Gate,
> London SW1E 6BE

▼ An early illustration of London's policemen from the 1870s.

CASE EXAMPLE

Metropolitan Police Archive

*Disciplinary books can be extremely interesting too, as the actor **Jeremy Irons** found out. Jeremy's great-great-grandfather Thomas Irons was a policeman with the Met, and a visit to the Metropolitan Police Archive uncovered records of Thomas joining the force in 1828, making him one of the first Peelers, but he was dismissed in disgrace in 1834 for being drunk and disorderly. Thomas went on to join the controversial Chartist movement and spent time in Newgate prison.*

The Metropolitan Police has some interesting pages about its history on the website www.met.police.uk/history.

The London Metropolitan Archives holds detailed records for the City of London Police from 1832, which are described on the Access to Archives database.

The Irish Constabulary was created as a national armed police force in 1822 and groups of part-time policemen were amalgamated to form the newly reorganized Irish Constabulary in 1836, known as the Royal Irish Constabulary from 1867. The Royal Irish Constabulary was disbanded in 1922 and The National Archives at Kew retains the records of around 84,000 men who saw service with the force between 1822 and 1922 in series HO 184, as described in the research guide Domestic Records Information 54: 'Royal Irish Constabulary Records'. Duplicates of Royal Irish Constabulary records can be found on microfilm at the National Archives of Ireland and in LDS Family History Centres.

Jim Herlihy has written *The Royal Irish Constabulary: A short history and genealogical guide with a select list of medal awards and casualties*, containing brief biographies of around 3,000 Irish policemen, as well as *The Royal Irish Constabulary: A complete alphabetical list of officers and men 1816–1922*.

The Ulster Historical Foundation has an online database of men who retired from the Irish Constabulary between 1836 and 1844 and a list of Irish Constabulary Sub-Inspectors in 1860 among its occupational databases at www.ancestryireland.com. *Police Casualties in Ireland 1919–1922* by Richard Abbott contains the names and biographical details of men in the police force who died in Ireland during this period.

Records of the Dublin Metropolitan Police, formed in 1786, are held at the National Archives of Ireland and Herlihy has also written *The Dublin Metropolitan Police: A short history and genealogical guide*, as well as *The Dublin Metropolitan Police: A complete alphabetical list of officers and men 1836–1925*.

Medical Professions

The medical profession covers a wide range of roles, from apothecaries, chemists, druggists, pharmacists and physicians who administered remedies to surgeons, doctors, nurses, midwives and dentists who treated and cared for patients. For a general guide to medical records useful to genealogists read Susan Bourne and Andrew H. Chicken's 1994 publication, *Records of the Medical*

Professions: A practical guide for the family historian. Most of the secondary sources, indexes and reference books mentioned in this section can be found in the Wellcome Historical Medical Library and any of the national libraries for England, Wales, Scotland and Ireland. The Wellcome Trust was established in 1936 with money left in the will of the pharmaceutical businessman Sir Henry Wellcome who had dreamed of creating a Museum of Man illustrating the medical past of mankind. The Trust's mission is to promote medical research and as such it has a fantastic library stocked full of books, manuscripts, archives, films and pictures on the history of medicine since the earliest times. You can search the Wellcome Library's collections online at http://library.wellcome.ac.uk.

A Medical Archives and Manuscripts Survey (MAMS) was carried out for more than 100 repositories in Greater London to establish the types of records they hold concerning the history of medicine between 1600 and 1945. The results of this survey can be located on the Wellcome Library website, giving descriptions of the archives of the General Medical Council, the Medical Society of London, the National Institute of Medical Research, the Royal College of Surgeons, the Royal College of Physicians, the Royal College of Midwives, the Royal College of Obstetricians and Gynaecologists, the Royal College of Psychiatrists, the Royal National Pension Fund for Nurses, the Royal Pharmaceutical Society, the Royal Society of Medicine and plenty of smaller institutions across the capital. The reports for each archive are found under the 'Databases' section of the 'Electronic resources' page on the website. You may want to contact some of these institutions if they cover your ancestor's occupation to seek their advice about your research.

If you manage to trace several generations of ancestors in the medical profession going further back in time, then *The Medical Practitioners in Medieval England: A Biographical Register* written by Eugene Ashby Hammond and Charles Hugh Talbot may be of use. The biographical register includes physicians and surgeons from England, Wales and Scotland, but very few from Ireland. The biographical detail has been taken from parliamentary rolls, the infirmarer's rolls of Westminster Abbey, charter witness lists, lay subsidy rolls, chancery and exchequer records, household accounts and many more sources.

In 1985 Wallis, Wallis, Whittet and Burnby compiled a register containing around 70,000 entries for *Eighteenth Century Medics (subscriptions, licences, apprenticeships)*, on behalf of the Project for Historical Biobibliography, covering a wide range of medical professions. It is a huge volume consisting of an alphabetical index of

▼ General Nursing Council certificate for Daphne Gillett, 1936. This is the kind of document you may find in your family's personal archive.

Nurses

Guides to researching nursing ancestors are on The National Archives website under various subject headings, including Domestic Records Information 79: 'Civilian Nurses and Nursing Services'; Military Records Information 55: 'British Army: Nurses and Nursing Services' (see also Chapter 9); Military Records Information 57: 'Royal Air Force: Nurses and Nursing Services' (see also Chapter 12); Military Records Information 56: 'Royal Navy: Nurses and Nursing Services' (see also

Chapter 10). If you know the name of a hospital your ancestor worked at the records of that hospital might be located using The National Archives' HOSPREC database at www.nationalarchives.gov.uk/ hospitalrecords. The Royal College of Nursing Archives has a website at www.rcn.org.uk/development/ library/archives describing their historical collections, which are open to the public and contain valuable material donated by nurses and their families.

The Central Midwives Board for England and Wales was established after the Midwives Act of 1902. The

Board's main function was to set up *The Midwives Roll* of certified midwives, which it published from 1904 until the 1980s. The Obstetrical Society of London had printed a *Midwife's Roll* dating back to 1872 containing the names of skilled midwives. The Central Midwives Board for Scotland was created by the Midwives (Scotland) Act of 1915 and has published *The Midwives Roll for Scotland 1917–1968*. Information about early midwives can be found in Joan E. Grundy's book *History's Midwives: Including a 17th Century and 18th Century Yorkshire Midwives Nominations Index*.

individuals compiled from The National Archives' apprenticeship records, subscriptions to the publication of medical treatises, alumni of UK medical schools, some membership lists of the medical Royal Colleges and a variety of other sources. Hard copies of the register can be consulted at the Wellcome Institute for the History of Medicine, the Pharmaceutical Society, the Royal College of Physicians, the Royal College of Surgeons, the Science Museum, the British Library, the National Library of Ireland, the National Library of Scotland, The National Archives at Kew and the Society of Genealogists, as well as at a select number of universities and regional libraries.

Lambeth Palace Library holds a directory of medical licences issued by the Archbishops of Canterbury between 1535 and 1775, which can be searched online at www.lambethpalacelibrary.org/holdings/ Catalogues/medics/medics_county.html by place name or by surname. Guildhall Library holds some records for apothecaries (most of which are due to be transferred to Apothecaries' Hall – see below) as well as some records for surgeons, physicians and other medical practitioners. An online guide to sources for tracing these occupations both at the Guildhall and elsewhere can be found at www.history.ac.uk/ gh/apoths.htm. The Guildhall holds membership and apprentice records for Barber-Surgeons who were members of the Barbers' Company of London from 1522 to the nineteenth century, and ecclesiastical licences for physicians, surgeons and midwives granted by the Bishop of London, and the Dean and Chapter of St Paul's Cathedral.

Apothecaries and Chemists

Apothecaries used to form part of the Grocers' Company, until in 1617 they separated to form the Society of Apothecaries, a Livery Company of the City of London (still in existence today). Apothecaries were an early form of chemist, keeping an open shop where they would give medical advice to customers and sell medicine to those who could not afford to visit a physician. In the eighteenth century apothecaries became either general practitioners or trading apothecaries who dispensed medicine. In 1775 the latter type of apothecary ceased becoming liverymen of the Society of Apothecaries.

Access to the archives at Apothecaries' Hall in Blackfriars Lane, London EC4V 6EJ can be arranged by writing to the archivist specifying what you are looking for. The Society of Genealogists has a microfilm copy of *A list of persons who have obtained certificates of fitness and qualification to practise as apothecaries from August 1, 1815 to July 31, 1840*. The National Library of Ireland holds material about apothecaries based there, including a list of admissions to the guilds of Dublin, 1792 to 1837, and records of Apothecaries' Hall, Dublin, for 1747 to 1833.

Chemists, who made their medicine from chemicals, and druggists who made their drugs from animal and vegetable products, were separate entities to the apothecaries, until the three merged into the Pharmaceutical Society of Great Britain in 1841. A comprehensive guide for records relating to these professions is *The Pharmaceutical Industry: A Guide to Historical Records* by Lesley Richmond, Julie Stevenson and Alison Turton. A register of chemists and druggists, including students, was established in 1852, requiring those on the register to have passed an examination. From 1868 a higher qualification of pharmaceutical chemist could also be obtained. The Pharmaceutical Society of Great Britain has published annual copies of *The Register of Pharmaceutical Chemists, Chemists and Druggists*. W. A. Jackson's book *The Victorian Chemist and Druggist* gives an illustrated insight into how Victorian chemist and drug stores worked, describing what they looked like, the types of medicines typically dispensed, how they were made and the containers and instruments used. Bryony Hudson wrote an article on 'Tracing People and Premises in Pharmacy' in 2005 that can be found in *Genealogists' Magazine*, vol. 28, no. 6 (2005), pp. 242–6.

CASE EXAMPLE

Apothecaries

Natasha Kaplinsky was astonished to discover that a fifth-generation great-grandfather on her mother's side, Benjamin Charlewood, was apothecary to the households of both George II and his son George III, the famously 'Mad King' who suffered from porphyria towards the end of his reign. The Worshipful Society of Apothecaries was able to reveal to Natasha that Benjamin became a master of the society in 1760, and although he died before George III's first bout of illness, his apprentices may have helped to treat the King.

Dentists

Prior to the Dentists Act of 1878 dentistry was almost completely unregulated. The Act called for a *Dentists Register* to be published annually, which became the duty of the General Council of Medical Education and Registration of the United Kingdom, after which it was published by the Dental Board of the United Kingdom. Registers of dental students were also made. The Dental Board of Ireland published its own *Dental Register* from the early twentieth century. Invariably the registers for each year are arranged alphabetically by surname and list details such as the dentist's address, their date of registration and a description of their qualifications. John Menzies Campbell was a dental historian who wrote many publications on the subject since the mid-twentieth century, and for those researching a Scottish dentist the History of Dentistry Research Group may be able to give guidance on the records available.

Physicians

Researchers looking for evidence of ancestors working as physicians should consult *The Roll of the Royal College of Physicians of London; comprising biographical sketches of all the eminent physicians ... 1518 to 1825* compiled in three volumes by William Munk and known as Munk's Roll. A further four volumes of Munk's Roll, compiled in the latter half of the twentieth century covering the years from 1826 to 1983, were written by George Hamilton Brown, Richard R. Trail and Gordon Wolstenholme under the title *Lives of the Fellows of the Royal College of Physicians of London*. Munk's Roll covers entries since the founding of the Royal College of Physicians in 1518. People researching early physicians may also find helpful John H. Raach's *A Directory of English County Physicians 1603–1643*.

Medical Directories have been published annually since 1845, and from 1866 onwards they contain the names of dentists as well as medical practitioners. The directories contain notes about dates of qualification, the types of qualifications obtained and any posts held in hospitals and the armed services. The General Council of Medicine has also published an annual list of qualified practitioners in *The Medical Register* since 1859. Both publications cover the whole of the British Isles and are available at the national libraries, the Wellcome Library, Guildhall Library and other major reference libraries.

Information about the careers of medical practitioners who were university graduates may be found in published university membership lists. The University of Edinburgh published *List of Graduates in Medicine 1705–1866* listing the names of graduates, when they graduated, the qualification they received and their specialism. The Royal College of General Practitioners has an online guide to 'Tracing Your Medical Ancestors' under the archives section of their website at www.rcgp.org.uk, and Alex Glendinning's article 'Was Your Ancestor a Doctor?' can be found online at http://user.itl.net/~glen/doctors.html.

Teachers

Early schools were set up and run by religious institutions, philanthropic and charitable organizations, principally for the benefit of the poor who could not afford to pay for their children to be educated by a governess or private schoolmaster. In 1603 schoolteachers were required to be licensed by bishops, and issues of such licences can be found in county record offices among Act books in the diocesan records. Unlicensed teachers could be prosecuted in the church courts. In England and Wales, Parliament began granting annual sums of money to help charities set up schools from 1833, namely the National Society, which was Anglican, and the British and Foreign School Society. Records of these may survive at the local record office, as might records of very small schools run by private individuals from their homes. A search of the Access to Archives database should uncover their whereabouts; if not, contact the record office local to where the school was based.

▲ Students at the Ragged School, Edinburgh, are given a lesson by a Reverend teacher, 1850.

Teachers in England and Wales

In England and Wales a pupil-teacher system was established from 1846 to train bright pupils up as teachers for three years at their elementary school where the headmaster would supervise them. At the age of 18 they needed to pass a King's/Queen's Scholarship Examination, later known as the Preliminary Examination for the Certificate. Those who passed the exam could then attend a training college for a further couple of years, usually a residential college set up by the Church of England and run by voluntary organizations.

- Records of teacher training colleges can be found at The National Archives dating back to the 1840s in series ED 17, ED 103 and ED 40, though the involvement of central government in teaching has largely been related to matters of supply, qualification and conduct, so many of the ED files relate to general business such as building applications rather than students.

In 1870 an Elementary Education Act was passed that required pupil-teachers to be trained at pupil-teacher centres instead of at the school in which they were taught.

- Some pupil-teacher centre files have survived for the period 1884 to 1911 and those are kept at The National Archives among records of the Department of Education and Science in series ED 57. A detailed memorandum about the pupil-teacher system written in 1902 can be found in ED 24/76.

This system flourished for the last two decades of the Victorian period, until the Education Act of 1902 set up the first national system of secondary school education under which many pupil-teacher centres turned into secondary schools. In 1902 Local Education Authorities (LEAs) were established with powers to train pupil-teachers at secondary schools that were emerging as a new form of higher education. From 1907 the pupil-teacher system was replaced with one that saw pupils who intended to become teachers studying at school until the age of 17 or 18 and then either acting as a student teacher at a public elementary school or attending a teacher training college.

- There are LEA files at The National Archives in ED 67 concerning the supply of teachers with some staff returns of teachers and students training in colleges, mainly for the period 1912 to 1915.
- From 1904 municipal training colleges were set up to replace the ones run by voluntary societies. Records for training colleges run by the LEAs are in ED 87 and ED 86, while ED 78 contains files relating to LEA colleges, those run by voluntary bodies and university colleges providing courses for teachers. Few records about staff will be located in those series, but the reports filed by HM Inspectorate concerning the premises, staff and curricula at training colleges in ED 115 may be of interest.
- Universities first began running teacher training courses in 1890, but in 1911 a four-year course was introduced with the final year devoted to teacher training. Information on university teacher training courses is in ED 81 and ED 119.

From 1926 teachers who trained in any type of training college needed to sit a final examination, which qualified students for certified recognition as teachers, conducted by the Joint Examination Boards and HM Inspectorate.

❛ Teachers had to be licensed, and evidence will be found among

In 1943 there was concern about how the nation would meet the demands of post-war teaching, and a need for more teachers was recognized. Secondary school education was reorganized by the Education Act of 1944. Fifty-five emergency teacher-training colleges had been set up by 1947 and representative files have been kept for one college in Alnwick, Northumberland, in ED 143/33–34 and for Borthwick Training College for Women in London in series ED 143/35–36.

The Education Act of 1899 put in place a register of teachers, but the National Union of Teachers protested at the manner in which the register was kept and so it was withdrawn in 1907. The Teachers' Registration Council, which had been responsible for compiling the register, was re-formed in 1912 and recommended compiling voluntary registration lists of teachers in alphabetical order. The Council was disbanded in 1948, at which point registration ceased, but the British Origins website has digital copies of registers from 1914 to 1947 with details of teachers who began their career as early as the 1870s, at www.originsnetwork.com. The registers provide details of around 10,000 people who taught in England and Wales, giving names and maiden names in the cases of married women, dates of registration, register numbers, addresses, and details of attainments, training in teaching and their experience, listing the schools they had worked at. There is a note to say whether the teacher was retired or had died. Two of the 162 original volumes are missing, so all names starting A to ALD are absent and a small percentage of names starting ALE to BL are not there either.

Teachers in Scotland

In Scotland local parish authorities or burghs ran schools for centuries, and the records of these are likely to be at the local record office. An introduction to early education in Scotland can be found in James Craigie's *A Bibliography of Scottish Education before 1872* (Scottish Council for Research in Education, 1984) and thereafter the chapter on schools in Cecil Sinclair's *Tracing Scottish Local History* is worth consulting. The Education (Scotland) Act of 1872 made formal education accessible to all children, supplying local school boards with sufficient funding to open new schools and train more teachers. Older records of local authority schools are usually deposited with the county archive, though some schools have retained their records. The National Archives of Scotland has records for some schools and reports

SUMMARY

The National Archives has a number of research guides on the website under the subject heading 'Education' detailing the records they hold concerning teachers and schools, including:

- *Domestic Records Information 67: 'Education: Elementary (Primary) Schools'*

- *Domestic Records Information 127: 'Education: Inspectorate and HMI Reports'*

- *Domestic Records Information 65: 'Education: Secondary Schools'*

- *Domestic Records Information 119: 'Sources for the History of Education'*

- *Domestic Records Information 23: 'Education: Records of Special Services'*

- *Domestic Records Information 63: 'Education: Records of Teachers'*

- *Domestic Records Information 24: 'Education: Technical and Further Education'* by inspectors. A research guide to education records at the NAS is online at www.nas.gov.uk/guides/education.asp.

Teachers in Ireland

The National Library of Ireland holds the *Irish Education Enquiry, 1826, 2nd Report*, listing all the parochial schools in Ireland in 1824 including the names and details of teachers, and Dingfelder's *Schoolmasters and Mistresses in Ireland* contains an index to the report. The Ulster Historical Foundation has compiled a database of those teachers and schools listed in Antrim, Armagh, Cavan, Donegal and Fermanagh in 1826-7, which can be found among the occupations databases at www.ancestryireland.com. The National Archives of Ireland holds records for National Schools set up in Ireland from 1831 onwards, and has produced a research guide describing their holdings at www. nationalarchives.ie/topics/Nat_Schools/natschs.html. Of particular interest are the registers for each school dating from 1832 until 1963 and salary books from 1834 to 1918. The Public Record Office of Northern Ireland has corresponding files for National Schools set up in Ulster Province and has also produced an online leaflet for tracing national school records - leaflet number 5 in the Local History Series.

Clergymen

Like most other professions, clergymen can be found listed in specialist directories published from the nineteenth century onwards, which are widely available in major reference libraries and give details such as the clergyman's benefice, any positions previously held and his education. Since 1858 *Crockford's Clerical Directory* has published biographical information about clergymen of the Church of England, the Church of Wales, the Episcopal Church of Scotland and, until 1985, the Church of Ireland. It was originally an annual publication but later was published every few years. The *Clerical Guide* published details of clergymen from 1819, and Cox's *Clergy List* was published from 1841.

The biographies of around 41,000 senior clergymen between 1066 and 1857 were published in the 1857 edition of *Fasti Ecclesiae Anglicanae* written by T. D. Hardy in continuation of John Le Neve's 1715 work. This work has been released in several volumes many times since, most recently edited by Joyce M. Horn who between the 1960s and recent years has had the volumes republished by dioceses. The

Society of Genealogists holds the Fawcett card index of clergymen, providing references to sources where they are mentioned, including university registers, parish records, *Gentleman's Magazine, Musgrave's Obituary* and probate records. The Arts and Humanities Research Council is funding the compilation of The Clergy of the Church of England Database, a biographical register of all clergymen between 1540 and 1835, though the careers of men on the database may extend past 1835. This is an on-going project that can be found online at www.theclergydatabase.org.uk.

From the thirteenth century bishops' registers record the ordination of deacons and priests and any subsequent appointments. Until the eighteenth century the documents are usually written in Latin. They can be found among diocesan records in county record offices. If your ancestor worked in the diocese of London then records about him should be found at the Guildhall Library, which has produced an online research guide for tracing clergymen at www.history.ac.uk/gh/clergy.htm. Prior to the mid-nineteenth century most clergymen studied at Oxford or Cambridge University, after which time theological colleges were set up to train them, so records of these may mention your ancestor.

Information about the daily toil of parish clergy can be gleaned from reading the parish records of the church from the time they worked there, such as vestry meeting minutes and parish registers signed by them. Clergymen were one of the many groups of society who were required to take an oath of allegiance to the Crown and Church of England from the late seventeenth century until the nineteenth century. The Common Pleas series at The National Archives contains Rolls of Oaths of Allegiance for clergymen between 1789 and 1836 arranged chronologically in CP 37, which can be searched in addition to the other Oath Roll series already mentioned.

Lambeth Palace Library, where records of the Archbishop of Canterbury's peculiar courts and jurisdictions are held, has produced an online research guide to tracing clergymen at www.lambethpalace library.org/holdings/Guides/clergyman.html, with tips for everyone researching clergy heritage back to the seventeenth century. For an in-depth guide to tracing Anglican clergymen read the Society of Genealogists' *My Ancestor Was an Anglican Clergyman* by Peter Towey.

USEFUL INFO

Suggestions for further reading:

• My Ancestor Was a Policeman: How Can I Find out More about Him? *by Antony Shearman (Society of Genealogists, 2000)*

• Records of the Medical Professions: A practical guide for the family historian *by Susan Bourne and Andrew H. Chicken (S. Bourne and A. H. Chicken, 1994)*

• A Bibliography of Scottish Education before 1872 *by Jaimie Craigie (Scottish Council for Research in Education, 1984)*

• My Ancestor Was an Anglican Clergyman *by Peter Towey (Society of Genealogists Enterprises Ltd, 2006)*

CHAPTER 21

Occupations: Trades and Crafts

This chapter will describe the way our ancestors worked as craftsmen, often in professions passed down the generations from father to son, or mother to daughter. Although the records are more scattered than in other lines of work, bearing in mind the aesthetic or prosaic qualities sometimes required in a particular field, some of the main sources are described, along with the historical context that gave rise to them.

Apprenticeship

The system of apprenticeship was practised widely in many different occupations from around the fourteenth century. The earliest apprentices were controlled by the guilds (see below) but the system soon developed independently. Apprenticeship was formalized in 1563 with the passing of the Statute of Artificers and Apprentices. The statute made it illegal to practise any trade without serving a period of apprenticeship for seven years, and this statute remained law until 1814. Those who chose to practise a trade without having first served an apprenticeship would be subject to fines and such cases would be recorded in quarter session records. Additionally, in 1601 the reforms of the poor relief system enabled parish overseers to send pauper children as apprentices, thereby increasing apprenticeship's popularity (see Chapter 24 for further details). It remained a vital system of providing skilled workers up until the beginning of the nineteenth century. However, increasingly during the eighteenth century, many apprentices would enter into more informal agreements with masters, especially

> *Apprenticeships taught young boys a trade by binding them to a master craftsman.*

◄ Apprentices, including two young boys, in a carriage works in Buckinghamshire, 1903.

for common trades. Moreover, as the Statute of 1563 only related to trades in existence at that date, many modern skills were bypassing the system altogether. By the nineteenth century, the vast changes brought about by the industrial and agricultural revolutions along with a large population increase meant that the system was no longer sustainable and it was no longer practised widely.

The scheme of apprenticeships was to teach young boys (and occasionally girls) a profession or trade by binding them to serve a master (sometimes their father) for a number of years (usually seven), or until the individual reached the age of 21. Indeed, during an era of shorter life spans an individual could serve as an apprentice for almost a quarter of their entire life. Female apprentices would learn from the master's wife or be trained in traditionally female trades such as seamstress. The child's parents would pay a sum of money to the master in return for the master's training. The master would be responsible for clothing, housing and feeding the apprentice during his or her time serving him. An agreement, or contract, between the two parties (the child's guardian and the master) would be formalized in the form of a written document known as an indenture. These indentures would list the name of the apprentice (occasionally the age and place of his or her birth) and his or her father (occasionally including his occupation and

▼ Apprentice records on the Origins website.

View records - Apprentices of Great Britain

Search criteria: Last name: BERRY + Close variants
First name: JOHN + All variants
Year range: 1442 to 1850

Viewing Images

Please note: Without a plug-in you will be unable to view the TIFF images in your browser. If you are unable to view images, or if your default image viewer is Quicktime, please go to www.alternatiff.com and download the free image viewer. This viewer will allow you to zoom, rotate, save and print images. ▶ Click Here for Help Viewing Images

Click column heading links to change order of results

Series	Top Last Name	Top First Name	Bottom Last Name	Bottom First Name	Image
1	BAILES	THOM	BOLFOUR	JAS	image
1	BASS	ROB	BEZLEY	DAN	image
1	BEAVER	ISAAC	BEVIS	JAS	image
1	BERRETT	THOM	BERRYMAN	JERE	image
1	BERNARD	THOM	BERTHET	FRAN	image
1	BERRY	JANE	BERRY	JOS	image
1	BERRY	JOS	BERRY	WM	image
1	BERRY	WILLIAM	BESBROWN	ABRA	image
1	BEETENSON	THOM	BETHAM	WM	image
1	BARRADALE	WILLIAM	BOOSEY	HEN	image
1	BERRELL	ROB	BURFORD	ISAAC	image

Apprenticeship Records

Surviving records for apprentices are scattered amongst The National Archives, local archives and the Society of Genealogists, and locating the appropriate record depends upon when and where your ancestor may have served.

Records at The National Archives

Between the years 1710 and 1804 stamp duty was placed upon all indentures of apprenticeship (except those of pauper children). Hence, for this period only, there is a central index of apprentices as the Inland Revenue would record the process of collecting stamp duty. The Commissioners of Stamps compiled these records as they kept registers detailing the amount of income received from placing a duty on indentures. The deadline for payment was relatively flexible, being one year after the expiration of the indenture. Hence, your ancestor's record could be in a large time period, spanning the entire duration of their apprenticeship. The records themselves can be found in TNA series IR 1.

The registers should provide the names of the apprentices, dates of their indentures and names and addresses of the masters along with their trades. The registers also included the names of the apprentices' fathers until 1752.

Thereafter this information was seldom recorded. There are indexes for the masters' and apprentices' names covering 1710 to 1774 available to consult in The National Archives, the Society of Genealogists and the Guildhall Library (the index of apprentices can also be searched online at www.britishorigins.net).

The National Archives also has additional apprenticeship records as follows:

● *War Office Records:* During the eighteenth and nineteenth centuries the War Office was attempting to settle the dilemma of whether to allow apprentices who had been recruited to the Army without permission of their masters to remain in service. These discussions have been referenced in *The Alphabetical Guide to War Office and Other Materials* (PRO Lists and Indexes). The piece WO 25/2962 has a list of all recruits who had joined the Army under such pretences and were subsequently returned to their masters from 1806 to 1835.

● *Admiralty Records:* The Admiralty also recruited apprentices to work in the dockyards and other occupations. Their recruitment can be found in the Admiralty Digests in ADM 12. Further registers of Admiralty apprentices can be found in pieces ADM 73/421 and 448 (miscellaneous registers of Greenwich Hospital).

● *Merchant Navy Apprentices:* The Merchant Navy also started recruiting apprentices to work on ships over 80 tons after the Merchant Seamen Act of 1823. Further details of these can be obtained in Chapter 13, which discusses Merchant Navy records in details.

● *Civil Service Records:* The Civil Service Commission contains records for those being examined as apprentices from 1876 onwards in CSC 10.

● *Board of Trade:* The Board of Trade has records for those apprentices working for this department in BT 19.

● *Poor Law Union Records:* The main series relating to Poor Law Union records (discussed in Chapter 24) may also include details of pauper apprentices.

The Society of Genealogists

The Society has compiled an index of approximately 1,500 private indentures, dating from the seventeenth to the nineteenth centuries. This index is known as *Crisp's Apprentices' Indentures.* The actual indentures if they survive may be at local archives.

The Society of Genealogists has also placed another one of its datasets online, available to search by name, at www.britishorigins.net.

This is the London Apprenticeship Abstracts, containing approximately 300,000 entries from 1442 to 1850. The entries relate to the apprentices, parents and masters of the numerous livery companies of London. The entries should also give the parish of residence of the apprentice's father.

Records in Other Archives

As indentures were private documents, many have not survived, and those that do may still be in private hands or in business archives. A certain number may be held in local archives, although this varies considerably for each archive. Those that do survive may also have been published by the local archive.

Additionally, local record offices should have other collections relating to local apprentices. It may be possible to find indentures of paupers in the parish overseers' records. As disputes between master and apprentice were settled in quarter session records, these documents may also contain relevant information. Most local record offices should also have registers of apprentices, covering a variety of date ranges and trades. It is worthwhile checking on the website of your local archive to ascertain the extent of their holdings.

residence), along with the master's name and trade. The apprentice was not allowed to marry or establish his or her own trade during the time of his indenture to his or her master. After completion of the period of apprenticeship the apprentice became a freeman of his company and could establish his own business in his learnt trade. These freemen were permitted to train their children in their trade without the children having to enter the formal apprenticeship system. Many apprentices did not become freemen after the completion of their service. Rather, these young men chose to continue working for their masters and were known as 'journeymen'. Additionally, it wasn't uncommon for apprentices not to finish their term of service and, therefore, not enter the trade of their master. If an apprentice absconded or ran away his sponsor/parent would have to pay the master a surety fee for the time the master had, up until that point, spent on training the runaway. Often children living in villages in close proximity to large urban areas would be apprenticed to masters living in such towns and cities.

The choice of which trade to learn was rarely made by the apprentice but, more often, depended on what connections his family had. Fathers would often be relatives or friends with the craftsmen whom they chose as their sons' masters. If the apprentice's family was relatively wealthy, they would be apprenticed into a trade that would also be well paid. The masters themselves were often surrogate parents of the apprentice, as the apprentice would be living with the master and his family for a number of years. In certain cases apprentices would end up marrying the daughters of their masters (sometimes as a means of accelerating their career prospects with their masters), or would form close emotional bonds with the master and his family. At other times the system could be open to abuse; either by the master using the apprentice as little more then a menial servant and not teaching the adolescent his trade in earnest, or by young apprentices stealing from the master.

Guilds and Livery Companies

Guilds were another key component in the history of British trades and occupations. The very first guilds or 'gilds' were formed in the twelfth century as religious groups but soon developed into specific trading organizations and came to be organized into three distinct bodies during the medieval period:

▶ An apprenticeship indenture for William Hamilton of Carlisle for seven years as a cabinet maker and joiner, 1839.

- *Merchant Guilds:* These groups of merchants originated in the twelfth century and were umbrella organizations for a variety of crafts and trades in urban areas.
- *Religious Guilds:* These guilds were founded on religious grounds and aimed to provide charitable support to the local community by operating schools and hospitals. They operated until the Reformation, and were subsequently abolished.
- *Craft Guilds:* These guilds were formed during the thirteenth and fourteenth centuries and were specific to individual trades or crafts (such as tailors, cobblers, blacksmiths, etc.). They controlled and regulated such crafts or trades.

By the sixteenth century the craft guilds became the main type of guild and absorbed merchant guilds into themselves. They became the dominant forces in the manufacturing sectors of the economy, each guild becoming an effective trade union for that particular industry, sometimes even monopolizing the craft. Each individual who joined the guild would be placed into one of three different categories depending on his skill level: apprentices (discussed above, those learning the trade), journeymen (newly qualified in his chosen trade, but without enough experience to set up his own business and, therefore, still working for the master) and masters (those fully qualified and established in the trade). Guilds operated as monopolies in that they were given exclusive rights to practise their trade and craft and could also control who entered their guild. As mentioned above, initially, apprenticeships were also solely controlled by guilds. Guilds also often provided welfare support to members, giving pensions to their elderly or infirm members, or their widows and children.

London became a particularly important centre for guilds. Indeed, different guilds in London each had their own distinguishing uniforms (or livery) for those in senior positions and, as such, they came to be known as 'livery companies'. There were numerous different livery companies operating in London, numbering almost 80 in the nineteenth century. All sorts of occupations were regulated by these liveries, including bakers, butchers, drapers, haberdashers, wax chandlers, carpenters, masons, musicians, plumbers, fletchers and weavers, to name but a few. The 48 companies in existence in 1515 were organized into an order of preference by the City's Court of Aldermen. This ordering was based upon each company's political or economic power and the twelve most important were known as the 'great twelve'. The most important of these was the 'Worshipful Company of Mercers'. Each company had halls where members could entertain their guests or conduct any relevant business, although many were destroyed either in the Great Fire of London or in the Blitz. The livery companies could monopolize their trade in the City of London by obtaining royal grants or charters. Sometimes certain livery companies' monopolies extended outside the area of the City. Although most guilds throughout the country declined in importance after the eighteenth century, the London livery companies were still influential during the nineteenth century. Indeed, many of these livery companies survive to the present day in London, although not as regulatory bodies.

Most guilds would include the leading masters of the trade, a few wardens who would gain their position by election, assistants, freemen, journeymen and apprentices. It was possible to become a freeman (i.e. to gain membership) of a guild by the following means: membership was granted if the individual's father belonged to the guild (patrimony), if the individual had served the appropriate time as an apprentice (servitude) or the individual would buy his membership (redemption). As mentioned, most newly qualified apprentices worked as journeymen. This title derived from the French word for day (*jour*) and referred to the fact that such men were day labourers who were paid daily. Only after a number of years would a journeyman become a master.

> *Many London livery companies survive to this day.*

Records of Guilds and Livery Companies

The surviving archives of guilds and livery companies are very useful for the family historian. Firstly, they provide useful details documenting the career progression of your ancestor. Additionally, they may assist in tracing your ancestors back for several generations, as many members would have become so through patronage and your family may have been working in a particular trade for generations. The surviving records may be found in local archives or, for the City of London, in the Guildhall Library. Certain livery companies may also retain their archives, in which case you will need to contact the clerk of the company directly. Many books have also been published recording the general history of individual livery companies or guilds. It is advisable to see if such a book has been published for the company you're interested in, as it may be easier to consult this source first before turning to the original records.

Records Held at the London Guildhall Library

The Guildhall Library has a large selection of records for the many livery companies operating in London, from the fourteenth to the twentieth centuries. This includes details for 85 livery companies and related organizations. The records fall into the following categories:

- *Freedom Admission Registers:* These list when an individual was given the freedom of the livery company.
- *Alphabetical lists of freemen:* These were recorded only for lists over long date ranges and not for single years.
- *Apprentice binding books:* These books would record when an apprentice was bound to a particular company. The registers end in the nineteenth century as the system of binding apprentices was no longer common practice after that date. Alternative sources such as court minutes and warden accounts may have details for those still being bound after that date.
- *Alphabetical list of apprentices:* Similar to the lists for freemen, these were only recorded for lists covering many years.
- *Quarterage books*: These annual lists record quarterly membership dues; they are not indexed.
- *Court minutes:* These documents record the meetings of the courts of the livery companies. Some books may have indexes to individuals mentioned in the minutes and most should include details of when freemen were admitted or apprentices were bound.
- *Accounts:* These record the financial affairs of the company.

◀ Corporation of London invitation to lunch for Mr and Mrs Fred Gillett at the Guildhall. Fred Gillett had his own tailor's shop in Fleet Street. He later became the Mayor of Bromley and a Freeman of the City of London. This invite was kept as an heirloom by his family.

The records described above are for individual livery companies and it is only feasible to use these sources if you know which company your ancestor was involved with. There is no comprehensive name index for members and apprentices of all companies. If you don't know what livery company your ancestor was part of, then the following may help in identifying the company.

- As most masters and apprentices would have become freemen of the City, check the records of this. They cover the years 1681 to 1923 and are held at the London Metropolitan Archives (LMA), although the name index is in the Guildhall Library. The LMA also has records for individuals purchasing freedom of the city for various dates from the fourteenth to the seventeenth centuries.
- Boyd's *Inhabitants of London* (available to search online at www.britishorigins.net) also includes residents of the city and details on liverymen, mainly for the sixteenth and seventeenth centuries.
- If your ancestor worked during the eighteenth century then the indexes to the apprenticeship records held at The National Archives may also give information on which livery company was involved.

▼ Another special family heirloom – a photo of Queen Mary being escorted by Fred Gillett through the City of London.

CASE STUDY

Natasha Kaplinsky

Natasha Kaplinsky knew a great deal about her maternal roots because a relative had already undertaken extensive investigations, drawing up a comprehensive family tree. After verifying the data it contained from certificates, census returns and parish registers – plus the occasional will at The National Archives – Natasha found an intriguing ancestor lurking in the eighteenth century, Benjamin Charlewood, who, according to family legend, was reputed to have been appointed as apothecary, or 'medical practitioner', in the household of George III,

the king who was famed for losing the American colonies and suffering from bouts of madness, thought now to be porphyria.

Included amongst family heirlooms was a silver meat dish, allegedly presented to Charlewood in recognition of his services and bearing a coat of arms to prove its authenticity. Natasha took the dish to Christie's for an evaluation, but to her disappointment it was made *after* the death of George III, casting doubt on the family story. Undeterred, Natasha decided to investigate the life and times of Charlewood in official documents, and started with the Worshipful Society of Apothecaries, a London livery company that represented the professional interests of all qualified apothecaries who successfully applied to become a member. The Guildhall Library, London, held all relevant records for membership, and to Natasha's surprise not only was Charlewood listed but he had actually been appointed Master of the Society in 1760 – the year George III ascended the throne. Furthermore, the records held there confirmed that he had indeed held the post of apothecary in the royal household.

Given his royal position, Natasha was able to follow up his appointment in official records, and found an entry for him in 'Office Holders in Great Britain', drawn from records in the Lord Chamberlain's office deposited at The National Archives. His original appointment took place in 1738, a position he held until his death in 1766 – therefore serving not only George III but also his predecessor, George II, as well! This enabled Natasha to order his will, also held at The National Archives, which revealed further details about his life, status, personal property and wealth.

- Probate documents may also detail which company your ancestor was involved in.
- Many freemen of the companies would also be given voting rights in the City of London and should be recorded in the City of London poll books and electoral registers, surviving from the late seventeenth century to 1872.

Records for Outside London

As mentioned, most surviving guild records should still be with local archives and record offices. Many general histories or actual records themselves may also have been published of the guild of interest, which will also be of help. The surviving records will be of similar nature to the ones detailed above relating to the London livery companies.

Freemasonry

Freemasonry developed from craft guilds. The term freemason originated from masons who used to work with freestone, a particular craft of the fourteenth century. Lodges were built to support the trade. Members involved in this craft were said to have secret signs to acknowledge fellow workers. However, during the seventeenth century the skills of the freemasons saw a decline as fewer cathedrals were being built. During this period the freemasons became general social clubs more involved in charitable work and not specific guilds dedicated to the craft of masonry. Lodges were now used for the purpose of social gatherings with regular meetings being held. Membership was increasingly given on an honorary basis to keep up the numbers. After 1691 membership was opened to a wider social spectrum, further increasing the membership.

The growth of the freemason movement culminated in the opening of the Premier Grand Lodge of 1717. The movement steadily grew during the eighteenth century and became a national organization in 1802. In 1813 the two separate Grand Lodges, the Premier Grand Lodge and the Atholl Grand Lodge, were amalgamated to form the United Grand Lodge of Antient, Free and Accepted Masons of England (this body included Wales too). This Lodge is still the main governing body of the movement today. The movement was also active in Scotland with the establishment of the Grand Lodge of Antient Free and Accepted Masons of Scotland. The Grand Lodge of Ireland was constituted in 1726.

Records for freemasons can be found in various places including the Grand Lodges themselves.

Freemasonry developed out of craft guilds.

Records for Apprenticeships and Guilds in Scotland and Ireland

Scotland

Scotland had a similar system of guilds and apprentices as south of the border. Guilds were first established in the Middle Ages in Scotland in various burghs throughout the country. These institutes would also have monopolies for their trade, provide financial support for their members and their families and have local political power and influence.

Surviving records of these guilds will either be at the National Archives of Scotland (NAS) or in the burgh records deposited in local archives. The NAS has listed the extent of its holdings (detailing the individual crafts they have records for, in which area and the dates covered) on its website, at www.nas.gov.uk/guides/crafts.asp. Local archives should hold burgess rolls or court books, which will contain relevant details. The latter may also have details of the regulations of crafts. Otherwise the Scottish Record Society has published numerous lists of men being admitted into burgesses (a right enjoyed by apprentices after completing their training) of various towns and cities of Scotland which may include your ancestor.

Surviving apprenticeship records may also be found in a variety of places. As indentures were private documents many have not survived to the present day, or will be in private hands. The NAS has some collections including records of the apprentices of Edinburgh from 1613 to 1783 (in NAS reference RH9/17/272–326). Another main collection can be found within the archives of George Heriot's Trust, a charitable organization providing apprenticeships for pauper children (in NAS reference GD421/10). A list of Edinburgh apprentices from 1583 to 1755 has

- *Masonic publications:* The Grand Lodges of England, Scotland and Ireland all published lists of freemasons. Such records will be found either at local libraries or at the respective Grand Lodges. Their contact details are as follows:

  England (and Wales): The United Grand Lodge of England
  Freemasons' Hall, 60 Great Queen Street, London WC2B 5AZ
  www.ugle.org.uk

  The library at the Grand Lodge should also have yearbooks for its members, published since 1908. If your ancestor was a senior member then consult *A Masonic Yearbook Historical Supplement (1969)* as it gives details of all holders of Grand Rank from 1717 to 1968.

  Scotland: Grand Lodge of Scotland
  Freemasons' Hall, 96 George Street, Edinburgh EH2 3DH
  www.grandlodgescotland.com

  Ireland: Grand Lodge of Ireland
  Freemasons' Hall, 17 Molesworth Street, Dublin 2
  Republic of Ireland
  www.irish-freemasons.org

also been published by the Scottish Record Society.

Scottish apprentices were also subject to the taxation imposed on English apprentices during the eighteenth century and hence the records held in The National Archives at Kew (discussed earlier) will be of relevance.

Ireland

The system of guilds and apprentices was similar in Ireland to other parts. Those who were apprentices and subsequently members of guilds would also be entitled to become freemen of the city. The guild system was very important in Dublin, dominating the economic and social history of the city for many centuries. As such many tradesmen and guild members would be included in the list of free citizens for Irish cities. Dublin City Archives has a list of the free citizens of Dublin from 1225 to 1918. Similar records may be found in other local archives.

The guilds and livery companies hold particular relevance for the history of Ireland as they organized the 'plantation of Ulster' during the seventeenth century. Ulster remained the only non-colonized part of the island during the seventeenth century and James I chose to redress this by encouraging Protestant Scottish and English migrants to settle there. The livery companies of London established a settlement near to the hill of Doire (Derry) on the River Foyle. The area they had settled was granted county status and renamed 'Londonderry' in honour of the participation of the London liveries in establishing this settlement. The apprentices were also intertwined with the history of Derry when the city was laid to siege by the deposed Catholic King James II and his forces in 1688–89. The Protestant settlers were besieged by these forces inside the city walls and the apprentice boys were instrumental in lifting the siege. The part played by the apprentice boys was subsequently honoured by annual parades, although these marches would become a symbol of sectarian strife during the late twentieth century.

- *Quarter session records:* Political uncertainty during the end of the eighteenth century (after the French Revolution) meant that the British government became increasingly suspicious of secretive organizations such as the freemasons. In 1799 the government passed the Seditious Societies Act making any society that had oaths not sanctioned by law illegal. Freemasonry was not made illegal but each lodge had to provide certificates listing names and addresses of its members to the Clerk of the Peace on an annual basis. Lodges could also be closed by quarter sessions if a complaint was upheld against them. Hence quarter session records have records for local freemason movements from that period onwards.
- *Published histories:* There have been many published histories of local lodges that may be of relevance. They may also include biographies of senior members.

If you have a strong interest in freemasonry then it would be worthwhile visiting the Freemasons' Hall in London. It has exhibitions, a museum and a library open to the general public.

CHAPTER 22

Migration: Immigration

Migrants have been coming to settle in the British Isles for many centuries. Some came as aggressive invaders, such as the Romans, Vikings and Normans, whilst other groups of people have arrived for economic reasons, or as refugees fleeing religious or political persecution. Whatever their reasons for coming here, tracking migrant ancestors can reveal fascinating stories for family historians.

> Britain has been home to countless waves of migrants from around the world.

The Jewish community, Huguenots and many other foreign Protestant groups are important examples of those seeking asylum from violent repression in their own homelands. Huguenot records have been described previously in Chapter 7, but records relating to the Jewish community will be discussed in detail below.

As Britain's empire expanded in the eighteenth and nineteenth centuries, people living in its dependent colonies increasingly became entitled to British citizenship. In the decades after the end of the Second World War, large-scale immigration from the former colonies of the West Indies and South Asia was actively encouraged by the government, as labour was in short supply. However, there has been a black presence in England far earlier than the mid-twentieth century. For example, many former African slaves lived in Britain in the eighteenth century until the trade in slaves was finally abolished by Britain in 1807, and after slavery was abolished in the British Empire in 1834 growing numbers moved here. More information about the black and Asian presence in this country prior to the twentieth century, from 1500 to 1850, can be obtained by accessing an online exhibition on The National Archives website (www.nationalarchives.gov.uk/pathways/blackhistory). The most recent wave of immigration has been from

Eastern Europe after many of these countries joined the European Union in 2004.

As there have been so many periods of immigration to the country it would not be uncommon to find at least one ancestor born outside the UK. Depending upon when someone arrived in the UK, and from whence they came, there may be surviving documents recording their arrival and subsequent life in the country. As a second stage it may also be possible to trace the migrant's family in their country of origin, depending which country it was and at what period. However, for a variety of reasons, this second-stage research can be time consuming and problematic and it is best to seek advice from suitable organizations on the feasibility of conducting such research. A good place to begin finding out further information is on the individual country pages on the website www.cyndislist.com.

▲ The *Empire Windrush* at Tilbury Docks after its journey from Jamaica, 1948.

Records for Immigrants Arriving in the United Kingdom

The British State has been recording the entry of foreigners through many methods for many centuries. Historically, foreign immigrants would be referred to as 'aliens' and were subject to various additional regulations such as extra taxation and were entitled to fewer rights than British citizens. These extra regulations placed upon the lives of immigrants led to extra documentation that has become good source material for anyone researching their family tree. However, anyone researching Irish migrants in this period will not find any mention of these individuals as they were not seen as foreign but as Britons. Systematic record keeping of aliens began from 1793 onwards with the Aliens Act, but there are certain relevant records (mainly related to tax) dating back much further.

Records Prior to 1793

The majority of records for this period are held at The National Archives, mainly in taxation, chancery and custom record series.

Information will be scattered amongst these series and there is no single index to all these sets of records.

- The very first records note foreign merchants and clergy from the thirteenth to sixteenth centuries in C 47. E 106 lists foreign laymen's possessions in England and what fines were placed upon alien clergy during the thirteenth and fourteenth centuries.
- The series E 179 holds the Exchequer Subsidy Rolls, which record taxes raised between the twelfth and seventeenth centuries. Foreigners had to pay twice the rate of tax on movable property (known as subsidies) as English citizens. These returns are arranged geographically (by county) then annually, and it is only feasible to search these records if a time and place for your ancestor's residence here is known (as no index is available).
- Port records list how much custom was payable by ships entering various ports throughout the country. They are found in series E 122 and are known as Particulars of Customs' Accounts, dating back to 1272. Along with the amount of custom payable is a record of the name of the ship, the master and the merchant who owned the goods that were entering the country. The record would also detail whether the merchant was a foreigner or not. Port books from 1565 to 1799 can be found in E 190 with similar information being found as the earlier records.
- The State Paper collections from Edward VI to James I also list returns of strangers in London and other parts of the country in 1571 and 1618. Some of these lists appear in the printed calendar of State Papers, which are indexed. Later calendars of State Papers also include the issuing of passes by Secretaries of State to allow Britons and foreigners to enter or leave the country.
- Treasury records also have records of foreigners and refugees who received pensions or annuities from the Crown as methods of payment for a variety of reasons. These payments are recorded in Treasury In-Letters in T 1 from 1557 to 1728, which are indexed in calendars.

Immigration Records Post 1793

A new wave of migrants started to enter the country from the late eighteenth century and early nineteenth century onwards, mainly arriving from France to escape the turmoil following the French Revolution and Napoleonic Wars. This new wave of refugees led to the passing of legislation in attempts to monitor and control the new influx. The first of these was the Aliens Act of 1793. From now on everyone entering the country had to register with the Justice of Peace and provide personal details about themselves. The JP would then forward this information in the form of certificates to the Aliens Office. Although the certificates held by the Aliens Office have not survived (apart from an index to the records in HO 5/25–32 for 1826 to 1836), local county record offices may have surviving documentation created by the JPs. Further records may be found in series HO 1. This correspondence series contains some passes given to aliens between 1793 and 1836. These passes contain the names of aliens, their entry port, religion, country of origin and occupation, amongst other details.

▲ Immigrants arriving at Dover, including Poles, Jews, Italians and people from the Baltic, 1910.

Another Aliens Act was passed in 1836 in response to the continued political turmoil in Europe and the threat of political revolution in many places, including the UK. Indeed, aliens came to be regarded increasingly with a certain degree of suspicion. Henceforth each immigrant arriving had to place his or her signature on a certificate of arrival. The surviving certificates, for arrivals into England and Scotland, are held in series HO 2 and cover the years 1836 to 1852. They are indexed in HO 5/25–32 (the indexes end in 1849). Each certificate provides similar details to the passes issued to aliens in HO 1. A specific index to certificates issued to some German, Polish and Prussian immigrants for 1847 to 1852 is also available at The National Archives (called the Metzner Index). HO 3 also contains records created in response to the 1836 Act.

Twentieth-century Alien Records

The status of new arrivals arriving into the UK continued to be monitored and regulated during the twentieth century, in part due to the great political upheavals, conflicts and changes that occurred in this century. The UK saw large numbers of arrivals of refugees and economic migrants in this time period and many of these arrivals would become naturalized citizens in due course.

The first Act passed in this period was the Aliens Act of 1905. This ensured that alien paupers or criminals could be expelled. Moreover, foreign nationals could also be deported if they were convicted of crimes. Trial and criminal records can be traced for such individuals in the same manner as other criminal records, discussed in detail in

Chapter 27. The National Archives series HO 372 contains registers of deportees. These registers should give the deportee's name, nationality and conviction details.

The start of the First World War in 1914 resulted in the Aliens Registration Act, an attempt to monitor in more detail the Germans living in the UK at this time (many of whom were interned during the war years). The Act required that all aliens aged over 16 register with the police, and also to re-register if their personal circumstances changed. Unfortunately, not many registration cards have survived for this period. A small number (approximately 1,000) are held in The National Archives in series MEPO 35. Depending upon where the individual registered, the cards may be with the appropriate record office, although again the survival rate is far from complete. Bedford Record Office has a large collection (approximately 25,000) from 1919 to 1980. Further details about what is available in local record offices can be found in Kershaw and Pearsall's *Immigrants and Aliens* (published by The National Archives).

The German invasion of Belgium in 1914 resulted in approximately 2,000 refugees fleeing their home country and arriving in the UK. They were housed in refugee camps throughout Britain and were registered by the Ministry of Health and the Local Government Board. Their entry was recorded on cards, known as 'history cards', which are now held in The National Archives series MH 8. Cards were created for every family arriving and give detailed information about the personal details of these families. Some local record offices may also contain information about refugees who were housed locally.

Communities Arriving Due to the Second World War

The menace of the Nazis in Europe during the 1930s and 1940s threatened many countries and communities in addition to the Jewish community (records of Jewish refugees will be discussed along with general Jewish records). Many of those under threat chose to escape their homes and find refuge in Britain and other safe countries.

- *Czechoslovakian refugees:* Many Czechoslovakians chose to leave their country after the Munich Pact in 1938. These refugees established the Czechoslovak Refugee Trust a year later. The Trust's administration files and a sample of personal files can be found in HO 294.
- *Polish refugees:* Around 160,000 Poles fought alongside the British against the Nazis during the Second World War. After their home country became a Communist state under Soviet influence many

USEFUL INFO

The main sets of archives relating to prisoners of war during the First and Second World War are held by the International Council of the Red Cross in Geneva. They are willing to conduct research at an hourly cost and can be contacted at the following address:

International Council of the
 Red Cross
Archives Division
19, Avenue de la Paix
CH-1202 Geneva
Switzerland

Poles were reluctant to return home and wished to stay in Britain. The British government allowed them to do so and allowed ex-soldiers to bring their wives over.

A Polish Resettlement Corps (PRC) was established to assist these ex-soldiers to settle into civilian life. Files found in WO 315 concern the affairs of the PRC and the Army Lists of the PRC found in this series name many ex-soldiers.

In 1947 the Polish Resettlement Act was passed to help Poles establish themselves in Britain. The National Archives has many records concerning this process in the Unemployment Assistance Boards (AST), in series AST 7, AST 11, AST 18 and AST 1, and also in ED 128.

- *Hungarian refugees:* Some Hungarian citizens also chose to flee the Soviet annexation of their country at the end of the Second World War and settle in Britain. A further influx arrived in 1956, after the failed Hungarian Uprising against Communist rule. Information concerning their admission and arrival can be located in HO 352, although there is very little on individual refugees themselves.

> *Many communities fled to Britain during the 1930s and 1940s.*

Internee Records

The two world conflicts during the twentieth century created not only a number of refugees who were welcomed into Britain, but also witnessed some foreign citizens being interned during the war years. There are very few personal records for those interned during the First World War, only lists of enemy aliens submitted by those running the internment camps in HO 45 and HO 44.

There are more details for those interned during the Second World War. The cards of those interned (approximately 8,000 aliens, including those who had fled Nazi persecution in Germany) can be found in HO 396. The cards have photographs of the internees and give personal details including nationality. HO 214 contains the few surviving personal files of internees. Further information can be found in HO 215, the series providing lists of internees in various camps. County record offices may have additional information on how the camps were administered. The Heritage Library of the Isle of Man is one such example, providing information on internment camps on the island during both conflicts.

Prisoner of War Records

A small number of foreigners would have been kept against their will in the UK, as prisoners of wars. A very limited amount of information can be found at The National Archives. WO 900/45 and 46 contain sample lists of German military personnel interned as prisoners during the First World

Arrival Records of Immigrants

Early records of immigrant arrivals have been discussed above in the port records section. From the late eighteenth century records began to be kept on a more systematic basis, recording the first moment an alien arrived on UK shores. Again, this was done as an attempt to monitor immigrants during the political uncertainties of the Napoleonic Wars and revolutionary fears.

The Aliens Act passed in 1816 required all ships' masters to list all aliens on their vessel and provide physical descriptions to the appropriate authorities. Surviving lists can be found in The National Archives series HO 3. The series is arranged chronologically but there is no name index to the series. There is little surviving material from 1860 to 1866. However, as this system was not thorough and many people arrived without being officially recorded, lists after 1869 were not retained.

Inward Passenger Lists

The National Archives series BT 26 holds inward passenger lists. These were compiled from 1878 to 1960, although the bulk of the records start in 1890. The series only contains lists for those journeys that originated outside Europe and the Mediterranean (although if the vessel made a stop en route to these parts then passengers would be included in the lists). The master of each vessel was legally required to detail the name of each passenger, their age and occupation and occasionally their intended residence within the United Kingdom. These lists would then be given to the appropriate customs officer. The records are stored by port of arrival, date of entry and ship name. As there is no passenger name index to date it is only feasible to search the series if the name of the ship is known. It is possible to search the series online by the name of the ship.

If you only know the port of arrival of your ancestor you may be able to find which ship he or she would have arrived on by using BT 32. The registers in this series state which ships arrived at each port and their date of arrival annually from 1906.

After 1960 no such inward passenger lists were kept as air travel had largely replaced travel by sea. However, it may be possible to find similar information after 1960 in the archives of private shipping companies.

War. General information can be found for prisoners of war in the Foreign Office records at The National Archives. WO 166 and WO 177 holds lists of names of prisoners scattered in war diaries and hospital records for prisoner of war camps during the Second World War, although no index is available for these records. Those held on a temporary basis in the Tower of London are listed in WO 94/105. Further information can be found in The National Archives' research guide no. 29.

Jewish Immigration

An important migrant community in the UK was the Jewish community, who initially came to the country with the Normans after 1066. The community was subsequently expelled by Edward I in 1290. In his Edict of Expulsion he demanded that Jews convert to Christianity, leave the country or be put to death. After this edict there was no Jewish presence in England until the sixteenth century, when a very small number arrived from Portugal and settled in London. Significant numbers of Jewish immigrants did not arrive until after 1656, when Oliver Cromwell rescinded the Edict of Expulsion. Since this time, there have been various waves of Jewish immigrants arriving from all over Europe up until the end of the Second World War.

As mentioned, the earliest Jewish community in England was expelled in medieval times. Hence, it will not be possible to trace any individual lineage back to this medieval community. Nevertheless, there is ample documentation recording the presence of this community in various contemporary records during the thirteenth century. These can be found in civil litigation and taxation records in the Exchequer series at The National Archives.

The earliest modern Jewish community settled in London from about 1655 onwards. However, over the next hundred years or so further communities established themselves in other major towns of England and Wales, and by the end of the eighteenth century there were almost 30,000 Jews living in Britain. There was another wave of Jewish immigration as they escaped persecution in Eastern Europe during the late nineteenth century. In fact the numbers arriving were large enough to cause concern to the British government, who passed the Aliens Act of 1905 to curtail this flow. The Jewish community was made up of two distinct branches – the Sephardic Jews (who arrived from Mediterranean countries such as Spain, Portugal and Italy) and Ashkenazi Jews (who were originally from Central and Eastern Europe).

If you find a Jewish ancestor during the civil registration period, they can be traced using the same sources as for other ancestors, by using the census and other civil registration records (see Chapters 5 and 6). However, this can be complicated by the fact that many Jews chose to anglicize their names, or Jewish names would often be spelt incorrectly. Prior to 1837, synagogues would record life events. These registers may still be at the appropriate synagogue as there was no requirement to submit the registers to the Registrar General in 1837 (as happened with non-conformist bodies).

Jews were exempt from the requirements of Hardwicke's Marriage Act of 1753 (see Chapter 7) and so marriages took place in synagogues. Synagogues would not necessarily keep a specific register, but may keep copies of the marriage contract (*Ketubah*). Some marriages in the early years of civil registration may not have been recorded due to ignorance of the law or language difficulties, or to avoid the expense.

USEFUL INFO

*Synagogue records can be complicated to use as many are in Hebrew and use the Jewish calendar. Additionally, there were no baptisms; rather new-born boys would have the date of their circumcision recorded (usually eight days after birth) by the circumciser (*mohel*). This information would be kept by the* mohel *and locating the surviving registers is very complicated. (Further assistance can be obtained from contacting the Board of Deputies of British Jews, at www.bod.org.uk.)*

As mentioned, many synagogues still retain their registers. The National Index of Parish Registers (NIPR, published by the Society of Genealogists) has a list of all synagogues in Britain prior to 1838 in its third volume. This volume also has guidance on searching for births, marriages and deaths of Anglo-Jews. The Court of the Chief Rabbi in Finchley, London, has a few early registers of some synagogues in its archive. Some registers held by this institute give the person's place of birth (which is very useful if the individual was an immigrant).

Other sources that may assist in searching Jewish ancestors include:

- *The Anglo-Jewish Archive:* This was originally held in the Mocatta Library, University College of London, and is now found at the Society of Genealogists or the Parkes Library at the University of Southampton. This includes pedigrees, wills and newspaper cuttings.
- *The Jewish Chronicle:* This is the longest-running and most important newspaper of the Jewish community in Britain. It has been printed on a regular weekly basis since 1844 and, similar to other newspapers, carries registers of births, marriages, deaths and obituaries. It has been partially indexed and is available at numerous Jewish archives and the British Library's Newspaper archive. The entire archive is also available online at www.jc.com. It is possible to search the newspaper collection free of charge but viewing the information incurs a cost.
- *The Jewish Year Book:* This has been published annually since 1896 and provides information on active synagogues.
- *The Jewish Museum* (www.jewishmuseum.org.uk): The Museum is in Camden Town, London, and has a variety of genealogical books and manuscripts including family histories.
- *The Jewish Historical Society of England* (www.jhse.org): This Society is centred at 33 Seymour Street, London. The Jewish Genealogical Society of Great Britain (www.jgsgb.org.uk) is also at the same location and publishes many references of Jewish genealogical interest.

The last great influx of Jews into the UK occurred in the 1930s and 1940s due to Nazi persecution. Their entry has been recorded in the following places:

- A very small number of records can be found in The National Archives series HO 382. It is possible to search this series by the individual's surname in the online catalogue, although a few records have not yet been opened.

- The London Metropolitan Archives retains the records for the Poor Jews' Temporary Shelter. The first such shelter was established in Whitechapel in 1886 (with another being established in Kilburn a little later) and many refugees fleeing from the Nazis would have first been housed at this shelter. The surviving records include files on individual immigrants and provide useful birth and country of origin details amongst other family details.

- The Jewish Refugees Committee was also instrumental in settling newly arrived refugees in the 1930s. The Committee retained approximately 400,000 personal files of refugees. The files are detailed and provide full information on birth, nationality, occupation and date of arrival. This collection is also with the London Metropolitan Archives. Some records may be on restricted access. This archive also has other useful collections for the London Jewish community, including records of the Jews Free School.

Immigrants from Commonwealth Countries

Britain faced a serious labour shortage in the years after the end of the Second World War. The British government sought to remedy this shortfall by encouraging migration from its Commonwealth countries, mainly from the West Indies and the Indian sub-continent. Additionally, migrants from these countries were eager to come to Britain in order to have a better quality of life.

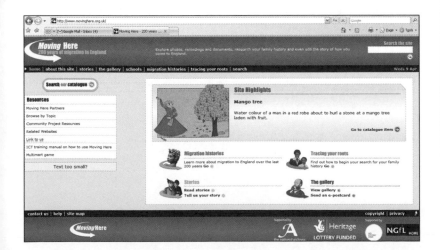

USEFUL INFO

www.movinghere.org.uk

• *This is a very useful website dedicated to the history of migration to the UK over the past two hundred years. The website has been funded by the Lottery Fund and is run jointly by The National Archives and 30 other libraries and archives in the country. It explores the motivations behind migration and the experiences felt by the new migrants upon arrival. It has separate sections on the migrant experiences of the following communities: Caribbean, Irish, Jewish and South Asian. The website also includes information on how to trace the family history of each of these communities.*

• *There are personal histories of migrants arriving in the twentieth century, recorded orally, in written stories and through pictures and photographs. Many public records from The National Archives and other archives have been digitized and can be seen on this website (such as the passenger list for SS Windrush).*

◀ Moving Here is the best place to seek advice about tracing an immigrant ancestor.

CASE STUDY

Colin Jackson

Colin Jackson was born and brought up in Wales, but knew that his parents came from Jamaica in the 1960s. After talking to them, Colin was able to piece together some of his history. His maternal grandfather, Dee, arrived in Cardiff in 1955 to find work, and brought his children with him. His wife Maria elected to stay behind, and returned to her native Panama to look after her sick father. Although she tried to keep in contact with her family, Dee did not pass on any of her letters to their children, fearing they might leave for Panama, and consequently the family lost contact with her. In the meantime, Dee bought a house in Cardiff, and rented out a room to other economic migrants from Jamaica, who in 1962 included Colin's father, Ossie.

Given the range of ethnic backgrounds associated with Jamaica, Colin wanted to investigate his mixed heritage. Colin was able to use archives in Jamaica to trace his father's family through various birth and marriage certificates to his mother, Marie Wilson, and her parents, Jacob Wilson and Eugenia Stewart. Still using certificates and parish records, Colin found that Jacob's father was Adam Wilson, an emancipated slave who was linked to the Greenmount plantation owned by Valentine Dwyer. Although Adam died a free man in 1849, he was born in slavery and lived to see the emancipation of his people in 1834.

Colin remained curious about the European side of his heritage and dug a bit deeper. From Panama he obtained his grandmother Maria's birth certificate, which showed her father was Richard Augustus Packer and her mother was Gladys McGowan Campbell. Working further back, Gladys's parents were Albertina Wallace and Duncan M. Campbell, and the trail led back to Jamaica. Duncan Campbell was part of a large Scottish community on the island, and Albertina was his black housemaid. It was not unusual for white 'gentlemen' to have children with their black staff. She was eventually given his house, and then worked as a prison warder – an important position at the time. Therefore her daughter Gladys would have been of mixed race, and at some time during her youth had moved to Panama.

There is a long history of West Indians heading to Panama during the various attempts to build the Panama Canal, first under the French and then under the American team in the first decades of the twentieth century. After some careful research in surviving employment records, Colin found that Richard Packer, Gladys's husband, worked on the canal in 1905 for six months, before finding employment in the hospital. He stayed in Panama until at least 1921, when Maria was born. She returned to look after Richard when he fell ill, leaving her own children to start a new life in Wales with her estranged husband Dee.

◄ Colin's parents, Ossie and Angela, on their wedding day in 1965.

Between 1948 and 1962 there was no restriction on migrants arriving from these countries, as the need for labour was so great. Indeed, the British Nationality Act of 1948 ensured that these new arrivals could obtain citizenship with relative ease, by simply sending applications to the Home Office. The Home Office would grant certificates of citizenship, each certificate having a number. The Home Office would retain a duplicate set of the certificates issued, and these can be found in HO 334 (see below for further details).

Various restrictions were placed from 1962 onwards, as it was felt that numbers now needed to be controlled. Hence, two Immigration Acts were passed in 1962 and 1968 to stem the flow. Commonwealth immigrants now had to apply for a work permit to be able to gain employment in the UK (unless they applied for a passport in the United Kingdom). Some specimen records of these work permits can be found in The National Archives series LAB 42.

Additionally, if your immigrant ancestor arrived before 1960 on a passenger ship it is possible to trace his or her arrival on the passenger lists in BT 26 (mentioned above).

Denization and Naturalization

Once immigrants had arrived in the UK some, depending on when they arrived, could seek to become British citizens. This could be granted through a process of either denization or naturalization. Denization was slightly different from naturalization in that although the individual would be protected by the Crown and law, he or she would not be able to vote, had to pay the same alien rate of tax as before, and could not apply for military and civil occupations. Some early immigrants chose denization over naturalization as it was the less expensive option. Naturalized citizens were awarded exactly the same rights as native subjects. It was not universal practice for immigrants to choose to do either, however, as both procedures entailed a cost and only minimal advantage was gained.

Denization Records

An alien was granted the status of a denizen by letters patent from the Crown. Once a denizen, the individual was entitled to purchase land only – the inheritance of land was still prohibited. Additionally, only children born after the individual became a denizen would be entitled

HOW TO...

...Trace Afro-Caribbean, African and Asian ancestors

The best resource is www.moving here.org.uk (see page 375). In addition, The National Archives provides an online exhibition of sources at www.nationalarchives. gov.uk/pathways/blackhistory.

Civil registration and parish records in the West Indies can be used to track ancestors back to the emancipation of slaves in 1833, whilst the slave registers and other sources at The National Archives and relevant local archives can be used to work further back in time.

Two good websites are The Institute of Commonwealth Studies, www.sas.ac.uk/ commonwealthstudies, and www.rootsweb.com, with links to the Immigrant Ships Transcribers Guild.

The Black Cultural Archives are at 378 Coldharbour Lane, London. Rrecords of the British in India are housed at the Asia, Pacific and Africa Collections at the British Library. You can also use www.ozemail.com.au/~clday for advice on how to research ancestors in India, South East Asia and parts of Africa.

▲ Naturalization certificate for
Joseph Conrad, 1886.

to British citizenship. Records of denization are to be found at The National Archives, in series C 66 enrolled within the patent rolls. A supplementary patent roll in C 67 also contains further records. Both series date to the thirteenth century and the early denization records, prior to 1509, have been transcribed in the *Calendar of Patent Rolls*. This publication does not index individuals by name, but searches should be conducted under the headings 'denizations', or for the early rolls, 'Indigenae'. Indexes are also available for the later period, from 1509 onwards to 1800, published by the Huguenot Society (see page 368). The Huguenot Society's indexes can be viewed at The National Archives and other large reference libraries.

Naturalization Records

Naturalization was a more costly option than denization and, therefore, even less popular. This was especially the case prior to the passing of the Naturalization Act of 1844 as prior to that date it was only possible to become naturalized through a private Act of Parliament. Additionally, anyone seeking naturalization had to take Holy Communion in the Anglican tradition (thereby excluding nonconformists and Jews). However, once naturalized the individual would enjoy the full rights awarded to natives.

The House of Lords' Record Office has the original acts from 1409. The National Archives has an incomplete set of Private Acts in its series C 65. Additionally, KB 24 and E 196/86 contain records of naturalization granted to all foreign Protestants who took the oaths of allegiance and supremacy in open courts. This enabled people to become naturalized without having the need of a Private Act of Parliament. Once more the Huguenot Society has published the information found in these oath rolls.

After 1844 naturalization became the responsibility of the Home Office and, as it was now a far cheaper option, it rose in popularity. The Home Office would only allow naturalization after the candidate had been satisfactorily investigated. The documentation produced by these investigations is part of the naturalization papers records that are held at The National Archives. They can be found in the following series:

- HO 1: 1844 to 1871
- HO 45: 1872 to 1878
- HO 144: 1879 to 1934
- HO 405: 1934 to 1948

USEFUL INFO

Naturalization records from 1922 onwards are subject to closure rules of a maximum of 100 years. However, due to the Freedom of Information Act the Home Office will review the closure status of each file upon request.

The above series refer to the naturalization papers and not the certificates themselves. The papers do contain a great deal of interesting personal information but researchers may also wish to view the certificates. The system of enrolling certificates in close rolls was used until 1873. The certificates were also kept in the series HO 334 for the years 1870 to 1969. The series is organized by certificate number. Up until 1934 the naturalization papers mentioned above should also give the certificate number, and finding the certificate is a straightforward task. Alternatively, there are indexes to the certificates in The National Archives. The Home Office chose not to keep any certificates after June 1969, although the indexes do continue after that date.

▲ Details from the Home Office case file for Conrad's naturalization application, includes a police statement about his suitability.

Further series may be of relevance. The scale of European immigration to the UK in the 1930s and 1940s led to a proportionate increase in those seeking naturalization. The papers for a few famous people can be found in HO 382. The remaining surviving documents are in HO 405. Unfortunately, this series does not include every individual who sought naturalization as the Home Office destroyed the majority of the records. A 40 per cent sample survives, some of which is still with the Home Office. The National Archives has all records surnamed with the initials A to N. As mentioned, these records are subject to closure rules.

Losing British Citizenship

As well as being granted British citizenship immigrants could also lose the same right, either voluntarily or involuntarily. The Crown had expelled aliens at various points in time. Additionally, the Aliens Act of 1905 gave the government the power to deport aliens if they did not comply with the regulations in the Act. Those who were deported between 1906 and 1913 are recorded in series HO 372. The series includes names of deportees, nationality, reason and date of deportation.

Additionally the 1870 Act of Naturalization gave the right to children born of naturalized parents to renounce their citizenship. Such a procedure was known as a 'declaration of alienage' and records can be found in series HO 45, HO 144 and HO 344/259.

USEFUL INFO

Suggestions for further reading:

• Immigrants and Aliens *by R. E. Kershaw and M. Pearsell (The National Archives, 2000)*

CHAPTER 23

Migration: Emigration

The previous chapter outlined the means by which you can trace any ancestors who moved to Britain from overseas countries – but the journey was not just one way. Millions of Britons left these shores both voluntarily and at the forcible insistence of the government (mainly convicted criminals), and so a sudden disappearance of an ancestor from your family tree might be explained by looking into the many emigration records that are described in this chapter.

Historical Background to Emigration

> *Hundreds of thousands emigrated in search of a better life.*

When researching your family tree you may come across an ancestor who migrated from Britain and Ireland to settle in another country. People elected to migrate abroad for a variety of reasons. Many went in search of a better life to escape poverty, others to flee religious persecution; some went unwillingly, convicted of a crime and transported to one of Britain's colonies. Although the Industrial Revolution increased employment in urban areas through the nineteenth century, many people in rural areas found their livelihood threatened as the mass production of textiles replaced many rural cottage industries. One way to escape this was to emigrate. Indeed, many poor emigrants may have been given state assistance when seeking a new life abroad.

The most popular locations where Britons chose to start a new life

▲ Emigrants to the US landing at Ellis Island, c.1900.

were North America, Australia, New Zealand and Southern Africa. After the end of the First World War the British government began officially to support migration. In 1919 a scheme was introduced to aid the migration of ex-servicemen and in 1922 the Empire Settlement Act was introduced, providing support for families to migrate to the dominions. During this period the government felt that the British race as a whole would benefit from sending people away from overcrowded cities to a better quality of life and helping single women to find husbands in dominions where women were a minority. Another less known policy of migration adopted by the British government was the child migration scheme to Australia, South Africa and Canada, popular from the late nineteenth century onwards. This will be discussed in further detail below.

It is estimated that many hundreds of thousands emigrated from the British Isles since the beginning of the seventeenth century. One of the first and most famous emigrant settlements was that of Jamestown, founded in Virginia, North America, in 1607. A group of 108 settlers left from London and established the first successful permanent English settlement in the New World. Soon after that date many more groups of English settlers began arriving and colonizing the New World and also the islands of the West Indies.

USEFUL INFO

Many of the early New World settlers were Puritans trying to find a home to practise their religion without fear of persecution. The most famous example of this was the arrival of the Mayflower *on Plymouth Rock, Massachusetts, in 1620, carrying a group of Puritan pilgrims.*

Thousands of emigrants arrived in the New World during the course of the seventeenth century, seeking a better life. The numbers did not decline after the American War of Independence in 1776 and the next century saw further large waves of migrants arriving from all parts of Europe, including Britain and Ireland. Indeed, the ties between the United States and Ireland have remained strong to the present day as many Americans are still aware of their Irish ancestry. Migration from Ireland was particularly strong in the nineteenth century as poverty was so rife and, in particular, many people left to escape the devastating effects of the Great Irish Famine in the 1840s. It is estimated that approximately 1.5 million Irish migrants moved to the United States and Canada in the ten years following the famine. The population of Ireland declined steadily from the 1840s due to migration, and this situation was only reversed in the 1960s. The success of the Industrial Revolution in Britain meant that the level of emigration was less severe but the numbers were still substantial. Some estimate that around 2.4 million English people emigrated between 1551 and 1851 and a further 3 million Britons migrated between 1870 and 1920 alone (although approximately 25 per cent of these later migrants returned to Britain for a variety of reasons).

The other popular destinations for British and Irish migrants were Australia, Tasmania (originally Van Diemen's Land) and New Zealand, although initially many arrived there against their will through the 'transportation' system. Indeed convicts were also transported to North America until the War of Independence in 1776, although the numbers were far smaller than those transported to Australia.

Settlement in Australia didn't begin until the late eighteenth century. The British government decided to use Australia as the new penal colony (after losing America) and the first fleet arrived on 26 January 1788 (the date now celebrated as Australia Day). The fleet contained 1,500 new settlers, half being convicts. They landed in Sydney Cove and on 7 February 1788 the British colony of New South Wales was formally declared. Large-scale migration to Australia began from this time and the numbers arriving increased rapidly after 1815, when government policy actively encouraged settlement by ensuring free settlers could arrive and purchase land at minimal costs. The discovery of gold in the 1850s was another important factor behind the decision to move to Australia. Convicts continued to be sent to Australia until the system was abolished in 1868, by which time over 150,000 had been sent to Australia and Tasmania (around 30 per cent of whom were Irish). The numbers of free settlers leaving Britain and Ireland were much larger, however, and continued into the twentieth century.

Emigrant Passenger Lists

The Board of Trade retained outgoing passenger lists on a similar basis to incoming lists discussed in Chapter 22. They are found in The National Archives series BT 27 and were kept from 1890 to 1960, detailing the name of each passenger, their age and occupation. The series is currently being transcribed and made available in digital format online at www.findmypast.com. At the time of publication all lists had been placed on the website from 1890 to 1939. The digitization of this series has greatly facilitated its use as it is now possible to search for an individual by name. Previously it was only possible to find an individual if the name of the ship he or she travelled on was known. It is possible to search the index free of charge, but viewing the data will incur a cost. The series covers the years of mass migration between 1890 and 1914. It was only after 1914 that migration began to be controlled on a stricter basis.

British emigration to New Zealand also began in the nineteenth century, after the Treaty of Waitangi was signed in 1840. Initially, emigration to New Zealand was less popular than to other parts of the world and the New Zealand Company had to actively encourage migration to the islands in rural England and Scotland by promoting New Zealand and offering free passage to some skilled workers. Economic assistance to encourage migration was also provided by the territorial governments of New Zealand from the 1850s onwards, resulting in a large increase in settlers arriving from this period onwards, although the numbers arriving were in tens of thousands not hundreds of thousands. The numbers dropped towards the end of the nineteenth century due to an economic depression in New Zealand, but they recovered at the beginning of the twentieth century when further groups of British and Irish migrants arrived with the assistance of the British government.

It may be possible to trace a migrant, depending on when they left and where they chose to go, through many British archives and the archives of the destination country.

Emigrant Records Found in the United Kingdom

The majority of records that list those emigrating through the centuries can be found in The National Archives. They are scattered amongst records found in the Treasury, Board of Trade, Colonial and Home

Records of Passports

Passports were not required as an official travel document in Britain until 1914. Hence, prior to that date, only a very small minority applied for such documentation even though the first records of those applying for passports begin in the sixteenth century. The earliest passports were used as licences to travel abroad and relevant records are found in The National Archives. Series E 157 contains registers of people applying for such licences from 1572 onwards; SP 25/111 has lists of passes issued during the Cromwellian Interregnum (between 1650 and 1653). Calendars of State Papers for the eighteenth century also include relevant records.

Office series. Unfortunately, there is no single index to these many records and reference may have to be made to a number of different records.

If you are researching an emigrant during the nineteenth century there are certain sources found within the Colonial Office records that may assist with research. A number of series specifically relate to emigration departments of the Colonial Office:

- Emigration Original Correspondence, 1817–96, in CO 384. This series includes letters and other documents from individuals who had settled or wished to settle in the popular destinations of Canada, the West Indies and Australia, along with other colonies.
- Emigration registers specific to North America are in CO 327 (1850–63) and CO 328 (1864–68).
- A Land and Emigration Commission was established in 1833 in order to provide assistance to would-be emigrants by free passages and land grants; the records are in CO 386 (1833–94).

Other records in The National Archives that may be of use are those created after the Poor Law Amendment Act of 1834. This law also allowed for parishes to assist their poor parishioners to emigrate. Records can be found in MH 12 (1834–90), arranged alphabetically under county and the new poor law unions. Hence it is only feasible to use this series if the name of the union is known. Additional information can be found in MH 13/252 (correspondence from 1853 to 1854 between the Colonial Land and Emigration Commissioners and the General Board of Health) and MH 19/22 (Poor Law Authorities and Emigration Commissioners correspondence between 1836 and 1876).

Transportation

As is well known, many British and Irish convicted of criminal offences were punished by being transported abroad, firstly to North America and later to Australia. At the time this was seen as a humane form of punishment, saving convicted criminals from the death penalty.

The starting date of this penal system was 1615, with criminals being transported to North America or the West Indies for a number of years to work on the new plantations, although many would not return. It is believed that approximately 150,000 British and Irish individuals were transported to North America, up until the American Revolution in 1776. If you find an ancestor who met this fate you may be able to trace his conviction and subsequent transportation. For details concerning researching in surviving criminal records turn to Chapter 27. However, it is also possible to ascertain details relating to the actual transportation. Fortunately, there is a large amount of information that has been published.

The best place to begin is by consulting Peter Wilson Coldham's book *The Complete Book of Emigrants in Bondage, 1614–1775*. Coldham has compiled lists of all those men and women who were transported using the available contemporary sources held at The National Archives. Most usefully he details at which court or session the individual was convicted, allowing further research into these sets of records. The name of the ship the convict was transported on is also listed, where known.

After 1776 an alternative destination for transporting convicts had to be found, and by 1787 Australia had become the chosen destination. As mentioned, the first group of convicts arrived in January 1788 and many more continued to arrive until the system was abolished in 1868 (although few arrived after 1857). During this time approximately 160,000 convicts were resettled from Britain and Ireland.

Records of those transported can be found in the UK and also in Australia. Unfortunately, unlike the records for North America, there is no uniform comprehensive published index for these sources. Rather, it is only feasible to search the original records if you are aware of the trial details (where and when the convict was tried) or the name of the ship they sailed on. However, there are useful published sources for specific transportation records:

- Lists of the very first convicts arriving in the first and second fleets have been published by P. G. Fidlon and R. J. Ryan, *The First Fleeters*

> *Transportation was seen as a humane alternative to the death penalty.*

(Sydney, 1981) and R. J. Ryan, *The Second Fleet Convicts* (Sydney, 1982) respectively.

- David T. Hawking's *Bound for Australia* (Chichester, 1987) is a useful overview of the subject and contains many transcriptions of relevant records.
- The Genealogical Society of Victoria has compiled an index to the New South Wales Convict Indents and Ships, detailing the names of the convicts who arrived in New South Wales and Van Diemen's Land between 1788 and 1842. This is available in microfiche form at The National Archives.

Various censuses and musters were taken in Australia by the British government during the early nineteenth century and many of these have been published and can be consulted at The National Archives and other large reference libraries. If you locate your ancestor amongst these sources the record should also provide conviction details and which ship your ancestor arrived on. These are the key details needed for consulting original records for transportation, such as the volumes of transportation registers found in The National Archives series HO 11 (arranged by ship name and date of travel) or the trial records themselves.

Trial records are very formal in nature and provide limited information of genealogical relevance. Often loved ones of the convict would petition

▼ A group of children at a Barnardo's orphanage in Liverpool have their luggage inspected before emigration to Canada, 1929.

the government for mercy. Additionally, some wives would request to join their convicted husbands in the colonies. Such records give very useful insights into the personal circumstances of the convict and can be invaluable in bringing the past into life. These petitions are also found in The National Archives, in the Home Office and Privy Council series.

Records for those being transported from Ireland are to be found in the National Archives of Ireland. These include transportation registers and petitions similar to those described above. Both these sets of records have been transcribed and can be searched online on the National Archives of Ireland website, www.nationalarchives.ie. Unfortunately, the records are incomplete due to the destruction of many Irish records in 1922 and transportation registers for before 1836 have not survived to the present day. Nevertheless, the database does contain records for those convicts who had petitions made for them before 1836. The database provides the name of the convict along with trial and crime details and can be used free of charge.

Another useful website for researching those transported to Australia can be found at www.convictcentral.com. The site gives guidance on how to research convict ancestors, with transcriptions of records of some ships transporting convicts to various parts of Australia and full information on the musters taken in Australia during the early nineteenth century, whether they have been published and where the original records survive. It is a very useful website and can be accessed free of charge.

Assisted Migration

Another major way people emigrated was by being 'assisted', usually financially. A large number went to North America and the West Indies as indentured labourers. Such individuals were given financial assistance in travelling and living in the New World, agreed to work on a plantation for a number of years and were given a plot of land after they had completed their agreed period of work. This was a popular system, with 200,000–300,000 indentured labourers arriving in the seventeenth and eighteenth centuries. A number of parishes would also sponsor such a system for their paupers as a means to relieve the burden of supporting the poor in the parish. The Poor Law Amendment Act of 1834 allowed poor law unions to give financial and material assistance to those emigrating, and similar assistance was also provided by the Commission of Land Emigration (both mentioned above).

> *Parishes would help paupers migrate to relieve the burden of supporting them in the parish.*

Child Migration

Another type of emigration was by poor or orphaned children to various parts of the colonial Empire. The origins of this system can be traced to the seventeenth century. The first group of 100 children from the Christ's Hospital School were sent by the City of London to assist the Virginia Company populate the new colony. The records for this school are now at the Guildhall Library and approximately 1,000 children were sent from this school between 1617 and 1775. Their names can be found in *The Complete Book of Emigrants in Bondage, 1614–1775*.

Child migration was used increasingly during the mid-nineteenth century onwards, thanks to the Poor Law Amendment Act of 1850 which allowed Boards of Guardians to send children abroad. Additionally, it was during this time that many charities, various church bodies and philanthropic organizations were established (such as Dr Barnardo's), and many of these institutions opted to send pauper children abroad to help them establish a better life. The process continued well into the twentieth century; between 1922 and 1967 approximately 150,000 children migrated to Canada, Australia, New Zealand and a small number to Rhodesia with the idea of populating the colonies with 'good British white stock'. Details of how to trace these migrants will be discussed under the individual destination countries below.

This policy was often very unsettling for these children, displaced from their familiar home country and sent to live in faraway places, where they would often be treated as little more than cheap labour. Some child migrants are still living in their new countries and an organization dedicated to helping migrants trace their roots was established in 1987 – the Child Migrants Trust. Further details of the aims of the trust can be found at their website, www.childmigrantstrust.com.

Records for those leaving under indentured labour can be found at The National Archives and also at the London Guildhall Library. A good book detailing assisted migration to North America and the West Indies is Coldham's *The Complete Book of Emigrants in Bondage, 1614–1775*, mentioned above. General sources relating to the colonial governance of North America and the West Indies, found in The National Archives (State Office, Colonial Office and Privy Council calendars), will contain information of relevance. Local parish records may also have information amongst their vestry minutes about individuals granted assistance to emigrate. These should be held at the local archive or record office.

Voluntary Migration/Free Settlers

As mentioned, many individuals chose to leave on a voluntary basis, attempting to establish a better life in far and distant places. Relevant records for such migration can be found in various archives in the UK and also in many different repositories in destination countries. Below is a guide to the key sources that may be consulted for the most popular destination countries for such migrants.

Records for Migrants to North America

Several states in the present-day United States and Canada made up one of the earliest colonies acquired for the British Empire. As such they were the earliest destination countries of both voluntary and involuntary migrants. Some general sources have been discussed earlier. The earliest colony to be inhabited was that of Virginia, where thousands of people were required to work on the plantations. Emigration was encouraged by using the 'headright' system, whereby people migrating would be given 50 acres of land if they funded their own arrival, and a further 50 acres if they paid for another migrant to arrive. This system enabled plantation owners to increase the size of their estates as most of the migrants would have had their travel paid for by the plantation owners. As such they would be indentured emigrants (see above). Nevertheless, below is a list of various records to assist in tracing free settlers during the colonial period.

- The calendar of State Papers Colonial, America and West Indies, 1574-1738, relates to all matters of administration of the colonies. It is available on CD ROM at The National Archives and other large reference libraries. It is indexed by name, making it possible to search the collections for an individual migrant.
- Various publications also provide lists of emigrants to North America and are worth viewing:
  - ➤ C. Boyer has edited four books that include lists of migrants: *Ships' Passenger Lists: The South (1538–1825); National and New England (1600–1825); New York and New Jersey (1600–1825); Pennsylvania and Delaware (1641–1825).*
  - ➤ P. W. Filby and M. K. Meyer (eds.), *Passenger and Immigration Lists Index*, 13 volumes. A thorough list, including approximately 2.5 million names of migrants arriving in the United States and Canada from the sixteenth to mid-twentieth centuries.
  - ➤ J. C. Hotten, *Original Lists of Persons Emigrating to America, 1600–1700.*
  - ➤ David Dobson, *Directory of Scottish Settlers in North America 1625–1825* was drawn from a number of original sources held at The National Archives.

- The National Archives holds a unique and detailed list of emigrants to North America between 1773 and 1776 in T47/9-12. The list provides names, ages, employment, residence and departure information for all those leaving.

Emigrants who left during the later nineteenth and early twentieth centuries can be traced using the passenger lists mentioned above. Additionally, many archives in the United States, such as the National Archives in Washington, contain information on their new settlers.

British colonization of Canada began at the same time as that of the rest of North America, although French settlers had been arriving in small numbers since the sixteenth century. After many conflicts between France and Britain over control of the territory, the French ceded the eastern part of Canada to the British in 1763 and thereafter it became the sole possession of Britain. The Hudson Bay Company was the main British organization involved in Canada, establishing small outposts since its inception in 1670.

Once Britain gained control of Canada, large numbers of emigrants arrived from the British Isles, particularly from Scotland and Ireland, and especially after America gained its independence in 1783. Such migration was actively encouraged by the Canadian authorities up until the twentieth century as a way of populating the vast land.

- The Hudson Bay Company's records include journals of early settlers. Microfilm copies of these records can be found at The National Archives, in series BH 1. The original records can be found in Canada, as are the majority of records relating to immigration to the country.
- The National Archives of Canada has an online exhibition detailing the immigrant experience, titled 'Moving Here, Staying Here: The Canadian Immigrant Experience'. It can be found at http://www.collectionscanada.ca/immigrants/index-e.html and has an online database searchable by name for people arriving between 1925 and 1935. The website also includes a database of passenger lists from 1865 to 1922 that can be searched at http://www.collectionscanada.ca/archivianet/passenger/001045-100.01-e.php. However, there is no name index of this database and it is only possible to search by the name of the ship.
- Another website which has a wealth of information concerning the Canadian immigration experience worth viewing is http://www.ist.uwaterloo.ca/~marj/genealogy/thevoyage.

Canada was also a major destination for child migrants from Britain, with approximately 100,000 such children arriving in Canada in the nineteenth and twentieth centuries. Their arrival may be traced in the Canadian sources mentioned above. Additionally, The National Archives also contains some records relating to this scheme, although they mainly concern general policy and do not contain specific lists of child migrants. They can be found in MH 102, covering the years 1910 to 1962.

Records for Migrants to Australia

Although convict settlement in Australia is infamous, the vast island was actually populated in greater numbers by free settlers. Many free settlers came due to economic factors and the hope of a better standard of life, and the numbers arriving swelled after the discovery of gold in the 1850s. Indeed migration to the country, including the celebrated 'ten pound poms' who were offered assisted passage by the Australian government, was very much encouraged well into the twentieth century, with the 'White Australia Policy' active from the 1890s to the late 1950s.

Along with the general passenger lists that record people leaving the UK and Ireland discussed above, there are other sources specific to Australia that may be of relevance. These can be found in the UK and in various Australian archives.

▲ Emigrating to Australia, 1942 – the celebrated 'ten pound poms'.

- The National Archives has a number of important records for newly arrived Australian migrants. The Colonial Office papers relating to the governance of New South Wales during the nineteenth century all contain names of emigrants. They can be found in series CO 201, CO 202, CO 360 and CO 369.
- The government took various censuses of convicts in New South Wales and Tasmania at various points from 1788 onwards. Although primarily concerned with recording convicts, they would also include those who were not transported. It is possible to find other genealogical data, such as age and occupation of these individuals, in these censuses, found in series HO 10.
- The Society of Genealogists also has a good number of records on new emigrants to Australia, including indexes to birth, marriages and deaths that occurred in the individual states.
- The National Archives of Australia has only a very limited amount of information on migration, as matters of immigration were not federal policy until 1901. Prior to that each of the six states would

> *Australia was populated in greater numbers by free settlers than by convicts.*

control immigration and their archives contain a variety of information on the new arrivals. It is best to contact the relevant state archive to discover the exact nature of the records held.

- The National Archives of Australia (www.naa.gov.au) does hold information for twentieth-century migration, including passenger lists form 1924 onwards, and some of their holdings can be searched online.

Records for Migrants to New Zealand

Another important destination for those choosing to emigrate was New Zealand. Unlike Australia, New Zealand was never used as a penal colony, and those who arrived came by their own free will. The present-day government of New Zealand has placed a very good history of immigration to the country on the website http://www.teara.govt.nz/ NewZealanders/NewZealandPeoples/HistoryOfImmigration/en. This is a very good introduction for anyone interested in finding out the history of immigration to the country. The same economic push-and-pull factors of searching for a better life and the active encouragement of the New Zealand government lay behind this migration. The very first Britons arrived in 1790 but large-scale migration only occurred from 1840 onwards, after New Zealand was declared a crown colony.

A large number of emigrants would have first arrived in New South Wales and it may be possible to trace your ancestor using the same Australian sources as above. After 1839 the New Zealand Company began an active policy of recruiting migrants. The company was in existence until 1858 and its archives are now at The National Archives in series CO 208. The Society of Genealogists also contains microfilm of birth, marriage and death indexes of New Zealand, where civil registration became compulsory from 1848 onwards.

The National Archives of New Zealand (www.archives.govt.nz) also contains records of immigration to the country. These include lists of assisted passengers arriving up to 1890 and passenger arrival lists.

Records for Migrants to South Africa

British and Irish migrants started to arrive in South Africa from 1806 onwards, when the Cape of Good Hope was officially ceded by Holland to Great Britain. Numbers arriving continued as further territories were obtained by the British from the Boers, culminating in the Union of South Africa in 1910. Other than the passenger lists in BT 27 (described above) there is only a very limited amount of information in the UK, found in The National Archives (payments made to Army and Navy pensioners settling there in the mid-nineteenth century held in the WO series, and correspondence files relating to settlers in the CO series), and also in some records at the Society of Genealogists. Although civil registration became compulsory for the entire country in 1923 (and earlier for some different provinces) the registers are not currently open to members of the public. It is best to contact the South African Society of Genealogists for further guidance if you are interested in tracing an emigrant ancestor in South Africa, at www.gensa.info.

Migration to Other Parts of the Empire

The countries that received large numbers of British and Irish migrants have been discussed above. However, many other parts of the British Empire also had people settling in them, many of those individuals involved in the direct governance of the many colonies. One guide for records for those who settled abroad and retained links with the 'homeland' is the book, *The British Overseas: A guide to records of their births, baptisms, marriages, deaths and burials, available in the United Kingdom* (3rd edition, 1994) published by Guildhall Library.

India

One of these territories was the subcontinent of India. There had been a European presence in India since the sixteenth century and the East India Company, established by Royal Charter in 1600, eventually became the effective ruling body of India by the mid-eighteenth century. After the failed Indian Rebellion of 1857, India came under direct rule by the British Crown until the end of the colonial era in 1947. During this period many hundreds of thousands of individuals went to India, employed firstly by the East India Company and later in the employment of the India Office.

A large number of records were created relating to the governance of India, including personnel files of employees and parish records. The East India Company maintained a separate army, which later became the Indian Army, and records for these have been discussed in Chapter 9. Similarly, the births, marriages and deaths of those residing in India

> *Migrants to other parts of the British Empire were often involved in the governance of the colonies.*

CASE STUDY

Alistair McGowan

Alistair McGowan was always curious about his origins, suspecting a European bloodline but uncertain exactly where it appeared in his family tree. He started by asking relatives about his background, knowing already that his father, George, was born and brought up in India. George's birth certificate listed him as 'Anglo-Indian', which Alistair found out from research at the British Library's Asia, Pacific and Africa reading room meant that one of the women in his family was Indian by birth. His quest was therefore to determine when his 'Anglo' family emigrated, and which one married an 'Indian' girl.

Sources revealed that generations of McGowans were born in India, part of the colonial administration of the Raj and, prior to 1857, part of the civil service established by the East India Company that controlled British interest on the sub-continent. A key document was the baptism of Ralph McGowan, which revealed that his father was a magistrate called Suetonius; no mother's name was listed on the document.

Alistair continued his research in India, and sought the advice of local historians who helped him work further back in time. He found a religious pamphlet that indicated Suetonius had married a noble-born Muslim lady, but because she refused to convert to Christianity her name was omitted from the baptism record that Alistair found in London.

Returning to the records at the British Library, Alistair was able to work even further back in time, and discovered that Suetonius's father was also called Suetonius, and was baptized in Bengal in 1775. His parents were John McGowan and Mary de Cruz, and Alistair established from military records, including pension funds, army gazetteers and service papers, that John first arrived in India as a private with the East India Company's army, and worked his way up to the rank of major – quite a rise. But the final surprise lay in the muster books for Fort St George, which indicated that John McGowan had sailed to India from Ireland – not Scotland as Alistair had always assumed. Was John McGowan born in Ireland? The records were inconclusive ...

◀ The McGowan family, including Alistair's great-great-grandfather Ralph McGowan (seated middle row, second from left).

are detailed in Chapter 5. Additionally, the large archive of the East India Company at the British Library contains various documents for individual employees. These can be accessed by using the many indexes available for the various series of the collection, and there is also a card name index held in the Library. The British Library has also placed the card index as a searchable database online at http://india family.bl.uk/UI/. However, this index only contains a limited amount of biographical information and it is also worth visiting the Library in person as many people would not have been included in the card index.

The Society of Genealogists also holds a selection of records for India. In addition, the 'Families in British India Society' is dedicated to British and Anglo-Indian families living in India and can provide assistance to anyone researching their ancestors. They have placed a number of their records online available to search by name. Further details about the society can be found on their website, www.fibis.org.

The Caribbean

The Caribbean was another important colony in the British Empire, and emigration proved popular during the height of the plantation era in the seventeenth and eighteenth centuries. As most emigrants settled prior to the late nineteenth century it is not possible to track these individuals using passenger lists in The National Archives series BT 27. Alternative sources have to be used, many of which are found in The National Archives.

The best starting point for detailed guidance is *Tracing Your West Indian Ancestors* by G. Grannum. The series E 157 has some registers of passengers travelling to Barbados during the seventeenth century. Further information about emigrants can be found in the calendar of State Papers Colonial mentioned above. The records of the governments of the various islands of the Caribbean may also contain relevant information, and these can be found in the CO series in The National Archives.

It is beyond the scope of this chapter to detail migration records for every country that an individual may have migrated to. However, it is possible to find out further details about specific countries online by visiting the country pages found on www.cyndislist.com or www. worldgenweb.org. Both websites contain country-specific guidance. The Society of Genealogists also has collections for a number of different countries and details of their holdings can be found in their online catalogues at www.sog.org.uk.

USEFUL INFO

Suggestions for further reading:

• Emigrants and Expats: A guide to sources on UK emigration and residents overseas, *by Roger Kershaw (Public Records Office, 2002)*

• Britannia's Children: Emigration from England, Scotland, Wales and Ireland since 1600, *by E. Richards (Hambledon, 2004)*

CHAPTER 24

Family Secrets: Poverty and Lunacy

One of the saddest and most poignant discoveries you can make is that one of your ancestors fell into extreme poverty, and spent time in the workhouse; or, worse still, was incarcerated in a mental asylum. When someone fell on hard times, their daily existence was grim. This chapter shows you how to spot the clues that suggest all was not well with the family finances, and follow them into document sources to find out what life was like in the workhouse or asylum.

Poverty was, and still is (though to a much lesser extent) an ever-present factor in the social history of the UK and Ireland. Because of this there were also always means of helping the very poor and destitute throughout the country for many centuries, initially through charitable relief through the Church and the monasteries, later through parish relief and the large-scale introduction of workhouses in the nineteenth century to the beginnings of welfare state in the twentieth century. If your ancestor was unfortunate enough to have suffered poverty, there may well be a record of the support provided to him or her in various sources, depending on the period concerned.

> *If your ancestors suffered poverty, there may be a record of support provided to them.*

Historical Background to Poverty and Poor Relief

The early methods of poor relief during the Middle Ages were provided by the Church as part of the religious obligations of the parish. The monasteries were also involved in providing food and shelter to the poor until their dissolution by Henry VIII in the sixteenth century. The dissolution came at a time when poverty in general was increasing. There was a large increase in population during the sixteenth and early seventeenth centuries. However, this was not paralleled by an equal growth in the economy and, therefore, there was not enough work to sustain this population increase, resulting in a rise in poverty levels.

The State had to provide solutions to this and did so by passing a series of laws from 1597 to 1601, culminating in the Elizabethan Poor Law Act of 1601. The measures introduced in this legislation became the backbone of poor law relief until the mid-nineteenth century. The Act placed the responsibility for poor law relief on the parishes of England and Wales. Each parish now had to raise a sum of money from its property-owning parishioners by levying a 'poor rate'. Two overseers were appointed to collect this sum of money and redistribute it to the needy. They would enter their transactions annually in an accounts book, and these books also detailed the recipients after 1690.

Most relief was in the form of 'outdoor relief' whereby people were given support in the form of money, food or clothing. The system of 'indoor relief' (the precursor of the infamous workhouse, discussed below) was introduced in the eighteenth century as a means of housing the infirm, orphans and the elderly. In 1723 an Act was passed to allow local parishes to build workhouses as desired. The Act also set the foundations of the 1834 Poor Law Amendment Act by allowing overseers to penalize those able-bodied individuals unwilling to enter the workhouse by denying them any other relief (although this was rarely enforced until 1834).

During the seventeenth century a problem arose regarding the implementation of the Poor Law Act – namely parishes had to fund the relief of individuals who were not born in that parish, but who had settled there. These people were seen as a burden on the parish and the situation was addressed by the passing of the Settlement Acts of the late seventeenth century. These Acts allowed two Justices of the Peace to examine those newly arrived parishioners during quarter sessions and remove them if they threatened to be a burden on parish ratepayers. The legislation also laid down the conditions by which an individual could claim 'settlement' of that parish.

Another potential burden on parishes' poor relief funds came from illegitimate children and their mothers. Although many couples in that situation would succumb to pressure to marry from family or the parish itself, still a number of children were born out of wedlock. Overseers in the parishes were keen to establish who the fathers of illegitimate

USEFUL INFO

The 1601 Poor Law Act stipulated that only the 'deserving poor' should be entitled to support and it was up to the parish overseers to decide who the deserving poor were (usually the old, sick or disabled). Those who were able bodied were helped by being given employment, and those seen as 'idle' (such as beggars) were disciplined by whipping. Orphaned children were placed into apprenticeships.

▼ A picture of poverty: an elderly woman cares for a child, 1877.

USEFUL INFO

*Further details about life in the
workhouse and the conditions there
can be found on the following
website, www.workhouses.org.uk.
If you have a strong interest in
the workhouse then visiting one
of the workhouse museums is
recommended. Many have been
preserved throughout the country
and they can offer first-hand
insight into the conditions at
the time.*

children born in their parish were, to avoid the mother and child claiming poor relief. They did this by a number of means. In 1576 an Act was passed giving JPs the power to investigate who was the father of an illegitimate child that may need parish support and demand maintenance from the father. These were known as 'bastardy examinations' and JPs would quiz the mother to discover the name of the father. Once this information was retrieved the local authorities could issue a 'bastardy bond' demanding the father provide his child with financial support. Women were also put under pressure from parishes to name the father, sometimes under oath. In 1610 an Act gave parishes the authority to send women to a House of Correction for producing such offspring. Another Act was passed in 1732 forcing women who were not married to declare they were pregnant and to name the father.

Original Records for Poor Relief

As historically the burden of poor relief fell on local parishes and authorities, surviving records will mostly be at county record offices. The National Archives does have some material relating to the administration of the Poor Law Unions, although these documents are of limited use to the family historian.

Early Poor Relief Sources

Prior to the enactment of the Elizabethan Poor Law any records relating to the care of the poor would be found in the vestry minutes of the parish, which cover all aspects of local life, not just poor relief. Vestries were the equivalent of church or parish councils who met on a regular basis to discuss parish matters including poor relief, church repairs, schools and other local issues. The council would be made up from the male ratepayers of that parish. All surviving material of these meetings will be in the relevant county record office. Some vestry minutes survive as far back as the sixteenth century although this is not the case for every parish and it is best to check with the archive concerned for survival rates.

Elizabethan Poor Law Records

As stated, legislation passed in Elizabethan times created a system of poor relief by making this the responsibility of individual parishes. The 1601 Act allowed parishes to raise a poor rate tax on parishioners. Two overseers were appointed to manage the funds raised and distribute them accordingly. The system lasted until 1834 when it was replaced by new legislation. It is surviving documents of the overseers that form the bulk of the relevant records for family history purposes.

- The first set of records that are worth consulting is the *account books* of the overseers. As overseers raised the appropriate funds on an annual basis they were obliged to list all their transactions at the end of the year in an account book. After 1690 these accounts also had to include the names of those who was being given poor relief. The records would also indicate who had been deported from the parish to their original settlement parish. Additionally, they would list which parishioners were paying the poor rate. These accounts form a very useful indicator as to the social circumstances of your ancestor, whether they were poor enough to need financial support or, on the other hand, wealthy enough to pay poor relief and, therefore, have some social standing within their parish (the amount paid depended on the value of the property the individual occupied). The level of detail in the records varies from parish to parish depending on the fastidiousness of the overseers or the size of the parish. These books can now be found in county record offices and some may also have been published.
- Another revealing source worthy of consultation is the *bastardy examinations*, especially if you come across an illegitimate forbearer (not overly uncommon). As mentioned, illegitimate children and their mothers could become a financial burden on the parish as they would often need poor relief, and it was in the overseers' interest to ascertain who the father was. Justices of the Peace had the authority to question such mothers and demand maintenance from the father by issuing 'bastardy bonds'. If the father refused to pay he could be sent to gaol until he did. The mother herself may have provided evidence as to who the father was so as to obtain such a bond. The surviving source records of this process are the bastardy bonds themselves, maintenance orders and entries in the vestry minutes or quarter sessions of the JPs. The records are particularly useful as they may mention the name of the father, which may have been omitted in the child's baptism record at the local church. Any surviving records will be found in the county record office.

Additional documentation may be found by using the surviving records of the 'settlement examination' system, introduced in the late

> *Historically the burden of poor relief fell on local parishes.*

The settlement system had detrimental effects for labour mobility. Workers would be reluctant to leave their own parish due to the difficulty in obtaining settlement in a new parish and the practice of home parishes not acknowledging their responsibility towards returned parishioners. This was altered in 1697 by parishes issuing certificates to those finding work outside their native parish. The certificates proved that the holders had legal settlement in the parish issuing the certificate and this parish would accept their responsibility to provide poor relief, if necessary.

seventeenth century. Poor relief was only granted to those were legally settled in the parish. Any newcomer entering the parish would only be granted legal settlement if they met certain criteria, and these rules were applied strictly. Parishes had the authority to expel 'illegal' settlers and JPs would examine newcomers in detailed interviews as to their eligibility for settlement and establish the home parish of poor newcomers in quarter sessions. Parish officials were particularly keen to ascertain the legal settlement of pregnant unmarried women so that they could be sent back to their home parish and not become burdens on their own parish. The JPs had the authority to expel individuals by means of issuing a 'removal order'.

The enforcement of this system has led to a great deal of useful sources for the family historian. Surviving settlement examination records may be found within the vestry minutes or the quarter session records, all found in your local records office. The records themselves include useful details of those seeking settlement, such as birthplace, age and occupation. If your ancestor was the examiner then that would also be an indicator of his social standing within the parish. The record offices may also have the removal orders or settlement certificates of your ancestors.

Pauper and orphaned children were often forcibly entered into apprenticeships by the relevant authorities so that they would not be

Poor Law Amendment Act at The National Archives

The National Archives has two main sets of records regarding this Act: staff records for those employed at the workhouse; and correspondence of the Poor Law Commission and Poor Law Board. These are to be found in The National Archives series MH (Ministry of Health).

● There were many people employed in running workhouses apart from the guardians, including cooks, teachers and nurses to name but a few. Their staff records can be found

in MH 9. The series comprises the paid registers of officers organized by county and then Poor Law Union for England and Wales. They should include date of appointment and resignation along with amount paid.
● If your ancestor was a senior officer then he or she should be included in the publication *Shaw's Union Officers; and Local Board of Health Manual* (published between 1846 and 1921).
● MH 12 is a large series of a whole range of documents (approximately

16,750 bound volumes of papers), including personal files for 1834 to 1900 (later files were destroyed by fire). However, the series is arranged by county and union and there is no index to individuals that may be mentioned in the series.
● MH 12 also contains the bulk of the documentation of the Poor Law Commission in general. Most of the material contains correspondence between the Commission and the Guardians discussing a range of matters for the administration of the workhouse. Although it is difficult to find information about individuals in this series, the

the responsibility of the parish. Children as young as seven were apprenticed for a minimum of seven years, often to be exploited as cheap labour by their masters. Indentures were signed by the parish overseers and the master formalizing the apprenticeship and the duty of the master. These indentures would then be sent to the JP for his approval and signature. Parish overseers would be eager to find a master outside the parish as an apprentice would be legally settled in the parish of his master after 40 days. These indenture records may also be found in the appropriate local archives (although many have not survived to the present day). The indentures would include the name and age of the apprentice (and occasionally his father), the trade and name of the master, along with the names of the relevant parish authorities. (There is more on the system of apprenticeships in Chapter 21.)

It should also be remembered that a minority of those seeking poor relief were encouraged to emigrate to seek a 'better life'. Parish authorities would pay for the passage of those encouraged to emigrate so that they would no longer be a burden on the parish funds. Emigration records have been discussed in detail in Chapter 23 but any decision to fund the emigration of poor parishioners would have been discussed in the vestry minutes or quarter session records and the payment should have been logged in the account books.

documents are of great use in researching the general conditions of workhouses. The series is indexed by MH 15 although the index is a subject and not a name index. Additionally many of the documents are in a poor state.

• MH 14 contains Poor Law Union plans and drawings of workhouse buildings, which may be of interest.

• MH 32 contains records of the reports made by inspectors of the workhouses. However, the series is not organized by the Poor Law Union but by the name of the inspector, making a search for a workhouse problematic unless you are aware of which inspector compiled the original report.

Another good source for researching the conditions of workhouses and the poor is the Parliamentary Papers, as this topic was often discussed in Parliament. Numerous royal commissions and parliamentary committees were established through the course of the nineteenth century and into the twentieth to ascertain the conditions of the poor. The most important of these were those carried out in 1832–4 and in 1905–8. These can be viewed along with the evidence supporting them. Additionally, Poor Law Commissioners were required to submit an annual report to Parliament explaining their activities in that year, and these reports can also be viewed.

Parliamentary Papers can be found in the Parliamentary Archives in Westminster. Alternatively, consult the CD ROM index available at The National Archives to lead you to the appropriate document order reference. Further details can be obtained in the appropriate research guide on The National Archives website.

Poor Relief in the Nineteenth Century

The beginning of the nineteenth century saw an increase in urban and rural poverty, with another large rise in population along with the rapid social changes brought on by the Agricultural and Industrial Revolutions. This was displayed by the increase in spending on poor relief: in 1800 parishes were spending approximately £4 million but this had risen to around £7 million by 1830. The poor relief system outlined above was becoming ineffective in dealing with this new pressure and changes were needed. Additionally, many people also believed that the system supported people unwilling to work as they could always claim poor relief. Hence, the existing system of poor relief was superseded in the

nineteenth century by the Poor Law Amendment Act of 1834.

The Act was passed following the recommendations of a royal commission, instigated by the government in 1832 to investigate the situation of poor law relief. This new Act drastically changed the nature of poor relief and, from this point forward, any able-bodied person seeking assistance would have to enter a workhouse, although they could also leave the institution at their own request. However, it was still possible for the elderly and sick to receive 'outdoor relief'. These workhouses were purposely designed to be as uncomfortable as possible to ensure that people would only enter them as a last resort and would,

therefore, be more willing to find work and not be a burden on the parish. Parishes were required to group together to form Poor Law Unions and provide workhouses. Each new union would elect a Board of Guardians who would now oversee poor relief for that particular union and would be elected by local ratepayers paying parish poor rates. Parish overseers were responsible for collecting the parish rates until 1865, when a new union rate was introduced. The Board of Guardians reported to the newly created Poor Law Commission based in London.

The workhouse was a prominent part of the Victorian social landscape until the introduction of state welfare benefits at the

beginning of the twentieth century by the Liberal Government (old age pensions were introduced in 1908 and unemployment benefit in 1911, thereby reducing the need for poor relief for the majority of the population). Conditions of workhouses would vary from relatively clean and adequate facilities to institutions serving as little more than prisons for the destitute. Along with the poor and those too ill to work were unmarried mothers or pregnant women (who may have been ostracized by their families) and the mentally ill (as separate asylums were only established after the mid-nineteenth century; see below).

Admission to a workhouse would be granted after the individual had been interviewed either by the Relieving Officer or, in cases of emergency, by the Workhouse Master and authorized by the Board of Guardians. Upon entry the individual would have a medical examination, be stripped and bathed and given a workhouse uniform. Men and women would be separated (including married couples) and had to live in separate quarters, sleeping in large dormitories. Their daily routine was strictly regimented by the Poor Law Commissioners, including waking and sleeping times, meal times and working hours. The work involved was mainly helping run the workhouse itself, with women doing more domestic tasks and men working on menial labouring. Punishments meted out to inmates who broke the strict rules varied in harshness; lesser offences were punished by withdrawing food luxuries (such as cheese or tea) while more serious offences could result in inmates being whipped or sent to solitary confinement or prison. Over the course of the nineteenth century, however, conditions in the workhouses became less harsh and regimented.

By the beginning of the twentieth century the use of workhouses declined. Guardians were more willing to give 'outdoor relief' as it was more cost effective and less cruel. In 1929 the institutions were abolished as new social legislation provided other means for assisting the poor and needy, although the Poor Law Amendment Act itself was not abolished until 1948 . The buildings themselves were sometimes used for other institutions including, in many cases, hospitals and schools.

Records of the 1834 Poor Law Amendment Act and the Workhouse

Workhouses became a more popular method for parishes to deal with their poor during the eighteenth century. The workhouse became institutionalized in 1834 with the passing of the Poor Law Amendment Act; henceforth, workhouses would be the only option for able-bodied individuals seeking poor relief. Parishes were grouped together to form Poor Law Unions and a new body was created to deal with the administration of the workhouse, the Board of Guardians.

You may discover that your ancestor was in a workhouse by obtaining their civil registration death certificate (or, worse, their birth certificate). Additionally, as workhouses were used mostly in the nineteenth century, census returns would also detail if your ancestor was residing in one. If you wish to trace an ancestor who was admitted into a workhouse you need to establish which Poor Law Union the parish your ancestor resided in belonged to. This can be done by referring to the

◄ People queuing at South Marylebone Workhouse, c.1900.

appropriate guide to their location published by the Federation of Family History Societies (and listed in the suggestions for further reading at the end of this chapter). These guides detail what records survive and for what dates for each union. The records themselves will be in local county record offices as part of the collection of the Board of Guardians of the workhouses. The local family history society may have produced indexes to some of these records and it is worthwhile checking this.

The main sets of workhouse records retained by local archives include:

- *Admission and Discharge Books:* This should include the name, age, residence (including parish of settlement) and occupation of the pauper. The books should also detail the circumstances surrounding the need of relief. Unfortunately, these books are rarely indexed.
- *Creed Registers:* These list the religious denomination of the individual. They are arranged alphabetically by first name and then by date of entry into the workhouse. The entry may also include date of discharge.
- *Registers of baptisms, births and deaths:* Guardians would note all these events in their registers, although recording burials was less common.
- *Records of Relieving Officers:* Relieving officers would interview and decide whether to admit an individual to the workhouse or to provide 'outdoor relief'. The surviving records for the latter can give a good background to the personal circumstances of that person.
- *Poor Law Union records:* Miscellaneous set of records that may include correspondence and minutes of the guardians of the workhouse. Administration records for the workhouses may also survive (including books detailing expenditure, rates and punishment of inmates). The system of making children apprentices continued, and records of such activities may survive. Additionally, the new guardians continued to sponsor the emigration of paupers, and documentation relating to this may also survive.

As with many other topics it is also worthwhile consulting local newspapers as they would often report the weekly meetings of the Board of Guardians and what the guardians recommended for individuals.

CASE STUDY

Jeremy Paxman

Jeremy Paxman stumbled across various examples of poverty whilst researching his family tree – not just among his Paxman forebears in Framlingham, in the heart of rural Suffolk, but also in his maternal lineage in turn-of-the-century Glasgow.

The Suffolk Paxmans were victims of the Industrial Revolution, as Jeremy discovered once he'd pushed his family tree back to the early nineteenth century by combining certificates, census returns and – eventually – parish records. Generation after generation had Yorkshire roots, until he found one family whose head, Thomas, was born in Framlingham, Suffolk, and from the places of birth listed on the 1851 census it seems he had then moved to the North West before migrating across to Yorkshire. Having obtained important geographical clues as to the rural origins of his family, Jeremy visited the county record office to discover more.

From census returns, it was apparent that Thomas Paxman had earned a living as a cobbler, but on reading about the social and economic conditions in Suffolk at the time Thomas was bringing up his family in the 1830s, Jeremy discovered that work was scarce and many people who could not support themselves on the land or through paid employment were facing starvation. Documents revealed that the local parish councils set up a repatriation scheme, and offered to pay for people to leave the area and find their fortune in the newly emerging industrial towns – a lifeline that Thomas gratefully seized.

Yet, as Jeremy found out when he researched his maternal line,

poverty was not just confined to the countryside. His maternal 2 x great-grandfather, John McKay, had served with the Royal Artillery – his Army records were located at The National Archives, Kew; but on discharge in 1891 he moved to Glasgow where he lived with his wife, Mary, and nine children, working as a school caretaker – but on checking the surviving school records, Jeremy discovered that John had died in 1894. Since John had not served long enough, Mary did not qualify for an Army pension, and without any further means to support the family she had to apply for poor relief.

Jeremy discovered documents at the Mitchell Library, Glasgow, which revealed that Mary's application was turned down because two children that she claimed were John's were actually born after he had died – and were therefore illegitimate. On these grounds, the parish council refused to grant her aid. To get a better idea of the conditions she had to face on a daily basis, Jeremy tracked down the tenement where Mary was forced to raise her family in cramped, squalid conditions and very little money. No wonder he was moved by what he saw.

◀ Jeremy's great-grandmother, Mary McKay.

Poor Relief in Scotland

Scotland passed a series of laws from the fourteenth to the sixteenth centuries to legislate for the relief of the poor, mainly as a means of dealing with beggars. Only a small number of relevant records survive to the present day. Records begin when parishes became responsible for helping the poor. This occurred following the Reformation and responsibility fell on the Kirk and the local landowners (known as heritors). The Kirk would raise revenue to support the poor by fines and charging for conducting its pastoral functions (baptisms and marriages, etc.). Records for these can be found in the general records of the meetings of the Kirks and Heritors. These can be found at the National Archives of Scotland under NAS references HR (Heritors) and CH 2 (Kirk).

A Poor Law Amendment Act was passed in Scotland too, in 1845, which significantly changed the nature of poor relief. The Act ensured that the government would now be responsible for poor relief and established parochial boards that would administer poor relief. These boards would create rolls listing the support they gave individual paupers and would provide significant details on these individuals (including age, place of birth and details of next of kin). These rolls are now held either in local county record offices or at the National Archives of Scotland. The National Archives holds records for some parishes in East Lothian, Midlothian and Wigtownshire. The rest will be found locally.

Poor Relief Records in Ireland

Until 1838 there was no systematic method of assisting the poor in Ireland. Poor relief was administered by religious institutes or private charities and was, therefore, dependent on donations. Such records as survive are scattered amongst a number of sources, mainly relating to the institutes or individuals that were donating. These may be found in church records or amongst the estate papers of donating landlords. Records may also survive of individual charitable organizations in local archives.

The workhouse system was introduced into Ireland in 1838 with the establishment of 137 Poor Law Unions in a similar fashion as in England. Initially only 'indoor relief' would be granted to those seeking help. However, the Great Famine of 1845 to 1851 created a huge increase in those seeking poor relief. The governing authorities did not have enough space in the workhouses to provide relief for everyone and therefore 'outdoor relief' would also be given if the local workhouse was full. Generally, outdoor relief was given to able-bodied individuals who did not need accommodation. The majority of those entering the workhouse would be the elderly, the infirm and children. The scale of poverty in Ireland as compared to England meant that a much larger proportion of Ireland's population sought assistance in this fashion.

Many records survive for the unions in either the National Archives of Ireland or, for the six counties of Northern Ireland, in the Public Record Office of Northern Ireland. The unions kept similar records to their English counterparts, including registers of entry and discharge, minute books, financial account books and records for those granted out-relief.

It should also be remembered that a very large number of people chose to emigrate from Ireland in the decades following the Great Famine to seek a better standard of living. Ireland suffered a steady decline in its population until well into the twentieth century. It is worthwhile checking the relevant emigration records in case your ancestor was one of the many Irish emigrants in this period (see Chapter 23).

The Mentally Ill

Specific care of the mentally ill was not provided by central government for all sections of society until the nineteenth century. Prior to that the mentally ill were only taken care of if their families could afford to do so. Otherwise, they would have been treated like any other poor person unable to work. The State itself only became involved when there was a dispute over the property of the mentally ill individual or 'lunatic' and when such cases reached the court of Chancery.

The Lord Chancellor would be responsible in cases of Chancery, and these individuals would be known as 'Chancery lunatics'. Indeed, this method of dealing with lunatics with property had been in existence since the thirteenth century through the *De Praerogativa Regis* ('on the King's Prerogative'), which gave the Crown the custody of the property of 'natural fools' (or 'idiots') or 'lunatics' – the difference being that the former would be deemed to be of unsound mind from birth and the latter only temporarily. The Crown would be responsible for the estate of an 'idiot' for the duration of their lifetime, but would only take over a lunatic's property during the periods of their lunacy. In practice it

> *Specific care for the mentally ill was not given by the State until the nineteenth century.*

▶ An engraving of scenes inside an asylum, 1735.

would not be the Crown who would actually administer the property but, rather, a 'committee' (individuals who were responsible for taking care of the lunatic or idiot, sometimes the next of kin).

An individual would be declared an idiot or lunatic only after an inquisition had been conducted by commissioners at the Lord Chancellor's request. The individual would have been brought to the attention of the Lord Chancellor by a number of means: through concerned relatives; by solicitors acting in circumstances where the supposed idiot/lunatic was involved; by creditors of such an individual; or by the Lunacy Commissioners themselves. The claim had to be supported by two sworn affidavits for the Lord Chancellor to allow an inquisition. He would then place the alleged idiot/lunatic and his property in the care of appropriate individuals – the above 'committee' – and would then investigate the accounts of such committees, which would be submitted to the Chancery Master. All these surviving records can be researched at The National Archives.

If your ancestor was a lunatic without property, then there would be no record in the Chancery files. However, if the individual had some financial support, he or she may have been admitted to a private asylum. These were administered locally by the JPs. The state provision of mental asylums only became compulsory in 1845 with the establishment of the Lunacy Commission for England and Wales. Prior to that pauper lunatics would be helped under the general poor relief system.

The private asylums of the eighteenth and nineteenth centuries started to come under increased State control by the passing of various pieces of legislation:

- The 1774 Madhouse Act introduced a licensing system for the operation of asylums. Henceforth it would be illegal to house more than one lunatic without such a licence. These licences would be issued by JPs in quarter sessions and the authorities would also inspect the asylums they were licensing. This licensing system remained in operation until it was superseded by the Mental Health Act of 1959.
- In 1808 the County Asylums Act encouraged local authorities to build separate lunatic asylums to house pauper lunatics.
- In 1845 the Lunacy and County Asylum Act made the provision of care for the mentally disabled compulsory for local authorities by providing public asylums. These asylums were to be under the authority of the Lunacy Commission, reporting to the Home Secretary. The Commission would inspect the newly created asylums, and the vast majority were created in the following 25 years. As these asylums were administered locally, most records will be found in local record offices.

Records for Lunacy

Records for lunatics recorded in Chancery documents can be found at The National Archives. The documents fall into the following categories:

- *Petitions:* Anyone seeking an inquisition to prove an individual's mental incapability had to provide two sworn affidavits in a petition. The majority of affidavits have not survived (apart from approximately 1,000 in C 217/55 from 1719 to 1733). However, they have been surmised in the petitions' abstracts found in C 211.
- *Inquisition records:* Records for inquisitions by the Lord Chancellor into ascertaining the state of mind of the individual can be found in the following series, depending on the date:
  - ➤ C 132–142 for before 1540. These are held with the Inquisitions post mortem
  - ➤ WARD 7 for 1540 to 1648. From 1540 to 1646 the Court of Wards was responsible for lunatics and idiots, not the Lord Chancellor
  - ➤ C 211 for 1648 to 1932 (PL 5 for Lancashire). If you are researching a Chancery case from the late 1700s onwards remember to check in newspapers as the press often reported these cases
  - ➤ C 43, C 44 and C 206 contain records for inquisitions that were subject to dispute ('traverses'), dating from the time of Edward I to Queen Victoria

> *Lunacy was seen as temporary, whereas 'idiots' were of unsound mind from birth*

- *Records for the possessions of lunatics/idiots:* The Clerk of Custodies collected this information and also decided who would be responsible for the lunatic and his or her possessions. The clerks would issue letters patent to the committees. Many of these records have not survived, but registers of bonds by committees dating from the eighteenth century onwards can be found in series J 103, J 92 and J 117.
- *Records of the Chancery Master:* The Master would collect annual accounts and also make reports on individual cases. Their records, where they survive, can be very useful:
  - ➤ Committee Accounts were submitted annually and are in C 101 (indexed by IND 1)
  - ➤ The Masters would also compile reports and collect exhibits relevant to the case. The reports are in C 38 (indexed by IND 1). The surviving exhibits are in C 103–115 and J 90. C 103–114 has an index in the beginning of the C 103 list
- *Decrees and orders:* These can be found with the main body of Chancery decrees C 33, with a separate series in J 79.
- *Visitor reports:* Visits made by officials are in LCO 10, starting in 1879.

Asylum Records

As mentioned, asylums were not provided by the government until 1845 (with some local authorities building some towards the beginning of the nineteenth century). Prior to that, asylums were run privately although subject to licence after 1766. As such, any surviving admission and discharge registers, along with other administrative records, will be held with the local record office. It is possible to search for the locations of the records of a specific asylum or hospital by using the following sources:

- *Hospital Records Database (Hosprec):* The database contains entries of over 2,800 hospitals compiled by the Wellcome Institute and The National Archives. It lists the location of any surviving records and the exact nature of the material along with dates. It is possible to search by name of institute or by locality. It can be searched free of charge online at http://www.nationalarchives.gov.uk/hospital records/.
- *Access to Archives:* This useful website (www.a2a.org.uk) also contains information on the location of surviving records for asylums and madhouses.

The National Archives also has a small number of records for asylums:

- A register of those admitted to private asylums outside London in MH 51/735 from 1798 to 1812.
- A list of insane prisoners held in houses of correction and prisons in 1858 in MH 51/90-207.
- Returns of mentally unstable individuals held in workhouses and asylums, 1834 to 1909 in MH 12. As discussed earlier, this does not contain any name indexes, and records are organized by county and Poor Law Union.
- The majority of patient files have not survived. A very small number can be found in MH 85, MH 86 and MH 51/27-77. MH 94 has registers of patient files from 1846 to 1960.
- The Royal Navy also had provisions and hospitals for naval lunatics. Records for these can be found within the Admiralty series, specifically:
  - Hoxton House in ADM 102/415-420 (for 1755 to 1818)
  - Haslar in ADM 102/356-373 (for 1818 to 1854)
  - Reports on how naval lunatics were handled between 1812 and 1832 are in ADM 105/28

Lunacy Records in Ireland and Scotland

Ireland did not have many hospitals until the eighteenth century as people's needs were cared for at home. In 1821 the first provisions for the mentally ill were made when 23 asylums were created. Records for hospitals and asylums will either be retained by the institution itself or held at the National Archives of Ireland or the Public Record Office of Northern Ireland.

In Scotland, care for the sick and mentally ill became a responsibility of the heritors (see above) and church councils after 1560. Matters relating to lunatics were discussed in sheriff courts and their records will either be held locally or at the National Archives of Scotland. In 1857 the Board of Commissioners in Lunacy was established and records of this body are also held at the National Archives of Scotland, under the Mental Welfare Commission (MC). They include:

- Minute books, 1857 to 1914, in MC 1
- Admission books, 1858 to 1962, in MC 2 (subject to 75-year closure rules)
- Register of lunatics held in asylums, 1805 to 1978, in MC 7

CHAPTER 25

Family Secrets: Illegitimacy and Adoption

Perhaps even more shocking than the discovery of poverty in the family tree is the realization that not all our ancestors were legitimate. Actually, there was far more illegitimacy around than you might perhaps think, and you may well find that all was not as it might appear. This chapter describes the changing attitudes towards illegitimacy and adoption through the ages, and how records were generated by society's attempts to deal with the issue.

> *Social stigma and poverty meant illegitimacy and adoption often went hand in hand.*

Illegitimacy has been frowned upon and deemed a social evil for generations, yet most people will discover at least one illegitimate birth, if not several, in their family tree, regardless of their ancestors' social standing. The abusive connotations attached to the word 'bastard' in the modern language bring home to us just how awful this label must have been to our ancestors who bore it on their baptism certificates. At worst the illegitimate child was blamed for the sins of their parents, condemned as being of 'bad blood', ostracized from society and had limited prospects of finding employment. Records of the nineteenth-century assize courts show that an alarming number of murder cases were against women for killing their illegitimate babies, such was the shame of bringing a child up alone. Baby farming and informal adoption were preferable alternatives for many mothers, however, and a lot of us will discover that our predecessors covered up the existence of an illegitimate child within the family by pretending the mother and child

◀ Nurses with a pramful of small children from Hutchison House Babies Home, 1930s.

were siblings. Legally, illegitimacy hindered the likelihood of an inheritance being awarded if the parents died intestate, because an illegitimate child had no legal parents in the eyes of the law.

Historical Context

It is only recently that illegitimacy has started to lose the social stigma that has clung to it for centuries. Pressure from the Church and other religious institutions for couples to marry before having children was behind this, but a number of other factors came into play as well. Local communities begrudged having to financially support those children whose fathers could not be made accountable, and in aristocratic circles the continuation of the family's name and wealth depended on there being a legitimate heir.

Hundreds of ordinary families have been told stories about a distant ancestor who was the illegitimate lovechild of a Lord or Duke, and indeed there were many illegitimate offspring from upper-class affairs, especially as often aristocratic marriages were little more than arranged dynastic unions and wealthy husbands were almost expected to take mistresses. However, the majority of us will stumble across illegitimate ancestors whose origins lie in the misery of a workhouse as opposed to the luxurious surroundings of a courtesan's bedroom.

Up until the mid-eighteenth century young women were known to consent to sex with a suitor after a promise of marriage, as common law ruled that a verbal promise of marriage should be binding. However, an engagement that took place in private could be easily denied when there were no witnesses, which often happened if the woman fell pregnant before she managed to get her lover to the altar. Men could escape their obligations by joining the army or fleeing the parish, and many women and babies were left dependent on parish relief (see Chapter 24 for more details on the workings of poor law relief). Poor Law Administrators would investigate who the father of an illegitimate child was in an attempt to force him to pay maintenance, and from 1732 unmarried pregnant women were required to identify the father under oath. Hardwicke's Marriage Act of 1753 put an end to the ambiguity surrounding marriage and common law, requiring all legally upheld marriages to take place in the presence of a priest. However, the practice of sex before marriage continued to be widespread.

The Poor Law Amendment Act of 1834 reflected parish authorities' fears that the old poor law system, which allowed unmarried mothers to take maintenance money from the father of their child, was actually encouraging bastardy. Therefore the 1834 Act transferred all financial responsibility for an illegitimate child onto the mother. This new system caused more hardship for women because they were invariably unable to work while caring for a baby, but if they went into the workhouse the poor law authorities were of the opinion that it was better to separate the child from its morally irresponsible mother.

Adoption and Foundlings

This double bind of social stigma and poverty meant that illegitimacy and adoption often went hand in hand. Many children were given up for adoption if they had been born out of marriage for fear of bringing shame upon the mother's family or in the hope the child would have a better upbringing if the natural father could not be made accountable.

If you are tracing the line of an ancestor who was adopted it can be difficult to get any further back than that individual without knowing something about their life before the adoption. Prior to the late 1920s and early 1930s all adoptions were arranged privately and the only way

---

CASE EXAMPLE

Adoption and family secrets

The eccentric antiques dealer **David Dickinson** *found out by accident that he was adopted at the age of 12. David had always been led to believe by his adoptive parents that they found him in a Barnardo children's home, but in the course of the programme David's search of the Barnardo records proved inconclusive. He discovered by talking to other relatives that his adoption had been a private affair. David's adoptive mother Joyce had actually been the hairdresser and a friend of his natural mother, Jenny. Joyce offered to look after David when he was born because Jenny had become pregnant by a married man and her strict Armenian father would not tolerate any shame being brought on the family. Jenny later married and moved to Jersey, creating a new life for herself where nobody knew about David.*

you may have of finding out more is by asking living relatives what they know. However, home truths like this can be hard to come to terms with, so if you plan to speak to relatives about a suspected adoption, be wary and sensitive to their emotions even if a long time has passed since the event.

Foundling children, who were abandoned by their parents in the hope that someone would literally find and care for them, were usually suspected of being illegitimate, though some foundlings' parents were married but felt forced to abandon their children due to excruciating poverty. The births of foundling children are often listed in the birth indexes without any names, written simply as 'male' or 'female' after the letter Z, and it is extremely unlikely you will be able to find out anything about their origins. If a foundling child was lucky they may have been taken in by a local family and brought up as one of their own children, but most were looked after by the parish authorities and an apprenticeship found for them when they were old enough to work. In 1739 Captain Thomas Coram established the Foundling Hospital to look after children abandoned on the streets of London, which cared for over 27,000 children between 1739 and 1954. And Dr Barnardo established one of the most famous organizations dedicated to caring for orphaned and abandoned children, opening his first home in Stepney in 1867. When the last of the Barnardo's homes closed in 1981 around 300,000 children had been helped by Dr Barnardo's organization.

Whether like David Dickinson you were adopted and need to discover who your natural parents were before you can work back any further, or like Griff Rhys Jones there's an intriguing story of adoption further back in time, the sources here should help you to uncover the truth about your mysterious past.

Researching Illegitimacy

The term 'illegitimate' covers a multitude of situations – it could be that both parents were single, or one or both parents were married to another person, that the parents were unmarried when the child was born but they married later on, or a variety of more complex situations which could lead to the legitimacy of the child being questioned. The possibility of sexual exploitation should not be overlooked either. There are many stories of domestic servants being taken advantage of by their employer, and allegations of rape were very difficult to prove so few women were prepared to go through the courts. If you have reason

CASE EXAMPLE

Researching adoption

Griff Rhys Jones unravelled the truth behind the adoption of his maternal grandmother Louisa Price, who was not illegitimate but was raised by a second cousin and his family. Griff's mother had been told that Louisa was orphaned at a young age after her parents died in a train crash, but Griff discovered that in actual fact Louisa's father died from injuries resulting from a drunken brawl and her mother seemed to disappear after an unsuccessful attempt to get parish relief. Distant relatives who watched the show were able to fill Griff in on what happened to Louisa's mother after her husband's death. It emerged that Louisa's mother, Sarah Louisa Price, was eventually forced to enter the workhouse, but not wanting her children to be brought up in such surroundings she accepted the kind offer of her husband's cousin to care for her young daughter.

Records of Institutions

If you have the birth certificate of an illegitimate child, then take note of the address where the child was born. If it was a workhouse or a maternity home then records of the establishment might survive. Special homes were set up for 'fallen' women from the early nineteenth century. The National Register of Archives can help you to locate the repository for such institutions. In Ireland, where Catholicism had a strong hold, the birth of an illegitimate child was considered particularly problematic. Thousands of young Irish women were sent to convents run by the Sisters of Our Lady of Charity, the Sisters of Mercy and the Sisters of the Good Shepherd from the nineteenth century until the late twentieth century, where some were kept for long periods of time and forced to work unpaid in miserable conditions washing laundry, cleaning, and caring for elderly nuns.

The lives of the girls and women who spent years in institutions such as the Magdalene Laundries are

▼ Register of baptism of foundling children in the Foundling Hospital, 1777.

to suspect the mother of your illegitimate ancestor may have pressed charges against the father of the child for rape or sexual assault, then Chapter 27 will help you to trace the case.

Civil Registration Certificates

There are very few official records created by illegitimacy, and the majority of us will find out about an illegitimate ancestor by stumbling across a marriage or birth certificate where the father's name has been left blank. In this case the child will have usually taken their mother's surname, so finding out who the father was can be an impossible task. Up until 1875 if a mother registered the birth of her illegitimate child and told the registrar who the father was, the registrar could enter the father's name on the birth certificate. After 1875 this could only happen if the father was present at the registration and consented to his name being put on the birth certificate. The shame surrounding illegitimacy often produced a web of lies, making a search of the records even more difficult. In rare cases it is possible that an unmarried mother would tell the registrar that she was married and register her illegitimate child's birth under the surname of the father, given that proof of the parents' marriage was not a requirement of registering a child's birth. These factors should be taken into account when searching for the birth certificate of an illegitimate child.

Parish Records

Parish records can be more accurate than civil registration records, particularly if the child was born in the parish where the mother lived, shrouded in mystery and the abuse suffered by many inmates has only recently come to light when mass graves of unidentified women were unearthed at a Dublin convent. The last Magdalene convent closed down in the 1990s and support groups have since been set up for the survivors and relatives of those who served time for their 'penitence'. Justice for Magdalenes is a support group whose website can be found at www.magdalenelaundries.com and while records for such institutions are difficult to unearth, they may be able to give you some guidance and direct you to indexes of the names of some of the women who lived and died in the convents.

If you are researching the background of an illegitimate child from a poor or working-class family, your first port of call should be the records of the Poor Law Administrators. You may not have found any evidence of your ancestors entering the workhouse from census returns or certificates, but if they lived on the bread line and an illegitimate child was born it is worth checking the records of the local workhouse to find out if they were forced to enter it for a while.

because the local community would usually know the truth about the mother's situation. Baptism records sometimes state that the child is a 'bastard', or 'the base child of', or more politely, 'the natural child of' the mother, meaning that he or she was born out of wedlock. Occasionally the baptism record will state the father's name as well, so it is worth locating the child's baptism record if their birth certificate leaves you with the suspicion they were illegitimate.

Census Returns

Census returns can be used to provide substantiating material. Returns just before and after the child's birth should be looked at to find out whether the mother was listed as single. Clues such as an unusual middle name given to the child may also hint at what the father's surname was, so take another look at the names of neighbours, employers and visitors found on census returns with the mother.

Census returns, civil registration certificates and parish records are the best tools available to genealogists investigating illegitimate births. They can tell us a lot about the impact illegitimacy had upon the child's life. If the mother married at a later date you may find that the child's stepfather took the child in as his own. This might be assumed if the child started using the stepfather's surname, or if the stepfather is listed as the father on the child's marriage certificate. Equally, evidence like this can prompt more questions than answers – was the step-father the child's natural father after all? Some illegitimate children were raised by their grandparents as their own child, so that their mother might lead a normal life, particularly if she became pregnant while still quite young. If you find a census return where there is a big

age gap between the youngest child and the other children in the household it is often worth ordering their birth certificate to find out whether the mother's name could be one of the females listed as a sibling on the census.

Poor Law Records

Poor law records can help in the hunt for the missing father of a child. It's easier to uncover documentary evidence about paternity for illegitimate births prior to 1834 because the local parish was required to either pay for the upkeep of the illegitimate child or chase the father to arrange maintenance payments. This means that a lot of evidence was gathered by parish officers, who would try to obtain a sworn statement from the mother ascertaining the identity of the father. Parish officials drew up bastardy bonds, or affiliation orders, as a sworn statement by the father that he would pay maintenance for the child. When establishing which parish to look under you should be aware that while most people were chargeable to the parish in which they were deemed to have settled, usually where their parents had lived, illegitimate children were chargeable to the parish in which they were born, which may differ from that of their mother's parish of settlement.

The Poor Law Amendment Act of 1834 may have put more responsibility on the mother for the maintenance of her child, but the parish authorities could still chase the father for payment in the quarter sessions if the mother became chargeable to the parish and she could provide corroborated evidence that he was the father of the child. Therefore, if you find that your illegitimate ancestor and his or her mother received parish relief between 1834 and 1844, it is also worth searching records of quarter sessions in the local record office.

Mothers were given more power to claim maintenance payments directly from the father from 1844, as well as costs such as paying a midwife or seeking legal advice. These records may be found among affiliation orders in the petty sessions in county record offices, and the system was used to claim money by unmarried mothers well into the twentieth century. You may find reports about the case in local newspapers, some of which have been indexed by local family history societies. Between 1844 and 1858 each petty session and quarter session was required to make an annual return of bastardy cases to the clerk of the peace, who in turn should send a copy to the Home Office. The bastardy returns of this period are usually found in the local record office amongst records of quarter sessions and petty sessions, but unfortunately the copies sent to the Home Office are not known to have survived.

Tracing Offspring of the Aristocracy

If you suspect your illegitimate predecessor was the product of an aristocratic love affair, and you have a notion as to the identity of the father, then look for circumstantial evidence that might place your well-connected would-be ancestor in the right place at the time of the conception and birth of the child. The movements of the well-to-do classes were chronicled in contemporary newspapers and journals, and biographies can give you more clues about their whereabouts at certain times. Hopefully you will be more successful in finding the answers you want than John Hurt was when he tried to prove that his great-grandmother, Emma Stafford, was the daughter of the Marquess of Sligo. By comparing the date his great-grandmother was born with an extended Sligo family tree John found that the dates did not work for Emma to have been conceived by the Marquess.

Always try to find the will of the person you believe to have sired your ancestor because they may have provided for them given that legally an illegitimate child had no automatic right to their parents' legacy. The National Archives holds records concerning the transferral of property belonging to illegitimate people who died intestate. Usually their property would go to the Crown, but petitions can be found in series T 4 from 1680 to 1819 from next of kin requesting that letters of administration be granted to them instead. TS 17 contains papers relating to the administration of estate papers for people who died without leaving a lawful heir between 1698 and 1981. Estate records found in local record offices and private repositories may also list regular payments made to the mother of the child for maintenance.

Adoption Records

The majority of adoptions prior to the 1930s were arranged privately, either by the child's mother or family, or by a charitable organization. In this type of set-up the adoptive family would act as foster parents and the child may or may not have taken a new name, but there was no legal requirement to change the child's name officially even if they became known by a different surname. This highlights the main

HOW TO...

... trace an illegitimate ancestor

1. *If you are investigating an illegitimate line in England or Wales then Ruth Paley's* My Ancestor Was a Bastard *(Society of Genealogists) is worth consulting for an in-depth look at all the possible sources available to you, from Poor Law records to legal records.*

2. *If the illegitimate child was a pauper then apprenticeship records might mention them because the parish authorities sometimes organized for a bastard child to be apprenticed to keep them off the poor rates.*

3. *The records of quarter sessions and church courts for charges of immorality against the parents are also worth searching.*

4. *The repositories of ecclesiastical courts in England are described in D. M. Owen's* Records of the Established Church in England *and the types of records these courts produced are explained in* Sin, Sex and Probate: Ecclesiastical Courts, Officials and Records *by C. Chapman, and* Church Court Records: An Introduction for Family and Local Historians *by A. Tarver.*

struggle with researching an adoption at any point in time: knowing what name to look for in the records. Some adopted children kept the name they were registered with at birth, but most will have taken their adoptive parents' surname and some will have been given a new first name by their adoptive parents, though this is more likely if they were adopted at a very young age. If the child's first name was changed within 12 months of the original birth registration the entry should be indexed under both names, as long as the child was not baptized with the name originally entered on the civil register.

Adoption registers for adoptions arranged through the courts with the aid of adoption agencies and charitable organizations exist from 1927 for England and Wales, from 1931 for Scotland and Northern Ireland and from 1953 for the Republic of Ireland. For most cases it is possible to order a copy of the adoption certificate if you can provide the adoptee's adoptive name and date of birth. The certificate will tell you the names of the child's adoptive parents and the court in which the adoption was granted. Alternatively you can order the original birth certificate if you know the child's original name and date of birth, which should provide you with the natural mother's name and possibly the father's name if it was recorded, as well as the place and time of birth.

In order to protect the identity of the adopted person it is not possible to cross-reference the information given in the original birth indexes with that given in the adoption registers. Therefore, while an entry in the birth registers may note that a child was adopted, it will not give the child's adoptive name, and the adoption registers will only give the child's adoptive name and date of birth, but not their place of birth or original name. You are therefore required to know a certain amount about the child before you commence a search. If you know what the child was called before they were adopted you should be able to order their birth certificate in the normal manner and use the information on that to work backwards. However, if you think they were adopted but do not know the name they were adopted under you will struggle to find out anything about their later life. Equally, if you know the person's adoptive name you may be able to order their adoption certificate and find out a little about their adoptive parents, but without knowing the child's birth name or the names of their natural parents you will probably not be able to find out about their origins.

Official Adoptions

If you are conducting a search for your own adoption records or on behalf of a relative who has given you permission to find their adoption records, there are three main sources of information. The General Register Office will hold the original birth certificate and the Adopted Children Register. If the adoption was arranged through an agency or organization then they should have records on the case, and court records should contain records of the adoption proceedings.

The Adoption, Search & Reunion website was set up by the British Association for Adoption and Fostering to help those researching an adoption that took place in the UK, as well as anyone who is thinking about searching for and making contact with birth and adopted relatives. The website contains a database to aid researchers in locating the repositories of adoption records which can be searched by the name of a home (such as a maternity home, mother and baby centre or women's shelter that might be given on the birth certificate), the name of an organization or local authority involved in the birth or adoption, the name of a member of staff who worked in the home or for the organization (perhaps they were the informant who registered the child's birth), as well as by place name. You can access the Locating Adoption Records database from www.adoptionsearchreunion.org.uk/search.

Adoptions in England and Wales

The General Register Office for England and Wales has an Adopted Children Register, a register of all the adoptions granted by courts in England and Wales since 1927. When an entry is made in the register the child's original birth entry should be marked 'adopted' and the adopted child should use their adoption certificate in place of their birth certificate for legal and administrative purposes. The Adopted Children Register is no longer open for the public to search, but if you can provide the GRO with the adoptive name of the child and their date of birth a copy of their adoption certificate showing the names of their adoptive parents may be issued. Send your application to

> The General Register Office
> Adoptions Section
> Trafalgar Road
> Birkdale
> Southport PR8 2HH

If you know the name the child was registered with at birth before the adoption you should be able to locate their birth certificate using the

Official adoptions

Nicky Campbell had known he was adopted, and decided to explain the background involved when looking for natural parents in his episode of Who Do You Think You Are? *– even though he decided to investigate the background of his adoptive parents during the show.*

Nicky had been given up for adoption when only five days old, but only decided to search for his biological family in later life once he was 30. He vividly described how emotionally draining the process was, and how it opened up issues with both his adoptive and natural family concerning identity, changes to existing relationships and long-buried feelings that rose to the surface once again.

Nicky Campbell followed the steps many others have taken, namely to find out as much information as possible from his adoptive parents, and then pursuing a paper trail – applying for original birth details, searching for a birth certificate, and finding information about his adoption file.

ordinary civil registration indexes. If you are trying to find out about the circumstances of your own adoption and do not know anything about your natural parents, the GRO's adoption service should be able to provide you with your original birth certificate. Before November 1975 many parents were told that the adopted child would not be able to find out their original name or the names of their natural parents. Changes in legislation that took place with effect from 12 November 1975 meant that it would be easier for adopted children to find out about their origins once they reached 18 years of age. Therefore, children who were adopted prior to November 1975 and wish to find out about their natural parents are required to meet with an adoption advisor before information about their original birth entry will be released. There is more information about the adoption service on the GRO website at www.gro.gov.uk/gro/content/adoptions.

Adoptions in Scotland

Children adopted in Scotland after 1931 can apply for information about their original birth entry by writing to

> The Adoption Unit
> New Register House
> 3 West Register Street
> Edinburgh EH1 3YT

Provide details of their adoptive name, date of birth and full postal address. A copy of the adoption certificate can also be requested by writing to this address. The General Register Office for Scotland has advice about tracing adoption records on its website at www.gro-scotland.gov.uk/regscot/adoption.html.

  Birthlink is a charitable organization set up by the Family Care Adoption Society in Edinburgh to provide support to adopted people and their relatives in Scotland. The organization can help adopted people to trace birth relatives and also gives advice on locating records concerning the adoption. The organization has a website at www.birth-link.org.uk detailing their services, or alternatively you can speak to a member of staff by telephoning 0131 225 6441, or write to them at

> Birthlink
> 21 Castle Street
> Edinburgh EH2 3DN

When the Adoption of Children (Scotland) Act was passed in 1930, adoptions could be arranged by charitable organizations or local authorities, and then be ratified in a civil court, usually a local sheriff court, although a very small number are passed in the Court of Sessions in Edinburgh. The process papers generated by these adoptions remain with the local courthouse for up to 25 years, after which time they are passed to the National Archives of Scotland where they are held in the Legal Search Room, but are subject to a closure period of 100 years. The only circumstances under which this rule may be relaxed are if the adopted person is over 16 years old and wishes to read them, or a person authorized in writing by the adopted person applies to see them, and in both cases proof of the adopted person's birth and identity are required. To locate the correct legal records the NAS needs to be notified in advance of your visit of the adopted person's birth name, the date they were adopted and the court that dealt with the adoption. This information can be obtained from the General Register Office for Scotland.

The adoption process papers vary in content from case to case, but should contain a copy of the child's original birth certificate, an official report to the court at the time of the adoption, a petition by the adopting parents, the consent of the birth mother and sometimes the birth father, the name of any adoption agency involved, and confirmation from the court that the adoption may proceed. The papers will not always give background information explaining why the birth parents wanted to give the child up for adoption; however, they may reveal distressing information about the circumstances. The NAS has produced an online guide to adoption records in Scotland found at www.nas.gov.uk/guides/adoptions.asp.

> *If you are trying to find out about your own adoption, the GRO's adoption service will be able to give you your original birth certificate.*

Adoptions in Ireland

There is a separate Adopted Children Register for Northern Ireland covering adoptions since 1931, and adopted people can apply to the GRO for Northern Ireland for a copy of their original birth certificate. The General Register Office for Northern Ireland has an adoption section on its website at www.groni.gov.uk where there is information about how adopted people can go about tracing their origins. Applications for copies of original birth certificates or adoption certificates should be sent to

The Registrar General
Oxford House
49–55 Chichester Street
Belfast BT1 4HL

Legal adoption in the Republic of Ireland did not begin until 1953. People researching adoptions in Ireland since the 1952 Adoption Act should contact the Adoption Board to seek advice by telephoning +353 (0)1 230 9300 or writing to

> Shelbourne House
> Shelbourne Road
> Dublin 4

The Adoption Board holds a file on each adoption effected in the Republic of Ireland since 1953 and its website gives further information about its services at www.adoptionboard.ie. This is unlikely to be your last port of call, but the Adoption Board should hopefully be able to give you contact details for other organizations and agencies involved in the adoption process.

Some children born in the Republic of Ireland were sent for adoption in England or the United States, particularly before the 1952 Adoption Act. The General Register Office for the Republic of Ireland will hold the original birth certificate, but the country of adoption should hold any other paperwork regarding the adoption.

Other Sources of Information

- The British Association for Adoption and Fostering (BAAF) has useful information on its website for birth relatives wishing to find out more about an adoption that took place in England, Wales, Scotland or Northern Ireland. They have experts on all areas of adoption and fostering and may be able to help you establish the records available for researching a UK adoption, depending on your relationship with the adopted child. The BAAF website at www.baaf.org.uk has contact addresses and telephone numbers for their numerous offices in central, northern and southern England, Cardiff, Rhyl, Edinburgh and Belfast.

- The Adopted People's Association (APA) is a similar organization established for those who want to find out more about Irish adoptions and for adopted people who have Irish roots. The APA actively encourages research into adoptions and may be able to help you locate the records you are looking for. More information can be found on their website at www.adoptionireland.com, or you can contact them by telephone on +353 (0)1 679 0011 or by writing to

> The Adopted People's Association Ltd.
> 14 Exchequer Street
> Dublin 2

Informal adoptions will be more challenging to research.

- If you were adopted and would like to try to contact your birth parents, or if you are the birth relative of a person who was adopted and would like to try to get in touch with them, you can join the Adoption Contact Register. The aim of the register is to match up the details of registered members, and so you will only be able to find your birth parents for example if they have also joined the register in an attempt to find you. Information about how to join the Adoption Contact Register for England and Wales can be found at www.gro.gov.uk/gro/content/adoptions/adoptioncontactregister and for Northern Ireland there is information on the GRO website at www.groni.gov.uk/adoption.htm, or alternatively write to the relevant GRO for more information. Birthlink control the Adoption Contact Register for Scotland and information about this can be found at www.birthlink.org.uk/adoption_contact_register.htm. The Irish Adoption Contact Register is administered by the Adopted People's Association and is free to use either online or by post. The database includes people who were adopted abroad but are looking for birth relatives in Ireland. The APA Irish Adoption Contact Register can be found at www.adoptionireland.com/register/index.html, but there is another contact register recently set up by the Irish Adoption Board known as the National Adoption Contact Preference Register (NACPR), information about which can be found at www.adoptionboard.ie/preferenceRegister/index.php. If you are seeking information about an adopted or birth relative with links to Ireland then it is worth joining both registers.

- There is an online contact register that has been running since 2004 at www.ukbirth-adoptionregister.com, which you can join for £10 by entering as much information as you know about the person you are searching for. You can search the database of members and email the organization if you think you have found a match for your relative. The UK Birth Adoption Register will then check that the information you have submitted corresponds with the data they have about the other member and will advise you how to go about contacting that person. The Adoption Contact Registers are a safer way of contacting adopted and birth relatives than tracing them using other means because you know that if their details are registered then they are presumably happy to be contacted and talk about the past. If they are not happy to be contacted then this information will be on their registration form.

Unofficial Adoptions

Informal adoptions that took place before the Adoption Acts introduced from the late 1920s will be more challenging to research. Court records can sometimes prove fruitful as the adoptive parents had no legal right to guardianship and from the late nineteenth century the birth mother was favoured for custody of the child even if the child had spent most of its life in the care of another family. Some mothers fought to retrieve their children through the courts from adoptive parents and if there was a dispute between the adoptive parents and the natural parents a record of this is likely to have appeared in the local newspaper. Unfortunately tracking down such evidence is not easy if you do not have a date to work with and newspapers for the area in question have not been indexed. Most of us will have to rely on circumstantial evidence extracted from census returns and birth and marriage certificates.

There is more hope for those researching foundling children taken in by a home or adoptions arranged by a charitable organization. The

▼ Children from Dr Barnardo's home in Stepney Causeway, East London, October 1904.

Thomas Coram Foundation still holds admission registers for children admitted to the Foundling Hospital from 1794. You can enquire about these by phoning 020 7520 0300, or by writing to

> 40 Brunswick Square
> London WC1N 1AZ

although access is restricted and a fee will be charged. The majority of records in the Foundling Hospital Archive can be found at the London Metropolitan Archives, whose collections can be searched using the Access to Archives database. The LMA has produced a leaflet to help genealogists, entitled 'Finding Your Foundling', and its records include petitions from parents for the admission of their children, apprenticeship registers, and tokens of affection left by mothers with their children.

The National Archives holds duplicates of baptism and burial registers for the Foundling Hospital in series RG 4, though the Thomas Coram Foundation is the only place where you can find details of parentage. If you believe your ancestor was a pupil at the Foundling Hospital then you may be interested in visiting the Foundling Museum at Brunswick Square, next to the original site of the hospital demolished in 1926.

Detailed records of children taken in by Barnardo's homes, including an extensive photographic archive of the children cared for, are held at Liverpool University but can only be searched by staff if a postal application is sent to

> The After-Care Department
> Barnardo's
> Tanners Lane
> Barkingside
> Ilford
> Essex IG6 1QG

detailing as much information as you know about the child. Telephone the Head Office in Barkingside on 020 8550 8822 to find out more.

USEFUL INFO

Suggestions for further reading:

- My Ancestor Was a Bastard *by Ruth Paley (Society of Genealogists Enterprises Ltd, 2004)*

- Illegitimacy in Britain 1700–1920 *by Samantha Williams, Thomas Nutt and Alysa Levene (Palgrave Macmillan, 2005)*

- Illegitimacy *by Eve McLaughlin (McLaughlin Guides, 1995)*

- Illegitimacy, Sex and Society: Northeast Scotland, 1750–1900 *by Andrew Blaikie (Clarendon Press, 1994)*

- Where to Find Adoption Records: a guide for counsellors, adopted people and birth relatives *by Georgina Stafford (British Agencies for Adoption and Fostering, 2001)*

- Tracing the Natural Parents of Adopted Persons in England and Wales *by Colin D. Rogers (Federation of Family History Societies, 1992)*

- Search Guide for Adopted People in Scotland *by Birthlink Adoption Counselling Centre (Stationery Office Books, 1997)*

CHAPTER 26

Family Secrets: Bigamy and Divorce

Another of the great taboo areas surrounding family life concerns the break-up of a relationship, even more so in the days when marriage was sacrosanct and ''til death do us part' was taken literally. Many couples simply moved apart and started afresh, even though bigamy was, indeed is, illegal. Despite their best attempts to keep these relationships secret, there are clues you can follow to check up on a suspicious partnership, and this chapter describes how to spot the telltale signs and where to look for more information.

Historical Context

Divorce as we know it today did not become common practice until the mid-twentieth century, although in England, Wales and Scotland civil divorce was possible from the 1800s. There was a stigma attached to separation and divorce that deterred people from leaving an unhappy marriage. Women, particularly from the upper and middle classes, were discouraged from leaving their husbands because they would lose custody of their children, be worse off financially and their prospects of remarriage were less hopeful than if they waited until widowhood. Equally, the lower classes simply couldn't afford to pay for the legal proceedings of a divorce. Before legal aid and changes in the law made divorce widely available to everyone it was very wealthy men alone who could afford to seek a separation through the courts, and indeed did so to protect their financial assets against their estranged spouse.

> *There were many alternatives to divorce that people resorted to.*

Alternatives to Legal Divorce

You might have been told that one of your ancestors divorced or separated from their spouse but you cannot find a record of any legal proceedings. There were many alternatives to divorce that plenty of people resorted to. A private separation could be agreed, and desertion was relatively commonplace. However, these alternative options did not allow either partner to *legally* marry again, though that is not to say everybody abided by the law.

Common law and local custom allowed for peculiar matrimonial practices, particularly among close-knit communities. For example, on market day prearranged wife sales were sometimes held whereby a husband who wanted to separate from his wife could sell her off to the highest bidder. The wife was usually in collusion with the husband and had chosen a suitor who may have settled on a price with her husband prior to the public auction. The sale was a symbolic ritual to make the matrimonial separation known to the community, with the transfer of person, property and responsibility from one man to another. Wife sales were not recognized in law but a wronged husband believed the public nature of the sale would protect him from any liability for his wife's debts in the future and his wife's lover believed he would be protected against any suit against him for adultery.

Desertion of one's spouse was more commonplace among the poorer classes who did not have any property to squabble over. A deserted wife and her children would often become dependent on parish relief and applications by deserted wives for poor relief can be found among quarter session records kept at the local county record office. Once a spouse had been missing for seven years the parish church might allow the deserted husband or wife to remarry on the supposition that their first spouse was dead. A married man or woman who deserted their spouse was not legally free to marry again, but if they fled to a faraway parish where nobody knew their history it was easy enough to commit bigamy. The number of charges brought to court for bigamy was probably far fewer than the number of bigamous marriages that actually took place. Researching an informal separation will be practically impossible, but circumstantial evidence proving bigamy can shed light on how a first marriage ended.

▲ 'Striking the iron while it is hot' – a quick marriage in Gretna Green, engraved 1791.

Records exist of applications for separation and divorce in the ecclesiastical courts, civil courts, parliamentary records and Probate, Marriage and General Register Offices, though the location and detail of these papers will depend on when your ancestor divorced and may require some prior knowledge of the type of divorce they obtained. Searching for the records of separations and divorces granted in ecclesiastical courts before civil divorce became legal can be complicated and time-consuming. Indexes are available for civil divorces, the earliest of which date back to 1830 in Scotland, though complete records of the court case do not always survive.

England and Wales
Divorce Before 1858

England was the only Protestant European country not to have some form of divorce law in place by the end of the sixteenth century. In fact the first legal change in divorce law was not introduced until 1858 and divorce did not become common practice until the mid-twentieth century as attitudes slowly changed.

Before the mid-sixteenth century, the only way a couple could separate and remarry was for the Pope to declare a marriage 'null and void' by granting an annulment. This in effect was a declaration that the marriage had never taken place. Henry VIII paved the way for divorce in England by instigating the English Reformation after Pope Clement VII refused to annul his marriage with Catherine of Aragon. The establishment of the Church of England in 1534 gave the English monarch supreme rule over the English Church, allowing Henry to grant himself an annulment from his estranged wife. Unfortunately for most of our ancestors this was not a luxury they could afford, and so many either stuck it out in unhappy marriages or deserted their spouse in search of a happier life.

Divorce in its truest sense, whereby both partners were free to remarry, could be granted by an Act of Parliament. However, this was rare and expensive to pursue, with only around 300 such divorces being granted between 1670 and 1858. Moreover, if a wife sought a divorce she needed to prove both adultery and life-threatening cruelty (or incest) on the part of her husband, resulting in only four female petitioners successfully obtaining divorce by Act of Parliament. A husband only had to prove adultery on the part of his wife, and it was more common for a husband to petition for a divorce on the grounds of his wife's adultery in an attempt to protect his property from being inherited by any illegitimate offspring she might produce.

- Divorce Acts are kept at the Parliamentary Archives with a sample of cases held at The National Archives in series C 89 and C 204, and reports were often published by *The Times* which may be searched online.
- About Archives produced a separate index to the names of spouses and adulterous lovers listed in court reports by *The Times* between 1788 and 1910, which can be found on microfiche and in book form at the Society of Genealogists and Guildhall Library, entitled *Index to Divorces as Listed in Palmer's Indexes to 'The Times' Newspaper London, 1788–1910*.

To obtain a divorce by Act of Parliament the couple would usually have been through various ecclesiastical courts to first of all obtain an annulment, then a declaration of nullity and a divorce *a mensa et thoro* ('from bed and board'), before finally submitting a private Bill before Parliament for a full divorce.

USEFUL INFO

If the husband was seeking a separation from his wife on grounds of adultery he may also have sought compensation from her lover by pursuing a Criminal Conversation suit in the court of King's Bench or Common Pleas on a charge of the lover having 'trespassed' his wife's body, which was her husband's property in the eyes of the law. Sensational reports of these civil suits can be found in both local and national newspapers from the late eighteenth century onwards.

Annulments

Courts of the bishops, or consistory courts, could grant an annulment, a form of legal separation that protected the rights of the wife and any children but did not allow either party to remarry. Annulment could be granted if it was found one of the parties had committed bigamy or if the husband and wife were too closely related and should never have been allowed to marry. Other grounds for seeking an annulment or nullity could be for non-consummation, lunacy or, as of 1754, because one of the spouses had been under 21 years of age at the time of marriage and parental consent was not obtained.

If an annulment was granted the couple could then seek to have the marriage declared null and void, which meant the marriage should never have taken place from the outset and any children of the marriage were therefore made illegitimate. A wife also lost the right to a third of her husband's property on his death, but the parties were still not allowed to remarry.

Divorce *a mensa et thoro*

Prior to going to Parliament, the Church courts could grant a couple a divorce *a mensa et thoro*, translated as 'from bed and board', if adultery or life-threatening cruelty was proved, though again neither party was legally allowed to remarry. Records of the consistory courts are held in local record offices, but if an appeal was lodged at a higher court all the case papers from the earlier court hearing were transferred to the court of appeal, in which case the records may be held at The National Archives. There was a right of appeal firstly to the archbishop's court, and further appeal could be taken to the High Court of Delegates until 1834, and then to the Judicial Committee of the Privy Council from 1834 until 1858. The records of the High Court of Delegates are held at The National Archives in series DEL 1, DEL 2 and DEL 7 and for the Judicial Committee of the Privy Council in series PCAP 1 and PCAP 3.

Private Separations

If a propertied couple who were financially well off wished to separate without the expense and publicity of obtaining a divorce, a private deed could be drawn up with terms agreed by the husband and the wife's trustee (in common law a wife had no legal power because she was the property of her husband). Private separation became more common during the seventeenth-century Cromwellian Interregnum when Church courts did not operate, and were also preferable to couples with children because a declaration of nullity would render their heirs illegitimate. The deed might include provisions for their children, some legal safeguards for the wife, and an agreed maintenance allowance to be paid by the husband, but marriage to another person would have been illegal and bigamous. The Matrimonial Causes Act of 1857 made no provision for divorce on the grounds of incompatibility, therefore couples continued to use deeds of separation after the introduction of civil divorce.

Deeds of separation can be found at The National Archives among the Close Rolls in C 54, an index to which is kept on open shelf in the Map and Large Document Reading Room, arranged by the names of the parties. Some deeds will have been kept among family and estate papers if the couple were relatively wealthy, which may be searched for using the National Register of Archives. If one of the parties to the deed broke the agreed conditions of the separation then a petition could be presented to the Court of Chancery to enforce the deed's terms. Such Chancery cases became more common by the late seventeenth century, records of which are held at The National Archives in series C; however, they are difficult to find without knowing the names of the parties (Chancery cases were sometimes filed under the names of acting solicitors) and an approximate date of the case.

Divorce in England and Wales After 1858

In January 1858 civil divorce was made easier with a new Court for Divorce and Matrimonial Causes opening to hear all cases regarding marriage and divorce. In 1873 this court became the Probate, Divorce and Admiralty Division of the Supreme Court of Judicature and is now known as the Principal Registry of the Family Division. Nevertheless, divorce remained costly and the court proceedings were centred in London, making it unattainable for most ordinary couples who did not have the time or money to hire a London lawyer, file a petition in the London registry and make a court appearance in the capital.

Poorer women could apply to a local magistrate for a separation and maintenance order from 1878 against an aggressive husband if he was convicted of aggravated assault against her. This type of legal protection was built upon through the latter part of the nineteenth century to provide for virtuous wives and children abandoned by their neglectful husbands and fathers, but if the wife became involved with another man after the separation, or if she had committed adultery before, she was not entitled to seek maintenance from her husband for herself or her children. Magistrates' courts granted thousands of separation and maintenance orders every year until divorce became a more affordable option, and records of these orders will be found at local record offices if they survive.

Divorce did not become an option for the less well-off classes until the 1920s. From 1922 ten assize courts across England and Wales were allowed to hear certain types of divorce cases, making divorce more accessible to people living away from the capital. In the 1920s legal aid was extended to help some couples obtain a divorce, and in 1923 women could at last pursue a divorce on the grounds of her husband's adultery without needing to prove another offence such as life-threatening cruelty or desertion. In 1927 the number of assize towns where divorce proceedings could be heard was increased to eighteen, while petitions could be filed in twenty-three district registry offices in addition to the Principal Registry in London. It was not until the 1960s that county courts were allowed to hear the suits and 1971 that divorce by mutual consent was allowed, which is the most common way of filing for a divorce nowadays.

Case files were created for every divorce suit, but very few survive after 1938. Those that do are kept at The National Archives, but for divorces granted in the last 20 years you should apply to the relevant court, which will destroy the case papers after 20 years. Case files contain the petition, affidavits, copies of any certificates and, from the 1870s, a copy of the decree absolute and decree nisi.

USEFUL INFO

The Principal Registry in London holds a central index to all divorce suits since 1858. Divorce certificates issued in England and Wales can be obtained for a fee from

> *The Principal Registry of the Family Division*
> *First Avenue House*
> *42–49 High Holborn*
> *London WC1V 6NP*

A certified copy of the decree absolute can be provided, but if you would like to know the cause of the divorce then a copy of the decree nisi should also be ordered.

- The majority of case files for divorces between 1858 and 1927 have survived and can be found in J 77 arranged chronologically and by case number. There is an index to the surnames of the parties in J 78, but the names are arranged under the first letter of the surname and then according to the order the cases went to court, so if you are searching for the name Barnes you would have to search the whole of the index for 'B'. The National Archives has produced a research guide in Legal Records Information 44: 'Divorce Records in England and Wales After 1858', which gives a detailed description of how to use the indexes to locate a case file.

- Case files have survived for around 80 per cent of divorces that took place between 1928 and 1937, which are also kept in J 77, but files from the district registries were destroyed and these are not indexed in J 78.

- Divorce files since 1938 have mostly been destroyed, with the exception of a very small annual sample that can be searched by surname in the catalogue, restricting the department code to J 77. Another small sample of files for divorces where the Official Solicitor acted for one of the parties can be found in J 132 for later cases. Most files for divorces after the 1930s are subject to a 30-year closure.

If you are searching for records of a divorce after 1937 you could use the date and information about the court given on the divorce certificate to try searching local newspapers for a report of the case.

Scotland
Divorce Before 1830

The Scottish Reformation of 1560 established the right for a husband and wife to seek a civil divorce. Prior to this an annulment would have needed to be sought from the bishops' courts or a separation could be arranged, which meant that the couple were still legally married but did not have to live together as man and wife. The National Archives of Scotland has some records pertaining to early annulments and separations granted by some bishops' courts.

The Commissary Court of Edinburgh heard consistorial cases between husbands and wives from 1563, though the Court of Sessions had been doing so since 1560. Divorce was allowed on grounds of adultery, and from 1573 on grounds of desertion. Unlike English law, Scottish law did not allow a husband to pursue a Criminal Conversation suit against the

◀ A marriage chapel in Gretna Green, Scotland, where people could get married without parental consent.

lover of his adulterous wife for 'trespassing' her body, but in 1600 Parliament tried to outlaw the marriage of divorced adulterers to their lovers, though there were means of evading this Act. Divorce on the grounds of cruelty was controversial because, like elsewhere in the UK, it was generally accepted that husbands had a right to discipline their wives, and so separation was usually granted instead, until the Divorce (Scotland) Act of 1938. While divorce became increasingly accessible in Scotland through civil courts it was expensive to obtain and so many people continued to choose an informal separation.

An index to the names of people who pursued divorces in the Commissary Court of Edinburgh between 1658 and 1800 was compiled in 1909 by Francis J. Grant and published by the Scottish Record Society. *The Commissariot of Edinburgh – Consistorial Processes and Decreets, 1658–1800* gives the references and dates of cases so that you can find records in the relevant volume of judicial decisions (decreets) kept at the National Archives of Scotland in CC8/5, with corresponding case papers found in CC8/6. There are some earlier case papers that have not been indexed kept in CC8/6/1 from the period 1580 to 1624.

There is a separate index at the NAS to those decreets and case papers that were not indexed by Grant for the period 1801 to 1835. Use the index to persons found in CC8/6/176 to find records in the decreets also kept in CC8/5, and use the case lists in CC8/20/6 to locate the case files in CC8/6.

▶ A marriage certificate for William Scott and Frances Bell issued at Allison's Bank Toll House, Gretna Green, 1847.

Divorce After 1830

In 1830 the Court of Sessions became the sole court with responsibility for cases concerning marriage, divorce and bastardy, though the Commissary Court of Edinburgh continued to hear some cases until 1835. Changes were made to the law with the Conjugal Rights (Scotland) Act of 1861 and the Divorce (Scotland) Act of 1938, widening the grounds for divorce to include not only adultery and desertion but also cruelty, anti-social behaviour and non-cohabitation. The Divorce (Scotland) Act of 1976 made the crucial change of allowing divorce by mutual consent; however, it was not until 1984 that local sheriff courts could hear divorce cases so that people did not have to travel to Edinburgh.

Records of divorces that took place at the Court of Sessions between 1830 and 1971 can be found along with all other cases that were heard at that court. There are printed general minute books of the court at the National Archives of Scotland in CS17/1, which lead to card indexes and bound indexes to the cases. Cases of judicial separation, which was more common than divorce before the 1938 Act allowed cruelty to be a ground for divorce, can be found in the same way as divorce records in the Court of Sessions, up until 1907. The sheriff courts heard separation cases after 1907, and dealt with issues of spouse and child maintenance and child custody. These records are held at the NAS among the civil court processes and the registers of extract decrees in series SC.

Divorce cases heard in the Court of Sessions after 1971 have been indexed in a separate commissary index kept in CS17/3. The case papers are divided into two series: 'extracted' and 'unextracted' papers. The Register of Acts and Decreets in CS45 contain the judicial verdict for the extracted case papers. If an 'R' has been written next to an entry this usually means that children were involved in the case and the papers would have been retained by the court until the youngest child reached the age of 16, at which point the records would be transferred to the NAS.

The General Register Office for Scotland has kept

a Register of Divorces since 1 February 1984, indexed alphabetically. Divorce certificates can be obtained by writing to

The Extractor of the Court Session
2 Parliament Square
Edinburgh EH1 1RF

An extract from the register will tell you the court in which the case was heard so that you may locate the case papers. If it was heard in a sheriff court the case papers will still be with the sheriff clerk. The Court of Sessions retains case papers for divorces heard there for up to six years after the case (unless children were involved), before they are sent to the NAS.

Ireland
Divorce in Ireland Before 1922

Divorce has always been a controversial issue in Ireland because of the strength of both the Catholic Church and the Church of Ireland, and the religious conviction felt there. When civil divorce was introduced in England and Wales in 1858 the law was not extended to cover Ireland, and the definitions of divorce and separation in canon and civil law were conflicting and remained unchanged for a long time. Very few divorces were obtained in Ireland; only 39 judicial separations and suits for Criminal Conversation were won between 1857 and 1910, all of which were by Protestant petitioners, and as few as 12 were granted in the eighteenth century. Between 1700 and 1800 the Irish Parliament granted just nine divorces; however, the manuscript records of the Irish Parliament were destroyed by fire in 1922.

Any Acts of Parliament granting a divorce between 1800 and 1922 would have been dealt with in London, so a search of the English records is necessary. The National Archives of Ireland has some indexes to annulments granted in the 1860s and 1870s and for the very early 1900s, stating the reasons that were petitioned, but there are no unified indexes. If you believe your ancestor obtained some form of judicial or ecclesiastical separation from their spouse in Ireland you should contact the National Archives of Ireland explaining the time period you are researching and what you know of the background to the case. They will be able to point you in the direction of any records, if they survive.

Though divorce was extremely uncommon in Ireland, informal separation and desertion were inevitable. Private separation deeds could be legally enforced from 1848, but wives were not allowed to undertake such contracts until 1882. From 1886 magistrates' courts could issue maintenance orders against husbands who deserted their wives, and these continued to be issued throughout the twentieth century. Records of deserted wives and children dependent on parish relief may be found in local record offices. David Fitzpatrick has written an essay entitled 'Divorce and Separation in Modern Irish History' in *Past and Present* 114 (1987) in which he examines why divorce was so uncommon and explains how social mores prevented unhappy couples from separating and how such obstacles were dealt with in everyday life.

Divorce in Northern Ireland

Civil divorce was introduced in Northern Ireland in 1939, before which time only the Church courts or a private Act of Parliament could grant a divorce or separation. The Public Record Office of Northern Ireland holds some records for Church courts and of private divorces granted in the 1920s and 1930s, though these were as few as one a year. It is best to write to PRONI explaining what you know already about the annulment or divorce so that they can assess whether their records will be of use to you. If you believe your ancestors sought an annulment or divorce from the Church of Ireland you should contact the archivist at the Representative Church Body Library to find out where any relevant records might be found.

When civil divorce was introduced, the High Court began granting divorces in 1940, although it was not until 1982 that county courts could also hear divorce cases. The Probate and Matrimonial Office hold indexes for divorce certificates issued by both these types of court since 1940. To obtain a copy of the divorce absolute and nisi write to them at

> Probate and Matrimonial Office
> Family Division
> Royal Courts of Justice
> Chichester Street
> Belfast BT1 3JF

It is important to state that you want a copy of the decree nisi as well as the decree absolute because the nisi will give the reasons for the divorce. The court papers are not available to the general public, but once you have obtained a copy of the divorce certificate you can write to PRONI giving them all the information you have gathered so far, explaining your relationship to the divorcees and why you want access to the papers so that they can consult with the court service to see if access can be granted.

Divorce in the Republic of Ireland

Until very recently divorce or official separation could only be obtained in the Republic of Ireland by petitioning the Church courts or by obtaining a private Act of Parliament. When the Republic of Ireland was granted independence in 1922 the Constitution of the Irish Free State, which governed Ireland between 1922 and 1937, was undecided on its attitude towards divorce. A Bill was passed in the late 1920s that made it virtually impossible to remarry after divorce, though it was not until 1937 that remarriage was made illegal. Divorce itself was also made illegal by the new 1937 Constitution. It was not until 1995 that people voted for in a referendum for divorce to be legalized.

Even since civil divorce has been reintroduced, the grounds for obtaining a divorce are more stringent than in the UK and people who marry in a Catholic Church and subsequently obtain a civil divorce may not marry another partner in a Catholic church if their first spouse is still alive. The National Archives of Ireland does not hold any records relating to modern civil divorces because court material has to be at least 20 years old before it is transferred there, and the information is not generally accessible by the public.

Between 1937 and 1997 it was still possible to obtain an ecclesiastical separation or divorce *a mensa et thoro* through the Church, but remarriage was strictly forbidden and very few people applied for such a divorce. The Central Catholic Library in Dublin and the Catholic Archive Society of the United Kingdom and Republic of Ireland may be able to help you trace divorces granted by the Catholic Church prior to the 1990s.

Investigating Bigamy

The constraints of public moral attitudes, religious pressure and rigid and expensive legal procedures meant that few couples divorced if their marital bliss turned sour. Even those who could afford the cost of divorce were deterred by the time and effort required to convince a court to grant one, not to mention the publicity and scandal generated by the court hearings. The majority would simply opt for an informal separation and the unlucky ones found themselves deserted by their unhappy spouse. It is hardly surprising, then, that at least one of the partners in the marriage

> *It is hardly surprising that bigamous marriages were not uncommon.*

CASE STUDY

Vic Reeves

Vic Reeves had always suspected that his maternal grandfather, Simeon Leigh, had a secret – a previous marriage. This was fairly common knowledge within Vic's family, as the story had been passed down to Vic's mother; but rumours that he was a bigamist also abounded. Vic decided to have a look into his mysterious grandfather to see if there was any truth in the tale …

The first step was to ascertain when he married Vic's grandmother, and see what his marital status was. Simeon Leigh married Lillian Crowe in 1926, and indeed the marriage certificate stated that he was a widower, aged 39. Consequently, Vic searched for his first marriage – and found it recorded in Sculcoates, near Hull, in 1900 to Mary Jane Payne. A cursory check of the birth indexes after this date, though, provided a real shock – Vic found three children born to this marriage from 1901 to 1907.

Therefore, the strong possibility existed that Simeon Leigh was a bigamist. To prove the case one way or another, Vic started to look for evidence either that Mary had died – as Simeon claimed at the time of his second marriage – or that they had legally divorced. A search of the indexes to divorce case papers at The National Archives failed to find an entry under the Leigh or Payne surnames (in case she had reverted to her maiden name); similarly, an extensive trawl of death indexes from 1907 – the birth of Mary's last child – to 1926, when Simeon remarried – failed to find a match. Of course, there could be numerous explanations for why someone's death was not recorded in the civil registration indexes – Mary might have emigrated and died overseas, for example. So Vic tried another tack – forward reconstruction.

By following the lives of the three children from Simeon's first marriage, Vic hoped to find a living

relative from that side of the family who could fill in the gaps in his knowledge. He ordered marriage certificates for each of Simeon's children, and then looked for the offspring of these marriages – repeating the process until he'd tracked down a living descendant of his grandfather, completing the search by locating them via modern electoral lists. On visiting, Vic learned that Mary Jane Leigh had indeed survived, and was presented with a picture of her looking very much alive and well at the time of her eldest son's wedding in 1927 – one year after Simeon married Vic's grandmother bigamously.

would move away from their community and start a life elsewhere, and, given that cohabitation was so frowned upon, bigamous marriages were not uncommon prior to the mid-twentieth century.

Some people who had left their first spouse would marry a subsequent lover in an irregular marriage ceremony to avoid being caught by the authorities. Prior to 1753 some marriages were valid under English common law but were not recognized by ecclesiastical law. For bigamous men and women such marriages made it easier to escape attention because without the need for banns to be read for three weeks in advance the marriage could be arranged hastily and could take place away from their home parish. (There is more on irregular marriages in Chapter 7.) Mark Herber has transcribed and published the registers of some clandestine marriages in London in *Clandestine Marriages in the Chapel and Rules of the Fleet Prison 1680–1754*, which has an interesting introduction to the history of irregular marriages. Other similar registers can be found in local record offices.

Hardwicke's Marriage Act of 1753 put an end to these irregular ceremonies by decreeing that a marriage was only valid and lawful if performed by a priest in the parish church where one of the spouses lived, and the reading of banns gave anybody who was aware of an impediment to the marriage, such as one of the betrothed being married already, the opportunity to come forward. However, the Act did not extend to Scotland, Ireland or the Channel Islands, so couples could elope there to marry.

The 1753 Marriage Act, the outlawing of irregular marriages and the centralization of civil registration indexes should, in theory, have made it more difficult to commit bigamy, yet it was still easy enough to get away with telling lies until your past caught up with you. Bigamy had been a felony in England and Wales since 1603. Newspapers are full of the scandal of bigamy trials and local record offices will contain court proceedings against suspected bigamists (see Chapter 27 on tracing criminal ancestors). Trials were supported by evidence from former neighbours and friends and from relatives who could vouch for the validity of the bigamist's first marriage or could prove that they married again in the knowledge that their first spouse was still alive.

There were, of course, plenty of people who got away scot-free after committing bigamy, and so you may stumble across a bigamous marriage in your ancestry by accident. There are usually some giveaway clues. The bigamous person will probably have lied on civil registration certificates or parish register entries and the discrepancies between various documents might arouse your suspicion.

USEFUL INFO

Suggestions for further reading:

- Victorian Divorce *by Allen Horstman (St Martin's Press Inc., 1985)*

- Road to Divorce, England 1530–1987 *by Lawrence Stone (Oxford University Press, 1990)*

- Alienated Affections: The Scottish Experience of Divorce and Separation, 1684–1830 *by Leah Leneman (Edinburgh University Press, 1998)*

- Family Law *by Alan Shatter (Tottel Publishing, 4th edition, 1997)*

- 'Divorce and Separation in Modern Irish History' *by David Fitzpatrick,* Past and Present *114 (1987)*

- *The National Archives Research Guides, Legal Records Information 43: 'Divorce Records Before 1858', and Legal Records Information 44: 'Divorce Records in England and Wales After 1858'*

CHAPTER 27

Family Secrets: Criminal Ancestors

Whether we like it or not, there is a strong chance that at least one of our ancestors in the dim and distant past was involved in criminal activity. Sometimes these stories are passed down the family, leaving you with tantalizing clues about nefarious activity; in other instances you can stumble across the evidence, revealing a completely different side to relatives you thought you knew already. This chapter introduces you to the various sources available to investigate some of these rumours and begin to piece together the evidence for your ancestors' crimes and punishments.

> There is a strong chance at least one of your ancestors was a criminal.

Perhaps one of the more interesting revelations in your family tree (depending on the nature of the crime and when it took place) is the discovery of a criminal ancestor. Over time, hundreds of thousands of people were convicted of crimes, especially for theft or poaching when the only other option facing many poor people was to starve to death. The punishment meted out to such individuals was extremely harsh by modern standards (capital punishment being used in many cases). Indeed prisons were not used as a means of punishment until the nineteenth century, but simply to hold individuals awaiting trial. Until that date many crimes were punished by execution or transportation, the latter seen as a humane alternative despite the fact that a large percentage of transported convicts died en route to their penal colony.

As exciting as it may be to find a criminal ancestor in the family tree, the drawback is that criminal records are rather complicated to use. The individual may have been tried in any one of a number of different courts, depending on the nature of the crime, the time period and the location. The general rule is that the more serious crimes would be tried in the higher courts (the assizes or palatinate courts) and the less serious crimes in the lesser courts (quarter and petty sessions). However, it is important to remember that this was by no means always the case. Another complicating factor is that most cases were recorded in Latin until 1732 and thereafter often used legal abbreviations and terminology. Below is a background to the history of the various courts used to convict criminals in the country following the Norman Conquest.

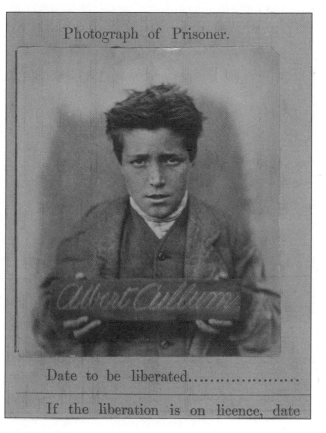

▲ A photograph of a young prisoner, Albert Cullum, taken on his incarceration.

Historical Background of the Legal System

Initially after the Norman Conquest, the manorial courts of individual lords of the manor were responsible for dealing with less serious criminal offences. By the fourteenth century, however, this system was replaced by using Justices of the Peace (or magistrates) who would settle matters such as petty theft, drunkenness and assault amongst others. These men were given authority by royal commission and were prominent members of the local community rather than professional lawyers or judges. They would meet four times a year in every county of England and Wales in what became known as 'quarter sessions' and their jurisdiction increased through the centuries. They could also give judgment in certain cases without the need of a jury. When judgment was made without a jury it was termed 'summary jurisdiction' and after the eighteenth century this became a popular method of dealing with minor crimes and was given the term 'petty sessions', as JPs would meet to decide these cases outside the quarter sessions. In urban areas such as towns and cities quarter sessions were known as 'borough sessions'.

The more serious crimes between the twelfth and fourteenth centuries were not tried at the manorial courts, but at the hearings of the General Eyre ('eyre' being derived from the Latin word for journey, *iter*). This court was made up of a group of judges appointed by the King who would travel around the country from their base in London

CASE STUDY

Lesley Garrett pt 2

Lesley Garrett had often wondered why little information about her paternal Garrett ancestors had passed down to her. Equally, her father, Derek Garrett, knew little about his great-grandfather Charles Garrett – so Lesley decided to investigate further. Putting together the family tree was relatively easy, by ordering certificates and cross-referencing with census returns. All these documents suggested that Charles had been a master butcher, with a shop in Finkle Street, Thorne; and this was confirmed by entries in trade directories, found in the local archives when Lesley visited to find out more about her elusive ancestor. So she was surprised to find that he had actually been one of the leading men of the town, rising to councillor and seen by his peers as one of the pillars of the community. Therefore why had no memory of such a prominent man passed down through the generations?

The answer came when Lesley did some further digging around Charles and his family. It became apparent that his wife, Mary Ann Garrett, had died in 1899; but the death certificate showed that it was not from natural causes! Indeed, she had died from poisoning by carboxylic acid, a chemical regularly used by butchers to swill away blood and other animal products from the workplace at the end of each day. The death certificate showed that a coroner's inquest had been held. By cross-referencing the date of the inquest recorded on the certificate, Lesley was able to search the local newspaper, where details of the inquest were reported.

According to the transcript, Charles Garrett had poured his wife her tonic drink but had mistakenly given her carboxylic acid instead of wine. The coroner absolved the 'distraught' Charles from any blame, but certain questions remained unanswered – such as how could a man, used to handling such a potent chemical on a daily basis, make a fatal mistake when carboxylic acid was clearly stored in a poison bottle, and why did he smash the offending bottle afterwards? Because no charges were ever brought against Charles, he continued with his business; but Lesley felt sure that members of his family suspected him of poisoning his wife on purpose – Charle's son, Tom, and his daughter-in-law, Mary, broke off contact with Charles. This may explain why so little information about him has filtered down to her today.

▲ Mary Garrett, Charles and Mary Ann's daughter-in-law, outside the family's butcher's shop.

▶ Lesley's great-great-grandfather Charles Garrett (far left bottom row) photographed with the Thorne Parish Councillors in 1901.

administering the King's justice in a wide range of civil and criminal matters. They would visit the various counties every few years to fulfil this duty, although the frequency and number of judges travelling would often vary.

The Eyre system was replaced by the Assize (the word 'Assize' is derived from the Norman-French for 'sitting') system during the course of the fourteenth century. Assizes would become the main courts for criminal trials until their abolition in 1971. The assize system was based on its predecessor, with judges from Westminster visiting the various counties of England. These judges would often travel in pairs on circuits for different parts of the country (the Western or Midland circuit, for example). The authority of these judges rested with the commissions of gaol delivery (to try prisoners) and peace and, from the 1530s onwards, of 'oyer and terminer'. The term 'oyer and terminer' literally means to 'hear and determine' and refers to felony cases that could not be tried in local courts (cases of treason, murder, rape, rebellion, burglary, etc.). The judges would hold court alongside a jury of 12 men.

By 1340 six assize circuits had been established: Home (the counties surrounding London), Western, Oxford, Norfolk, Midland and Northern. Certain parts of the country fell outside the jurisdiction of these circuits. The City of London (and later Middlesex), Wales, Bristol and the palatinates of Chester, Durham and Lancaster all had separate systems and will be discussed below.

The highest court in the land, with overriding jurisdiction over any of the above courts, was the King's Bench. This court was situated in Westminster, being established in the twelfth century, and dealt with both criminal and civil matters. Criminal cases were dealt with under the 'Crown side' and civil cases under the 'plea side'. It served as a court of appeal for any supposedly erroneous convictions made by the lesser courts and was reformed as the Queen's Bench Division of the High Court of Justice in 1875.

Searching for Trial Records

As has been seen, there were numerous courts dealing with criminal matters. In order to search the original records it is necessary to have a rough idea of where and when the trial occurred. Often the best course of action is to start searching through published sources including newspapers and pamphlets, as they provide the crucial pieces of information necessary to search the original records.

Published Sources

1.

Newspapers

Local newspapers would often report and give useful summaries of criminal trials occurring in the locality and may also list the exact date of the trial and which court the trial was held at. If the crime was particularly newsworthy it may also be featured in national newspapers such as *The Times* and it is always worthwhile searching these sources if the crime occurred the late eighteenth and nineteenth centuries (when newspapers began to be published).

2.

Law Reports

These were published summaries and judgments of key civil and criminal cases that were used to decide matters of law. As they were only compiled for those casing affecting the drafting of law they only cover a very small percentage of total cases. Nevertheless, if the case of interest is included, the report will also give the exact details of the trial to allow easy access to original sources. They can be found in large reference libraries and legal libraries and institutes.

3.

T. B. Howell's A Complete Collection of State Trials and Proceedings for High Treason *(Longman et al., 33 vols, 1816–26)*

This is a published collection of criminal trials from the twelfth to the nineteenth centuries, relating in the most part to those being tried for crimes against the state. The source is now available to search on CD-ROM.

4.

British Trials 1660–1900 *(Chadwyck-Healey)*

This publication is sourced from the numerous contemporary pamphlets that were printed reporting the more sensational criminal cases. The source is available in a microfiche format and is indexed by names of victims and defendants, location of the trial and the nature of the crime. The microfiche can be found in The National Archives, the British Library, London Guildhall Library and other large reference libraries. The original pamphlets themselves are scattered amongst national and local archives.

5.

Notable British Trials

This is an 83-volume collection of the more famous and interesting trials that occurred in England and Scotland, published from 1905 to 1959. It includes details of the trials themselves along with witness accounts, and therefore provides the required details needed to access the original records.

Original Trial Records

As mentioned above, the trial records themselves, where they do survive, can be complicated to use. Prior to 1732 they were recorded in Latin and there are no comprehensive name indexes to the records (although indexes exist for certain sections of the records). The records themselves can be found either in The National Archives or local repositories, dependent upon which court the individual was tried in. The records are also subject to closure rules, also dependent upon when the trial occurred, and some may not be in the public domain.

Manorial Courts

These were the oldest courts dealing with less serious criminal matters following the Norman Conquest. The records survive in the form of court rolls or books detailing the minutes of the courts. Three types of manorial courts existed:

1. *The Court Leet:* This court would try certain offences including assaults, poaching and other nuisances. Its judicial role came to be superseded by the Justices of the Peace after the fourteenth century (see below).
2. *The Court Baron:* This court was responsible for disputes between the lord of the manor and his freehold tenants.
3. *The Court Customary:* The Court had the same function as the court baron but for disputes between the lord and his customary tenants. The court baron and court customary would usually sit together and would be known as the court baron; they were not primarily interested in criminal matters.

The records for manorial courts are scattered in numerous archives. They may be at The National Archives, the British Library, the Bodleian Library, local record offices or even private collections. The collections begin from around the thirteenth century but many do not survive in their entirety. The best place to begin searching is by using the Manorial Documents Register held at The National Archives. The Register details exactly where records for each manor are located along with what exactly survives for a specific manor in England and Wales. It has been partly computerized for certain counties and can be searched online on The National Archives' database. However, not every county has been placed on the online database at the time of publication and it may only be possible to search the register onsite by visiting the archive.

Trial records can be complicated to use.

Quarter/Petty Sessions' Records

From about the fourteenth century onwards, the function of the manorial courts gradually became replaced by judicial sessions held by Justices of the Peace, known as quarter or petty sessions (detailed above). Indeed by the sixteenth and seventeenth centuries the JPs had a number of administrative functions aside from less serious criminal trials, including licensing certain trades and overseeing the enforcement of the Poor Law. The JPs continued to meet quarterly per annum until 1971 and surviving archives for such meetings originate from the records maintained by the Clerk of the Peace.

Surviving records for quarter or petty sessions are to be found in local record offices or archives. The type of information retained by archives includes order and minute books, indictments and sessions' rolls. As surviving records may be scattered amongst different archives of the county it is advisable to refer to Jeremy Gibson's book, *Quarter Sessions Records for Family Historians* (1995). Gibson provides a detailed list of the exact location of surviving records, including for which dates the material survives. Additionally, it is always worthwhile referring to the local county record office's website as many such institutes are placing their catalogues online.

Records of the General Eyre

The General Eyre was responsible for administering justice for the more serious crimes between the twelfth and fourteenth centuries in all counties of England apart from Durham and Chester. These records are now in The National Archives, mainly in the Eyre series in JUST 1–4. The nature of the records themselves is very formal and they are heavily abbreviated by the clerks who produced them. Hence, using these original records may be somewhat problematic. However, some sections of the records have been translated and published by the Pipe Roll Society and local history societies. A list of exactly what has been translated can be found in David Crook's *Records of the General Eyre* (Public Record Office Handbook No. 20, 1982). This publication offers a comprehensive overview of the records of the General

▼ Prisoners working on the tread-wheel in the Vagrants' Prison, Coldbath Fields, 1862.

Eyre and their level of survival for each county. Additional guidance can also be obtained from referring to the appropriate research guide on The National Archives website.

Assize Records

The majority of criminal trials were held in assize courts for England and Wales. As mentioned, these courts were the successor courts of the General Eyre and were responsible for trying capital cases or cases involving transportation. The majority of surviving records are held in The National Archives, although some records may still be with local record offices. The records that make the archives originate from the documents kept by the Clerks of the Assize at the time of the trials although many have not survived to the present day. The archives are organized on the original circuit system by which the judges would visit the counties, and then by the type of record. The nature of records for the assizes differed over time and the survival rate depends on which time period it was, but generally the main types of records are:

- *Crown and Gaol Books:* These were also known as minute books. They would list the prisoners and their alleged crimes for which they were to be tried. They may also contain notes of the subsequent verdicts of the trials.
- *Indictments:* These are possibly the most useful documents as they list the exact nature and details of the charge, the name of the accused (including his or her plea) and victim, the date of the offence and any other relevant detail. They may also contain later notes detailing the trial verdict. Unfortunately, very few indictments survive until the mid-seventeenth century apart from for the Home circuit.
- *Depositions:* These were the pre-trial statements taken from witnesses of the alleged crime and were written in English. They may form a separate series of records or be included with the indictments. After 1830 the depositions were only kept for the most serious offences (murder, riot), although the survival rate improves again from the mid-twentieth century.

The majority of the surviving records are in The National Archives in the ASSI series. In order to make a search feasible it is necessary to know approximately when the trial took place and in which county. It is then possible to refer to the key for the records to pinpoint the correct reference, which is also given in the appropriate research guide of The National Archives.

> *Additional details can be found by referring to prison and convict records.*

Records for Areas Outside the Remit of the Assize System

1. City of London/Middlesex

The City of London had a unique judicial system combining the quarter session and assize system and this system incorporated Middlesex after 1540. It was under the administration of the Lord Mayor, aldermen and recorder meeting at the Old Bailey, with certain sessions being held at the London Guildhall. These sessions were replaced in 1834 by the formation of the Central Criminal Court whose jurisdiction covered the same area, together with certain parts of Essex, Kent and Surrey (which came under the greater London area). The Central Criminal Court continued to sit in the Old Bailey.

The records for the Old Bailey prior to 1834 are now held at the London Metropolitan Archives. Additionally, a great deal of material has now been placed online at www.oldbaileyonline.org. The website offers online access to all proceedings from 1674 to 1834. It is possible to search this website free of charge and by individual surname. There are also plans to digitize later proceedings, from 1834, when the Old Bailey became the Central Criminal Court, to 1913. This project will add a further 100,000 cases onto the database and is due to be launched in 2008. Otherwise, the records of the Central Criminal Court can be accessed in person by visiting The National Archives. They are held in series CRIM for the following sets of documents (all subject to 75- or 100-year closure rules):

- After trial calendars in CRIM 9 (1855 to 1949) and HO 140 (1868 to 1968).
- Depositions are in CRIM 1 (1839 to 1971) and J 267 from 1972 onwards.
- Indictments are in CRIM 4 (1834 to 1871), indexed by CRIM 5 and J 268 (1972 to 1974), indexed by J 368.
- Court books are in CRIM 6 (1834 to 1949).

2. Bristol

Bristol also had an ancient right to hear criminal cases, until the right was abolished in 1832 and Bristol became incorporated into the assize system. Records for Bristol prior to 1832 are with the Bristol Record Office. The assize records can be found in the appropriate assize section at The National Archives.

3. Wales

The Great Session of Wales also had autonomous jurisdiction from 1542 to 1830; thereafter Wales was also incorporated into the assize system. The National Library of Wales will hold the surviving documentation of the Great Session of Wales and The National Archives has the relevant records for the assizes.

4. The Palatinates

The palatinates of Cheshire, Durham and Lancaster also had their autonomous criminal courts until their incorporation into the assize systems in 1830, 1876 and 1877 respectively. These records of the palatinate period are not held locally but with The National Archives in series CHES (Cheshire) DURH (Durham) and PL (Palatinate of Lancaster); the later records are with the assize records.

Records of the King's (or Queen's) Bench

The Court of the King's Bench, established in the twelfth century, became the court of appeal for criminal matters and could overturn judgments made in the assizes or any other courts. The surviving records are held at The National Archives. The court was abolished in 1875 and replaced by the Supreme Court of Judicature. The main sets of records worthy of consultation are as follows:

- Indictments: KB 10–12 from the mid-seventeenth century onwards.
- Depositions and affidavits: KB 1 and 2.
- Controlment rolls: KB 29 (detailing the progress of a case).
- Court orders: KB 21.
- Exhibits: KB 6, for the nineteenth century only.
- Final judgments: KB 28, although most cases did not reach that stage. There is a card index for this series for 1844 to 1859.

Searching for Prison and Convict (Transportation) Records

After tracing your ancestor in the appropriate trial records, you may also find additional details of such individuals by referring to prison and convict records. Indeed on occasion, as they can give details of trial, it may be more effective to search these documents first as they can point you in the right direction. Although prisons were built from the thirteenth century onwards, imprisonment was only used by central government as a wide-scale form of punishment from the nineteenth century onwards. Prior to that local counties would run and maintain prisons, usually as temporary means of holding those awaiting trial. Indeed most individuals convicted of a crime would be either fined or whipped (for less serious offences), executed, or from 1700 onwards transported to North America (the term 'convict' referred specifically to those sentenced to transportation or hard labour imprisonment and 'prisoner' to individuals simply imprisoned). Transportation to North America ceased after the American War of Independence and convicts would be imprisoned in 'hulks' (old ships used to house prisoners with the

▼ A chain gang, Hobart Town, Tasmania, 1831.

▼ Portraits of prisoners in Pentonville, 1876.

eventual expectation that they too would be transported). The use of the hulks constituted the first large-scale involvement of central government in using prisons as punishment.

As both local government and central government used prisons during different time periods, surviving records will be found in either local archives or The National Archives.

Local Archive Records

Initially, as we have said, local counties used gaols as places of temporary imprisonment for those awaiting trial. However, after the passing of Elizabethan poor laws, 'houses of correction' (or brideswells) were established to house criminals, vagrants and those unwilling to work. Sheriffs would provide lists of those housed in local gaols and houses of correction to the JPs and these surviving lists may be found with the quarter session records at county record offices. These records may also include reports on the conditions of gaols and journals of gaolers, as an Act in 1823 made JPs responsible for maintaining prisons.

Registers in The National Archives

The main body of records held in The National Archives are those created from the late eighteenth century onwards. The main sets of records worth consulting include:

- Criminal Registers in HO 26 and HO 27. HO 26 covers those convicted in Middlesex from 1791 to 1849. Thereafter Middlesex is included in HO 27, the series for all the other counties of England and Wales, from 1805 to 1892. Both registers provide details of where the crime was committed, where the individual was tried, age and sentence (including date of execution).
- Criminal Registers in HO 140. These are post-trial calendars for all those convicted in England and Wales in assize and quarter sessions from 1868 to 1971 (with closure rules).
- Prison registers and calendars. Many counties kept these from the late eighteenth century onwards to record the prisoner's name, crime, trial date, verdict and date of custody (in calendars published after the trial). The Home Office only become responsible for the administration of prisons in 1877 and, therefore, early documents may still be with the local archive. Nevertheless, The National Archives has a good representative number before 1877 in a variety

CASE STUDY

Jeremy Irons

Jeremy Irons already knew a fair amount about his family history, as various members of his family – including his cousin – had already done a great deal of research. The Irish side of his family traces back to Cork at the start of the eighteenth century via his grandfather, Henry Sharpe. One character, though, that stood out from another part of the family was his 2 x great-grandfather Thomas Irons, who was alleged to have been involved with the Chartist movement of the 1830s and 1840s, which sought political reform and an extension to the franchise. Jeremy investigated the biographical details of Thomas's life in more detail, finding his marriage certificate in 1840 that revealed he was a policeman – though a search of his police records at The National Archives revealed that he had actually been thrown out of the force in 1834 for being drunk and deserting his post.

Of particular interest was Thomas's involvement with the Chartists following the massive rally on Kennington Common of 10 April 1848. Jeremy's research into the history of the period in his local library showed that on this occasion a large petition of over 1 million signatures was handed to Parliament, but nothing came of the proposals. As a result, elements of the Chartist leadership adopted more radical tactics, including various attempts to start insurrections in London.

Two such plots were exposed on 16 August 1848, known as the Orange Tree and Powell Plots. The newspaper reports of the day showed that one of the men arrested at the Angel pub in Webber Street was – Thomas Irons. The conspirators were found with weapons, and the leaders tried, found guilty and transported to Australia. Although Thomas was tried and convicted too – Jeremy located the court papers in The National Archives – he was sentenced to 18 months in Newgate prison instead. Once again, Jeremy was able to use newspaper reports to find out what happened to his family during his incarceration; and it would appear that they were supported by donations from Chartist sympathizers.

◄ Jeremy's ancestors also included the Sharpe family (from left to right) Henry Curtis, William, Charles and Catherine (bottom left).

Transportation

Transportation arose in the seventeenth century as a humane alternative to capital punishment. It has already been touched upon in Chapter 23. Although the majority of records relating to transportation to North America and the West Indies were discussed, information for Australia concentrated on records for the physical transportation of convicts rather then actual criminal documents. The majority of these records are held in The National Archives. In the first instance it may be worthwhile consulting published works such as David T. Hawkings's *Criminal Ancestors* or *Bound for Australia* (Chichester, 1987), as they give transcriptions of many of the original sources containing information for transportation. Additionally, the CD-ROM compiled by the Genealogical Society of Victoria (see Chapter 23) may be worth checking as it contains an index to all convicts arriving to New South Wales between 1788 and 1842 (including the name of ship they were transported on, and therefore details of trial can be obtained).

● The transportation registers can be found in The National Archives series HO 11. They cover the entire period of transportation to Australia, from 1787 to 1868. This series is organized by name of ship and the date the ship left. If you have already found your ancestor in the trial records, details of the ship may also be included, allowing you to access these records. Otherwise, as there are no indexes to the name of the convict, the search may be unfeasible.

● It would not be uncommon for wives to wish to travel with their convicted husbands and they could petition the government for the of series, such as PCOM 2 (1774 to 1951, including photographs of Victorian prisoners in some sub-series), HO 23 (1847 to 1866 for a number of counties), HO 24 (London prisons from 1838 to 1875) and HO 16 (1815 to 1849, for those awaiting trial at the Old Bailey).

● Quarterly returns of prisoners on hulks and prisons in HO 8 (1848 to 1876). These are arranged by name of ship or prison but include details of trial, physical description, employment and marital status. Photographs may also be included.

● The central Habitual Criminal Registry was established after the passing of the Habitual Criminals Act in 1869. The Act was passed in response to concerns that criminals were no longer being transported to Australia and would, therefore, have to be released back into the community at some stage. Local prisons were asked to compile lists of 'habitual criminals' (anyone who had committed offences specified in the above Act) along with physical description, residence, age and other identifiable details, together with photographs, and send these to the central Registry where a national list was compiled. An incomplete set of these registers can be found in PCOM 2 and MEPO 6/1–52.

● The increase in costs by having to imprison criminals and not transport them led to the development of a parole system, known as right to do so. These surviving petitions can be found in The National Archives series PC 1/67–92 (1819 to 1844) and from 1849 in series HO 12 (accessed by using the registers in HO 14).

• Series TS 18 contains records of contracts made with agents who transported the convicts. The series provides full lists of the ships used and names of convicts (along with trial details). Registers for hulks specifically carrying convicts can be found in HO 9, TS 8, ADM 6/418–23 and PCOM 2/105 and 131–7.

• Death sentences were often commuted to transportation, sometimes after petitions of clemency were received by the government from the convict's loved ones. Some of these petitions also survive and can be accessed at The National Archives. They are often more useful than trial records for the family historian as they provide a great deal more detail of the convict, the circumstances around his or her crime and details about his or her family and occupation. They can be accessed in series HO 17 (1819 to 1839) and HO 18 (1839 to 1854), indexed by the registers in HO 19. HO 19 refers to petitions from 1797 onwards, although the actual petitions only survive from 1819 onwards. Further petitions can be found in HO 48, HO 49, HO 54 and HO 56 but there are no indexes available for easy access to these. Please note that these petitions were not solely for convicts but for all prisoners in general.

• Another useful group of documents relating to criminal records is the judges' reports. They often contain unofficial summaries of evidence provided in cases, comments on the case and witnesses made by the judges, and petitions compiled by the accused's loved ones. They are also in The National Archives, in series HO 47 (1784 to 1829) and HO 6 (1816 to 1840).

'tickets of leave'. It was introduced in 1853 and allowed convicts displaying good behaviour to be released into the community early by a licensing system. These licences were issued from 1853 to 1887 and can be found in PCOM 3 for male convicts and PCOM 4 for female convicts, indexed by PCOM 6.

The above is a brief summary of some collections held at The National Archives. A more thorough list can be found by consulting David T. Hawkings, *Criminal Ancestors: A guide to historical criminal records in England and Wales* (Alan Sutton, revised edn 1996).

Execution

Execution by various means was used as a punishment in the UK until its abolition in 1965. It was commonly used for many minor offences (forgery, pickpocketing, poaching and other forms of theft). Indeed, at its zenith over 200 crimes were punishable by death. Often juries would be unwilling to find defendants guilty as the punishment was so severe, which was the reason why transportation was seen as the 'humane' alternative.

The extensive use of capital punishment was reformed during the nineteenth century and by 1861 the Criminal Law Consolidations Act ensured that only five crimes would be punishable by death: murder, treason, espionage, arson in royal dockyards and piracy with violence. In 1868 the Prisons Act made public hangings illegal, and hanging itself was finally abolished in 1965.

Most records for capital punishment lie within the usual sets of criminal records. As people were often sentenced to death for murdering someone these trials would often attract the attention of the press and it is important to search local and national newspapers. Another published source that might be of use is an examination of people hanged for minor crimes in the eighteenth century in London – Peter Linebaugh's *The London Hanged: Crime and Civil Society in the Eighteenth Century*. The book has accounts of many individuals hanged at Tyburn for committing minor crimes.

There are also some records specific to capital punishment at The National Archives.

- General records for those hanged in the nineteenth and twentieth century can be found in HO 163, MEPO 3 and PCOM 9.
- There is a small sample of files for those hanged in the twentieth century in HO 336 to display the type of information collated for such individuals.
- Details of how capital punishment was conducted can be viewed in PCOM 8, HO 42 and HO 45.
- HO 324 contains the records of prison graves, including a register of those buried after being hanged in HO 324/1 (1834 to 1969).

Criminal Records in Scotland

Scotland has always had its own legal system for both criminal and civil cases. Most surviving records can now be found at the National Archives of Scotland. The records have been listed in detail by Sinclair in *Tracing Your Scottish Ancestors: A guide to ancestry research in the Scottish Record Office,* and a summary is provided below.

The least serious crimes, such as drunkenness, debt and other minor offences, were usually tried in burgh or franchise courts or, after 1609, by Justices of the Peace. Sheriff courts were responsible for administering justice in offences such as theft or assault. Surviving records can be found in the National Archives of Scotland, under series SC (sheriff court) and JP (Justice of the Peace). They begin in the eighteenth century but are not indexed and tracking a case may prove time consuming.

The highest court in the land was the High Court of Justiciary, established in 1672, which sat in Edinburgh and also travelled the country on circuit. The court dealt with the most serious offences (rape, robbery, murder) and also served as an appeal court with the authority to overrule the decisions of the minor courts. The surviving records include minute books, lists of cases and various indexes. There are also published guides which may be useful, such as *Ancient Criminal Trials in Scotland, Compiled from the original records and mss; with historical illus* by Robert Pitcairn (1833), and *Records of the Proceedings of the Justiciary Court, Edinburgh, 1661–1678*, 2 vols, ed. W.G. Scott-Moncrieff (Scottish History Society, 1905).

Other sources that may prove useful include:

- Statements made to the Lord Advocate (known as 'precognitions') when deciding if an alleged crime should be tried. These are found in the National Archives of Scotland under reference AD.
- Registers of prisoners, mainly for the nineteenth century, are also at the National Archives of Scotland.
- All those who were executed in Scotland between 1750 and 1963 have been listed in a book by **Alex Young**, *The Encyclopaedia of Scottish Executions.* The illustrated book has biographies of all of the 464 individuals executed in Scotland during that period.

Criminal Records in Ireland

Many legal records for Ireland were lost during the fire in the General Register Office in 1922. What little does survive can be found in the National Archives of Ireland. It has the surviving records for the Grand Jury Presentment sessions and the General Prison Board Collection (1836 to 1928). The Public Record Office of Northern Ireland (PRONI) has a significant collection of court records for Ulster.

As the survival of original records is limited, it is recommended that you consult published sources such as local and national newspapers. Another useful source is *The Irish Reports* by Butterworth Ireland Ltd, containing hundreds of volumes of reports of legal cases, starting in 1838. The National Library of Ireland has copies of the *Hue and Cry* (later renamed the *Police Gazette*), a weekly publication from 1822 listing all wanted criminals and escaped prisoners.

USEFUL INFO

Suggestions for further reading:

- Criminal Ancestors: A Guide to Historical Criminal Records in England and Wales *by David T. Hawkings (Alan Sutton, 1992, revised edn 1996)*

- Family Skeletons: Exploring the Lives of Our Disreputable Ancestors *by Ruth Paley and Simon Fowler (The National Archives, 2005)*

- Quarter Sessions Records for Family Historians *by Jeremy Gibson (1995)*

- A Guide to the Records of the Great Sessions in Wales *by Glynn Parry (Aberystwyth, 1995)*

- Tracing Your Scottish Ancestors: A guide to ancestry research in the Scottish Record Office *by Cecil Sinclair*

- Records of the General Eyre *by David Crook (Public Record Office Handbook No. 20, 1982)*

CHAPTER 28

Social History: Working Further Back in Time

The majority of this book has shown you how to research aspects of your ancestors' lives generally through the nineteenth century, or at a stretch the late eighteenth – simply because this is the period of history when you can trace most of your ancestors with the greatest degree of success. However, it is possible to research further back in time, particularly if you have ancestors of a higher social status.

> *There is a compelling desire to explore family histories further back in time.*

There are several good reasons for covering topics and associated record collections from the eighteenth and nineteenth centuries. First, the world we see around us is a product of the Industrial Revolution that grew in momentum as the century progressed, completely altering Britain's landscape, economy and way of life for most of its citizens in the process. Cities such as Manchester and Birmingham simply didn't exist in their current size and form prior to the mid-eighteenth century, when mechanized industry first saw the rise of factories, mass employment and urban expansion. Prior to the Industrial Revolution, port cities such as London and Bristol were the dominant centres of commerce, whilst even further back in time, when agriculture and the wool trade dominated the medieval economy, provincial centres and market towns were the prosperous places to be. For most of the time prior to the seventeenth century, Norwich was England's second city, drawing upon the fantastic wealth generated by the East Anglian wool trade. This is why there are so many large stone-built churches in Norfolk and Suffolk serving tiny parishes; the riches of local merchants funded their construction.

Second, the population was far smaller prior to the massive expansion witnessed throughout the nineteenth century. Today, we live in a country of 60 million souls, with 49 million in England. In 1801, there were only 8.5 million people in England. Within a hundred years, however, this had increased to 30.5 million, rising on average by 2 million each decade.

▲ A ploughman (14th century).

By way of contrast, the population on the eve of the Black Death in 1348 was 3.75 million. It had risen at the end of the Tudor period in 1603 to 5.8 million, a figure that didn't expand greatly over the next century or so. The population began to increase in the mid-eighteenth century, when the Empire brought trade, wealth and opportunity to Britain and fuelled the growth of provincial towns, particularly along the coast. The nineteenth-century expansion was due to new towns and cities rapidly growing, providing accommodation for families leaving the countryside to find regular work, with infant mortality gradually lessening over time to keep the population rising. Even more people arrived through some of the immigration patterns outlined in Chapter 22.

Closely linked to the Industrial Revolution and population expansion is the accompanying interest by the State in these events, which in turn generated a need to register, regulate and record. The main products of the Victorian obsession with statistics and paperwork have been described in Chapters 5 and 6 – civil registration and census returns – and these now form the bedrock of most people's family trees. Prior to these innovations, no central set of records exists that covers the entire population. Parish records provide a record of ecclesiastical events, but are not systematically kept, and the further back in time you go the sparser the information becomes. Furthermore, there are no complementary records for most sections of society that can be used to corroborate or flesh out details. Wills, as detailed in Chapter 8, were the preserve of the wealthier classes who had goods and possessions to pass on; the majority of the population were not of sufficient status to bother with or need wills.

Research Tips

People often assume that they can trace their ancestors back to the Norman Conquest, and their Holy Grail is to find a relative in the Domesday Book. Whilst there are a few families that have a legitimate claim, borne out by proof, that they came over to this country with William the Conqueror, most of the time there are large gaps in the family tree. For the majority of us, it simply isn't going to be possible to collect enough evidence, even though we clearly have a link to *someone* who was alive at that time, otherwise we wouldn't be here! The moral of the story is that you'll need to have some fairly realistic expectations before you start; it can

be tricky enough proving a link between generations in the early to mid-nineteenth century, when census records stop being useful and civil registration was a new innovation that many people treated suspiciously. Furthermore, you'll encounter problems with the evidence that you do collect. To help you navigate further back in time, here are some useful research tips and pointers:

● Before haring off into the distant past, you should first try to consolidate your family tree. Complete as many branches as you can with official sources, so that if you get stuck working back down one line, you can return to a known ancestor and start again.

● When working backwards prior to civil registration and census returns, it's better if you choose a branch of the family that has an unusual name as they will be easier to spot in earlier records such as parish registers, where there are fewer sources against which to corroborate.

● The social status or occupation of your chosen family will be important as well. The reason we can track back using certificates and censuses is that they included everyone; further back in time, it's much easier to trace people with land or money, or both. Skilled professions based in towns also help.

● Since the majority of our ancestors were largely country

Yet there's a compelling desire to work further back in time, to break through the nineteenth-century record barrier and explore the lost world of the past that lies beyond. Many thousands of people have done this, though it is tricky, and it will take far longer to construct a family tree prior to 1800 than after. Moreover, unless you are very lucky, it will be much harder to work out what these people were like, since records for occupation, residence and lifestyle rarely exist. Nevertheless, it is well worth carrying on regardless – you never know, you might strike lucky and make a connection to a branch of the family that were of sufficient status to leave records, or maybe even prove a link to aristocracy or blue blood!

Manorial Records

If you strip away the towns and cities of modern life, and return to the days of the early 1700s, you'll find that there are no major population centres apart from London, which, with its population of 1 million in dwellers and without much in the way of wealth, the only time they tend to appear in records is when they fall foul of authority, or are on the margins of society through poverty or illness. These were the exceptions rather than the norm, so don't be surprised if you can't find your relatives listed in official sources.

- Geography becomes even more important, since the majority of people didn't move around much outside their local area. On average, someone born in a rural community would never move more than ten miles away from their place of birth. Therefore if you are searching for records online, it becomes more important than ever to focus on local matches.

- The survival of records becomes much less likely the further back in time you go. Very few personal documents will exist in archives, aside from estate and family papers deposited by landed gentry and aristocracy, whilst prior to parish registers the records diminish dramatically. This means that you are unlikely to get much further back than the sixteenth century.

- Where documents do survive, official sources are likely to be written in Latin prior to 1732. Similarly, handwriting may also be hard to read – which is why you should follow the advice provided in Section One and enrol on a palaeography course, and invest in a Latin dictionary.

- Finally – and it should hardly need repeating at this stage of the book – never assume a link between sources.

With these points in mind, it's time to examine some of the key sources for tracing your family prior to 1800. They are grouped into four main areas, starting with the records of one of the most important medieval and early modern administrative units – the manor. After that, various government sources are examined, followed by a look at some of the civil courts in which our ancestors regularly disputed with one another, with a final look at the records that survive for higher ranks of society.

1800, was the largest city in Europe. Most cities were on the coast, or acted as provincial centres; throughout the medieval and early modern period, none could rival London in terms of population size. For example, by 1700 just over half a million people lived in the capital, which represented about 10 per cent of the overall population; the next largest city at the time was Bristol with about 30,000 inhabitants, having finally outgrown Norwich, previously England's second city in terms of population density. Provincial centres such as York, Nottingham, Exeter and Worcester made up the next tier, averaging about 10,000 residents. By way of contrast, Manchester was home to 6,000 people, Liverpool and Birmingham 4,000 each.

Essentially, England was a rural country until the commercial boom of the mid-eighteenth century, followed by the industrial expansion of the late eighteenth century onwards. The majority of its population lived in villages and hamlets clustered around market towns, with the provincial towns and cities acting as hubs for the local economy and providing a base for commerce and industry. Within these rural communities, there were two main administrative units. From an ecclesiastical perspective, the parish was the key binding force that kept everyone together, providing the spiritual heart to life. Yet equally important, and dating back to the late eleventh century and the Norman Conquest, was the secular manor; and the records generated by the manorial system can be used to trace your relatives, potentially even further back in time than the earliest parish registers.

The Manorial System

A full discussion of the medieval and early modern social hierarchy and methods of local government would fill a book in its own right, but a brief discussion of the manorial system is required to understand the records generated as a result, and how you can use them to trace your ancestors.

You may be familiar with the manorial system from history lessons, with simplified diagrams of the 'feudal pyramid' showing how all land was owned by the King, who granted it to a few tenants-in-chief, who in turn passed some of this land to their sub-tenants, and so on. Leaving the Church out of the equation, this picture is essentially accurate, with the process known as 'subinfeudation' leaving everyone who held land, apart from the Crown, as the tenant of an overlord.

The basic unit of land that was passed down via subinfeudation was a manor. The origins of manors are unclear, but the actual division of land would appear to be based on old Anglo-Saxon allocations or holdings prior to 1066. One common misconception is that a manor was a compact and easily defined geographical entity. This is not the case; land belonging to a manor was often scattered amongst the lands of other nearby manors, although most were centred on a core portion of land. What defined a manor was the unifying bonds of allegiance owed to a single lord, and as such it was more of a social and economic unit than a block of territory. This isn't easy to grasp, and it's made a bit more confusing by another common misconception: that manors shared the same boundaries as parishes. On occasion this may have been the case, but some parishes may contain more than one manor, whilst a large manor may straddle the borders of several parishes.

The way land was held in a manor was very important, and would vary from place to place depending upon the 'customs' of the manor. In theory, every manor had a 'lord', who would grant strips of land within the manor in return for service, which usually comprised any combination of rent, military obligation or work on the lord's land according to custom. Anything not granted to manorial tenants but held by the lord was known as his 'demesne', which could be farmed by him, or leased out to individuals on negotiable terms. The way in which land was granted by the lord and held in the manor was known as 'tenure', and it would determine whether a tenant was 'free' or 'unfree'. Free tenants were required to perform a fixed amount and type of work on their lord's land each year, whereas unfree or 'customary' tenants, often referred to as 'villeins', had only the amount of work set, and would be told the nature of the work at the time it was due. At the lower end of the social scale, 'cottagers' worked the lord's land but owned no land of their own, residing in a cottage with a small garden to cultivate; whilst 'slaves' did not even enjoy these privileges, working exclusively for the lord and eking out a living working 'waste' land that no one else wanted.

> *Records of the manorial system revolve around the manorial court.*

Court Rolls

The records generated by the manorial system revolve around the main source of administration – the manorial court. There were two main types that would be held during the course of a year. Twice a year, a 'view of frankpledge' and 'court leet' were held to try minor offences that had occurred within the manor, as well as to inspect 'tithings', groups of ten men who were mutually responsible for each other's behaviour – effectively a form of self-policing. However, the main type of manorial court was the 'court baron'.

The court baron was held on a regular basis, and was presided over by the lord of the manor's steward or deputy. The aim was to conduct the routine business of the manor, which would include financial penalties for offences against the customs and rules of the manor; general manorial administration; announcements of the deaths of any manorial tenants since the last court; and the admission of new customary tenants into copyhold land – so called because they were given a 'copy' of the record of the court to prove they had been admitted. All free tenants were expected to attend the court baron and acted as jurors; customary tenants were also required to attend, and unless they provided an 'essoin' or excuse (usually with an accompanying payment) they were given a financial penalty.

The records generated by the court baron are known as court rolls, even though the later documents were usually written in bound volumes. Hand-written by the steward, or most likely his clerk, they were in English after 1732 and Latin before that date. The composition may vary from manor to manor, but tended to contain the following items:

- Heading, including the type of court held, the name of the manor, the names of the lord of the manor and his steward, plus the date on which the session was held.
- The names of the jurors are provided next, followed by the essoins of the customary tenants who paid not to attend.
- The first section usually relates to routine administrative matters brought before the court, including offences against the customs of the manor, the judgment of the court and any ensuing financial penalties. Sometimes disputes between tenants are recorded here.
- The next section relates to the regulation of copyhold land. The names of deceased customary tenants were presented to court, followed by 'admissions' of new tenants. Many manors practised primogeniture, or the succession of the first male heir, though this was subject to the customs of the manor and the discretion of the lord.
- Sometimes tenants 'surrendered' land back to the lord, to be re-granted on more favourable terms. This would often include a named series of successors within the family, or a 'surrender to the use of a will' so that the tenant could pass the land freely to his nominated successor. The terms of the will would be recorded in the roll for reference.

Clearly, for those of us whose ancestors were customary tenants, court rolls are a potential goldmine of genealogical information. There are several other bonuses as well:

- Many court rolls contain internal alphabetical name indexes, particularly from the eighteenth century onwards; indeed, separate index books, registers of admission and surrender and other supplementary documentation survive.
- Once you've found an ancestor in a court roll, you will be able to follow a chain back in time, possibly over several generations. If a marriage agreement or will formed part of the initial admission, the relevant section will be entered in the court roll, providing you with a copy if the original no longer survives.
- Most court rolls continue into the nineteenth century, so you should be able to connect with some of the main official sources, parish registers and wills – any of which may indicate you have an ancestor who was a customary tenant. It is well worth checking the tithe apportionments for signs of landholding too.
- Since manorial courts were held since the inception of the system after the Norman Conquest, there may be court rolls extending back

into the medieval period. Some manors have surviving records as far back as the thirteenth century.

Court rolls are particularly useful if you have an ancestor who was a customary tenant, but there are other records generated by the regular business of the manorial court in which free tenants might appear.

- *Estreat rolls* contain fines levied during a court session, so may have a list of manorial tenants included, free and customary.
- *Minute books* and draft court rolls may have additional notes taken during the course of a court session, possibly relating to genealogical data for newly admitted tenants.
- *Suit rolls* and *call books*, when compiled and retained, contain the names of all tenants who were due to appear at court, and were often amended and dated when a tenant died, providing important genealogical information.

During the latter stages of the nineteenth century, land was gradually converted from copyhold to freehold, a process known as enfranchisement. Most of the official records are at The National Archives, though some returns can be found in county record offices.

Where to Find Manorial Records

There is no one single repository for manorial records, since manors were owned by individuals and institutions such as the Crown. Consequently you'll need to search a variety of archives, bearing in mind that material for Crown manors is likely to have been deposited in The National Archives – along with relevant records for privately owned manors that might, for whatever reason and however long ago, have fallen into Crown hands at some point in the past. There are several research guides available on The National Archives website to help you locate and interpret manorial records amongst their holdings, at www.nationalarchives.gov.uk.

The majority of material will be in county archives, though many manorial documents are incorporated in the private papers of large estate owners, much of which may not even be in the public domain. Many of the larger families owned manors in more than one county, so you'll need to establish where the central place of deposit is. To help you locate court rolls, you can search the Manorial Documents Register at The National Archives. The returns for several counties are available online at www.nationalarchives.gov.uk/mdr with plans to make more available over the coming years; the paper lists are on open access in the main search room. Alternatively, you can conduct a search of Access to Archives at www.a2a.org.uk by the name of the manor in the hope of locating other related sources.

Rentals, Surveys and Stewards' Accounts

The legal requirement to hold manorial courts was only one part of the steward's administrative duties, particularly if the manor formed part of a larger estate. To assist his work, various rentals, periodical surveys and regular accounts were prepared, many of which can be useful to a family historian. Some of the most useful are listed here.

- *Extents* often list all manorial tenants, free and customary, alongside the lands held by each, a brief description of the land and the rental value to the lord. The drawback is that they were a much earlier source, disappearing from most manors by the sixteenth century.
- *Rentals* replaced extents and were less detailed, but were created on a more regular basis. They often only list the tenant and their rent, but were frequently annotated on the death of a tenant.
- *Surveys* survive on a more regular basis from the seventeenth century onwards, and often link tenants to a particular piece or parcel of land. They comprised many subsidiary documents, including rent rolls – a list of rents owed per tenant – and often maps of the manor, with the inclusion of data about the historic customs that tenants were required to observe.
- *Accounts* were regularly compiled by stewards as they collected the rents due from the manorial land, not just from tenants but from individuals who leased the lord's demesne lands too. Most are complicated financial documents, but you will find names of manorial tenants listed, and the rents due.

In addition to these 'manorial' documents, you may also find miscellaneous information relating to the running of an estate.

Crown Records

Before the inception of civil registration and census returns in the nineteenth century, and the organization of central government in the late eighteenth century, various institutions had traditionally overseen the administration of Crown affairs and ensured that its rights were enforced. Some of these institutions date back to the time of the Norman Conquest, and the records they produced over the centuries are full of personal names and genealogical information. Here are some of the most important that you should consider using to trace your ancestors through time.

Feudal Rights of the Crown

As described above, the feudal system of landholding was essentially a pyramid, with the King at the top who granted estates to tenants-in-chief in return for service. However, the Crown also retained various rights over these lands, particularly at the death of a tenant-in-chief. If an heir wished to inherit, he (and it was generally the oldest living son) would pay the King a 'relief' to enter into possession of the land. However, they had to be 21 years old to do this, and if they were underage the King would be granted 'wardship' of the estate, running it as he saw fit and taking the profits. Often, though, the wardship would be granted to a third party, usually at a price. The ability of unmarried heiresses to choose a husband, or widows to remarry (or not, depending on the circumstances), was also governed by the Crown, and money was paid to the King to control the right of marriage. Together, these were known as the Crown's 'feudal rights' and were fiercely protected. The paperwork – or parchment – that was generated contains a great deal of information about medieval families up to the seventeenth century, when many of these rights were relinquished after the Civil War. These feudal rights generally only affected the top level of society, though with subinfeudation more and more people came under the Crown's sphere of operations.

Inquisitions Post Mortem

One of the most important sources of information for family historians are Inquisitions Post Mortem (IPMs), not to be confused with the modern post mortem examination to determine the cause of death. Medieval IPMs were conducted whenever a major tenant-in-chief or principal landowner died. The records start in the thirteenth century and continue right up to the start of the Civil War, and were created when a royal writ was sent to the sheriff, or more usually the royal escheator (an official in charge of looking after lands in the King's hands), for each county in which the deceased held lands, instructing them to gather together a jury of men of standing, usually knights or esquires, who would answer questions about the lands held. The aim of the inquisition was to determine:

- The name and date of death of the deceased
- What lands they held in the county, including a description
- Whether they were held directly in chief from the King, and if not from whom they were held
- What military service was owed on the land (the technical term being 'knight's fee')

> An important source of information is the Inquisition Post Mortem

- What rent was due from the land
- Whether the heir at law was sane, or of age to inherit (i.e. over 21 years old)
- If under age, who the guardian of the child was

If the heir was under age, then the lands would default to the Crown and the heir would become a ward; and if they were of age, then the level of relief would be set according to the amount of land held. Three copies were usually made, with one sent to the Royal Chancery, which had issued the original writ; another to the Exchequer (which dealt with financial aspects of the royal administration); and, with records dating from 1509, the last was sent to the newly established Master of the Wards to ascertain whether wardship was required. The surviving documents are written in abbreviated Latin, and – given their antiquity – many of them are now hard to read.

The records of IPMs are now held in The National Archives, with the period from 1236 to 1418 in series C 132–138 and E 149, between 1418 and 1485 in C 138–141 and E 149, from 1485 to 1509 in C 142 and E 150, and from 1509 to 1640 in series C 142, E 150 and WARD 7. There are various ways to access them; one of the best is to search the catalogue online by surname at www.nationalarchives.gov.uk, where there's also a research guide to explain more about the records and how they are arranged. In addition, there are printed 'calendars' in the reading rooms, which are indexed volumes that summarize the content of the records. Whilst they do not cover the entire period, it is well worth using them as many contain translations into English, allowing you to pick out the salient genealogical points.

Inquisitions Ad Quod Damnum

As well as IPMs, the Crown routinely conducted other surveys into its rights. Another similar source were Inquisitions Ad Quod Damnum, literally an inquisition to ascertain 'what damage' was done to Crown rights when private grants were made. Once again, they are stored at The National Archives, and can provide a fascinating insight into what our more lordly ancestors were doing to incur the interest of the Crown, such as endowing monasteries with land, requesting the right to hold a market or fair, and other such staples of medieval life. They can be found in series C 143 and cover the mid-thirteenth century to the late fifteenth, with calendars and indexes also available to help your research.

(Miscellaneous inquisitions are in series C 145, and contain references to forfeited estates, when your ancestors really stepped out of line.)

Wardship

To regulate and deal with the growing business of wardship handled by the Master of the Wards, first appointed in 1503, the Court of Wards and Liveries was created in 1540, surviving over a century before becoming another casualty of the Civil War. There are a wide range of documents relating to the work of the institution, aside from the copies of IPMs described above; grants of wardships to third parties, leases of wards' lands, deeds and evidence, accounts, valuation documents and – perhaps most importantly – judicial records. These are described below along with other equity court records, but were the material generated by disputes between rival claimants over rights to exploit the profits from wardships that had fallen into private hands. Together, the records of the Court of Wards and Liveries form a rich resource for the patient researcher, and can be found in The National Archives in series WARD 1–15. Further information can be found at www.national archives.gov.uk

Taxation

One of the principal ways that the Crown funded its operations was through regular grants of taxation from its subjects – an unpopular move since it was first tried on a large-scale basis in the late twelfth century. Although occasional levies on property and movable goods were attempted throughout the thirteenth century, regular grants only became established under Edward I (1272–1307) and became known as lay and clerical subsidies. The records for national taxation are stored in The National Archives in series E 179, but in many ways the records listed there are only the tip of the iceberg.

As well as taxes on property, individuals were assessed depending on the rate of tax granted. The assessments form the bulk of useful material in E 179, though a whole range of other documentation exists – poll tax returns, early income tax documents from the fifteenth century, clerical taxation, alien subsidy returns, as well as exemption certificates, accounts, arrears and inquisitions. Assessments tend to list every eligible taxpayer in a parish, with totals for the amount on which they were assessed, and the amount they were due to pay. These can be searched online by name of parish (so you'll need to know where your ancestor was living at the time) at www.nationalarchives.gov.uk, and there's a research guide on taxation to help you locate the most suitable records.

> *Tax assessments list every eligible taxpayer in a parish.*

Civil Court Records

Chapter 27 has dealt with records generated by the criminal courts, when our more wayward ancestors were caught by the long arm of the law. However, apart from nefarious activities that society frowned upon, our ancestors proved particularly adept at fighting one another in court, as well as rowing frequently with friends, neighbours, landlords and a variety of other people. The records generated by inter-personal lawsuits fall under two main branches of civil litigation – common law and equity.

A quick sketch of how civil law evolved is necessary before the records are examined, because it can be rather confusing. Common law was the older branch of legislation open to civil litigants, and its courts dispensed justice based on the ancient custom of the land and used past precedent to settle cases. However, in many instances this produced unfair resolutions, and so a separate legal code based on the principle of 'equity' developed, which used justice and conscience to reach a judgment rather than basing a decision solely upon what had happened in the past. This was the preferred route for families that were in dispute, often concerning inheritance or property.

Enrolment and State Papers

One of the main government departments was the Chancery. Its original function was to write all royal writs and correspondence, but it quickly compiled its own archives where copies of all correspondence were written in long rolls. The main two sets were known as Patent Rolls, so called because correspondence sent out by the King with the royal seal attached but with the letter open for all to see was known as 'Letters Patent'; and Close Rolls, which earned their name for recording letters that were sealed 'closed' so that only the recipient could see the contents. These two record series, dating

back to the late twelfth century, quickly became important places where people could 'enrol' information, on payment of money to the Crown; a record of the Crown's interaction with individuals was preserved permanently in written form.

The rolls are now in The National Archives in series C 66 and C 54 respectively, and indexes and calendars are available to help you search for the thousands of individuals whose names are recorded within them. In many instances, issues of contemporary genealogical importance are preserved, such as petitions from

the heirs of an executed traitor attempting to get their lands back; lists of eligible men who could be called upon to serve in the militia; copies of land grants to individuals; appointments to office; indeed, any aspect of royal administration in the shires that required correspondence to be issued. Further information on the history and content of the rolls can be found online in a research guide at www.nationalarchives.gov.uk. You may even find the original correspondence or grant within private collections in county archives.

Similar enrolments can be found in the Chancery series of records for financial agreements between

Before 1875, there were four main common law courts – King's Bench, Common Pleas, Exchequer of Pleas and Chancery (Plea Side) – with two principal equity courts, Chancery and Exchequer, with at various times other courts applying equity methods of settling suits, such as Star Chamber, Court of Requests, and Court of Wards and Liveries. In 1875 all these courts were abolished and a Supreme Court of Judicature established with five divisions, with either common law or equity applied where necessary. A further change in 1881 reduced the Supreme Court to three divisions – King's (Queen's) Bench; Chancery; and Probate, Admiralty and Divorce.

Equity Court Records

The system employed by Equity Courts such as Chancery was called 'bill pleading', whereby a plaintiff would bring a written 'bill of complaint' before the court, outlining their grievance against the defendant and justifying the recourse to an equity court rather than under common law. The defendant would then provide a written 'answer' explaining why the bill of complaint was unjust, to which the plaintiff would

Crown and individual (Fine Rolls, C 60); grants and inspections of charters (Charter Rolls, C 53); and expenditure authorized by the Crown and the issue of money to individuals (Liberate Rolls, C 62).

In addition to the official enrolments are the various pieces of miscellaneous correspondence, working papers, reports and documentation that were used to conduct the day-to-day business of government from the sixteenth century onwards. These have been brought together by reign and organized according to date, type and place, and form the State Papers collection in The National Archives. Although there are many stray individual documents that

predate the series, the earliest relate to the reign of Henry VIII (1509–47) and are usually broken down into Domestic and Foreign, with separate collections emerging for relations with Scotland and Ireland. The series ends in the reign of George III (1760–1820) when modern Departments of State were introduced, such as the Home Office and War Office. Most of the earliest documents relate to correspondence between the leading political figures of the age, such as Elizabeth I's Secretary of State William Cecil (Lord Burghley) and his spymaster, Sir Francis Walsingham.

Like the official enrolments, Calendars of State Papers provide

name and subject indexes to the documents, which are largely available on microfilm at The National Archives. These are fantastic resources, literally crammed full of information about your ancestors. For example, during the mid-seventeenth century there was a real problem with piracy on the south coast from Barbary slavers raiding villages and carrying off women and children to be sold in Africa. The State Papers are full of petitions from angered menfolk seeking permission to mount private expeditions to rescue their families, which provide personal accounts of the trauma and loss as well as detailed genealogical information.

respond with a 'replication'. This would be contested by a 'rejoinder', and the process would continue until the limits of the case were clearly defined. Together, this material is known as court 'proceedings'.

At this stage, the Masters of Chancery, who were in charge of each case, would step in and request evidence to assist with the task of reaching a verdict in the case. Depositions from witnesses on both sides were required, in response to a series of questions known as interrogatories. Sometimes, voluntary statements made under oath called affidavits were brought before the court. Other forms of evidence was brought in, such as title deeds, pedigrees or family trees. Usually this was sufficient information to reach a final verdict, issued in the form of a decree, though during long cases the court might issue an order of judgment to one party or another. All of these processes generated records, which can be viewed at The National Archives.

Similar practices were adopted in the courts of Exchequer, Requests, Star Chamber and Wards and Liveries, though Chancery was the most popular court and handled the bulk of business. Exchequer was meant to be reserved for cases where the plaintiff was a Crown debtor or, at a stretch, a tenant, but other material was heard too. It conducted business from 1558 until 1841, when it was superseded by Chancery. The Court of Requests was established in 1483 and was intended as a court for the poor, given the rising costs of hearing cases in Chancery; it was shut down after the Civil War in 1649. It was formerly an offshoot of the King's Council, as was Star Chamber, established two years later in 1485 and named after the patterned ceiling of the room in which the court sat; it was abolished during Charles I's turbulent reign. Finally, the Court of Wards and Liveries was set up in 1540 and dealt with cases relating to disputed rights to control profits of wardship, but also failed to survive the Civil War.

Aside from Chancery, for which many records are searchable by name online, searches in the other equity courts can be speculative and somewhat tricky, as there are few online name indexes available and therefore you are required to plough through index books in the hope of tracking down an ancestor. Even more frustrating is the fact that litigants often moved between equity courts, starting proceedings in one and then transferring them to another. Another annoying trend was for cases to be brought to court simply to force the other party into an out-of-court settlement, so there may not even be a complete set of records, judgment or overall conclusion. However, the good news is that, unlike most common law material, the records generated by equity cases were predominantly written in English.

Some other points to remember:

- Most records are indexed or arranged alphabetically by the name of the plaintiff.
- Multi-party cases could be brought to court, and it may be the case that only the first-named plaintiff appears in the index.
- Court cases could rumble on for years, often after the death of the original litigants; so you may have to search a long time span to gather all the necessary paperwork.
- If this was the case, you may find new names appearing, or a single-party case turning into a multi-party case as sons and daughters took over from a deceased parent.

> *Unlike common law material, the records generated by equity cases were usually written in English.*

Records for Proceedings

Records for Chancery are now largely searchable by name online via the catalogue at www.nationalarchives.gov.uk, with some series also containing detailed descriptions of the subject matter of the case that can be searched too. The earliest records, from about 1358 up to 1558, are in series C 1, with later material running to the middle of the seventeenth century in series C 2-4. By this stage, the business of the court had expanded to the extent that records from the seventeenth century to the court's abolition in 1875 were divided into six divisions, known as Six Clerks after the staff who originally worked on behalf of the various Masters of Chancery. These records are in series C 5-10 prior to 1714 and C 11-18 up to 1875. It may be necessary to search all six divisions, as cases could move between them – and frequently did. Contemporary index books are also available on the shelves in the reading rooms, and there is a microfilm index (known at Bernau's index after its compiler) of litigants in the Six Clerks series, and witnesses who prepared depositions, at the Society of Genealogists London.

It is less easy to search for Exchequer equity proceedings as the records are not searchable by name online. They are in series E 112, with some material (that is searchable) in E 111 and strays in E 193. To access them, you need to consult a series of index volumes in series IND 1; a full list is available via a research guide online at www.nationalarchives.gov.uk. Proceedings for the other courts can be found in:

- Court of Requests – series REQ 2
- Star Chamber – series STAC 1-9
- Wards and Liveries – series WARD 13

Common Law Court Records

In contrast with equity court records, Common Law courts generated different records since they used a different system, whereby pleas were entered into court and judgments were made on the basis of the plea and precedence. As with equity pleadings, cases could move between courts, making it trickier to research one particular dispute in full, and once again, many cases were dropped or private agreement reached so that a final judgment is never recorded. The records prior to 1732 are also in Latin, making it harder to understand the full legal nuances of a case given the technical language sometimes employed – although you may find recitals from English documents are kept in the original language. Sadly, it is not possible to search many of the records by name online, since this information is contained within the main body of the documents listed below.

Plea Rolls

The material contained on the plea rolls is similar for all courts, and records the formal processes conducted during the course of a case. Normally, they contain details of the action brought before the court, including the names of the litigants and a description of the cause of dispute; how the case proceeded within the court; and (if one was actually made) a final judgment. The records are at The National Archives in the following series:

- King's Bench – early cases in series KB 26 (1194–1276) with the remainder in KB 27 (1273–1702) and KB 122 (1702–1875); there are printed calendars for KB 26, with indexes in IND 1 for the remainder.
- Common Pleas – series CP 40 (1273–1874), with some material in CP 43 (1583–1837) for pleas relating to land; there are some docket books which serve as indexes to CP 40 in series CP 60, with indexes to CP 43 in IND 1.
- Exchequer of Pleas – series E 13, with a partial calendar index in IND 1 and repertory rolls in E 14.
- Chancery Pleas – records from 1272 to 1625 in series C 43–44, which are searchable online; and 1558–1901 in C 206.

Judgment Records

Although judgments were usually recorded on the plea rolls, many

Evidence: Depositions, Interrogatories and Affidavits

These are very useful records, because they contain personal accounts relating to salient points concerning the case – often from friends and family of the litigants. Chancery depositions are split into 'town' (meaning London and a few surrounding counties) and 'country' (meaning everywhere else), as well as chronologically, and can be found in a variety of series – C 1 and C 4 (prior to 1534), C 11–16, C 21–22 and C 24; related interrogatories are in series C 25. Indexes are available in series IND 1, most of which are on the shelves in the reading rooms, whilst names of witnesses have been transcribed and form part of Bernau's index, now available on microfilm at the Society of Genealogists, London. Some of the material is now searchable online via The National Archives' website. Affidavits are in series C 31 and C 41, again with indexes available in series IND 1.

Exchequer depositions for London are in E 133, whilst material for cases were simply not filed by the eighteenth century given the volume of business. In these instances, the only records that survive are the separate records for court judgments. These are also in The National Archives in the following series:

- King's Bench – series KB 168 (1699–1875), Entry Books of Judgments, with some indexes within this series that have to be ordered.
- Common Pleas – series CP 64 (1859–74), Entry Books of Judgments, with some indexes within this series that have to be ordered.
- Exchequer of Pleas – series E 45 (1830–75), Entry Books of Judgments.
- Chancery Pleas – series C 221 (1565–1785) and C 222 (1638–1729), Remembrance Rolls.

Civil Cases in Assize Trials

Aside from the common law courts, there was also the possibility of holding a civil court at one of the itinerant assizes, described in Chapter 27, if a writ of *nisi prius* (literally 'unless before' or 'unless sooner') had been served in one of the common law courts, allowing the case to be heard by the claimants' local assize circuit judges before the date set in the writ for a hearing in the London common law court. A list of the various series at The National Archives where the records are stored for England, and Wales after 1831, is available on their website www.nationalarchives.gov.uk. Records for Wales prior to 1831 can be found in the National Library of Wales, Aberystwyth. Further information can be found on the library's website at www.llgc.org.uk.

Supreme Court of Judicature

When the legal system was reorganized in 1875, a Supreme Court of Judicature was created. Although any one of the five divisions, later reduced to three in 1881, could hear cases and apply the principles of common law or equity when required, most common law suits tended to end up in King's (Queen's) Bench. Many of the records have been destroyed, and the best place to look for information about the names of parties, and a brief description of the case brought before court, is in the cause books, in The National Archives in series J 87 and J 168, with some indexes from 1935 in J 88.

the remainder of the country is in E 134, with interrogatories in E 178. The country depositions are available to search online and can be very detailed, based on indexes that are still available in the reading rooms. Special commissions, interrogatories and depositions are also in E 111, another area searchable online. Affidavits survive in series E 103, for which there are partial indexes, and E 128, for which there are not.

Depositions for the Court of Requests can be located in series REQ 1, along with affidavits. Only depositions survive for the Court of Wards and Liveries in WARD 3, and there are no records for Star Chamber.

Evidence: Masters Exhibits

Perhaps the richest source of information, but also one of the least used by researchers, is pieces of evidence brought before the courts but never reclaimed by the relevant parties. Chancery masters exhibits are fully searchable online, and are contained in series C 103–114, where you can find a wealth of genealogical information about the litigants and their families in the documentation that was brought to court. Deeds have been separated into C 115, whilst manorial court rolls are in C 116. Chancery masters documents are similar in terms of content and are in series C 117–129, though they are not as well indexed. Plans are ongoing to list the records more thoroughly, however, and put them online.

Similar records exist for the Exchequer, but far fewer survive compared to Chancery. The majority can be found in E 140, but there are a variety of miscellaneous series in which you are likely to find strays. No material survives for the courts of Request, Star Chamber or Wards and Liveries.

Judgment: Decrees and Orders

Chancery decrees and orders are a useful way to track the progress of a case, and entry books – the main means of reference – can be found in C 33. Confusingly, there are 'A' and 'B' books for each year, kept annually and arranged by the name of each plaintiff. Indexes are available on open access in the reading room. Similar material exists for the Exchequer, in a wide range of series – entry books, decrees and orders are in E 123–128, with material for the reign of Philip of Spain and Mary in document E 111/56; original decrees are in E 130–131. There are some decree and order books for the Court of Requests in REQ 1 and Wards and Liveries in WARD 2, but nothing survives for Star Chamber.

Supreme Court of Judicature Post 1875

Many of the Chancery Division equity records for the Supreme Court of Judicature are simply continuations of the earlier material. Proceedings are in series J 54, depositions in J 17 and affidavits in J 4. The decree and order entry books are in J 15, whilst exhibits are in J 90.

Tracing Aristocratic Roots

Chapter 4 has already examined the subject of pedigrees, blue-blooded ancestors and the work of royal heralds. This is a popular subject and more people than you might imagine are able to prove a connection to royalty, using standard genealogy to unearth an 'aristocratic' family name and then using this to fast-track back using pre-researched and well-established pedigrees to link into the royal family tree. Many people make this connection around the late fourteenth or early fifteenth century, mainly because Edward III (1327–77) had so many offspring – fourteen children, to be precise, though two died as infants. In particular, four sons – John of Gaunt, Duke of Lancaster; Lionel of Antwerp, Duke of Clarence; Edmund Duke of York; and Thomas Duke of Gloucester – had numerous children, both legitimate and illegitimate, and consequently they intermarried with the other leading noble families and gentry of the day.

Here are a few reminders about how and where you can check to see if you have a chance of making a similar connection.

1.
Find a gateway ancestor

Someone who has proven links to royalty or aristocracy. They are usually of higher social status, often with military connections or a role in the church – favoured professions for younger sons of gentry or nobility.

- *The Dictionary of National Biography* will provide information on their lineage and background
- Publications such as *Debrett's Peerage*, *Burke's Peerage* or the *Complete Peerage* may provide a pre-researched genealogy

2.
Find a famous surname in your family

There may be chance that you can prove a link to a more famous or well-to-do branch.

- Standard genealogy will help you to prove or disprove a match
- Surnames often appear through the female side through marriage, so ensure you check both sides of the family tree
- Look for family or estate papers for further clues

3.
Coat of arms

The right to armorial bearings or an inherited coat of arms is a strong signal that there is some connection with an ancient family line.

- Check the collections of the College of Arms, where they can verify the lineage and right to bear arms
- Heraldic visitations published by Harleian Society
- Check for pedigrees at The National Archives, British Library, Society of Genealogists

4.
Investigate stories of illegitimate links

Many children were fathered on household servants by sons of nobility or lesser royals, and were often financially helped in later life. In particular, look out for:

- Children receiving a better education than their social status would suggest
- Regular financial payments to 'parents'
- Expensive baptism gifts, engraved heirlooms or other presents from a noble family
- Purchase of a military commission or an army career for a boy, which provides social status
- Good marriage into gentry or wealth for a girl, which provides social status
- Investigate the credibility of the story – geography, time period and key players have to match

CASE STUDY

Matthew Pinsent

Matthew Pinsent was able to research both his paternal and maternal lines quite a way back in time, using standard resources such as census returns, certificates, wills and parish registers. However, on his mother's line a few interesting surnames kept cropping up, such as Landale – and by continuing to link certificates with census records, particularly through the female lines of the family, he discovered he was related to General Sir George Anson, a renowned and titled military commander who died in 1858 and was actually Matthew's 3 x great-grandfather.

Since General Sir George Anson was well known, it was relatively easy to track down his background in standard history textbooks and use him as a 'gateway' ancestor. Indeed, Anson warranted an entry in *Burke's Peerage* – a standard genealogical reference work that lists the nobility and their connections to the past – and from there it was fairly easy for Matthew to establish a link to the Howard family who were particularly prominent in the sixteenth century, mainly

because they often fell foul of the Crown. For example, Matthew was descended from Lord William Howard, who was the uncle of Catherine Howard, Henry VIII's fifth wife who was beheaded in 1542.

Matthew was then able to trace the Howard genealogy back through the fourteenth and fifteenth centuries, mainly through female descent and marriage, to prove a connection with William de Bohun, 1st Earl of Northampton, whose maternal grandparents were Edward I and Queen Eleanor of Castile. This journey led Matthew to the College of Arms, where a family pedigree confirmed the connection and even extended the royal lineage via William the Conqueror back to biblical times!

◄ David Landale (third from right) in a family photo from the 1920s.

In addition to the sources listed above, there are original documents relating to aristocratic, gentry and noble families in archives around the country. Most of these can be located online at websites such as www.nationalarchives.gov.uk/nra, www.a2a.org.uk, www.bl.uk and www.archiveshub.ac.uk amongst deposited family and estate papers, pedigrees and legal paperwork relating to marriage settlements and assignments of dower.

Official sources can also help, especially Crown surveys of its rights in relation to landholding and leading aristocratic families. The most famous is Domesday Book, available to view online at The National Archives, but other works can also help push your family tree even further back in time. For example, in 1166 the Domesday Book was found to be so out of date that a new survey was commissioned, with tenants-in-chief sending in data about their landholding and military service; the resulting data was compiled into a document now called the Cartae Baronum. Material from other surveys was copied by Exchequer staff in the early fourteenth century, covering material from the late twelfth century, into a book called the Book of Fees, or Testa de Nevill, whilst another composite volume, Feudal Aids, brings together similar information from a range of records dating from 1283 to 1431.

▼ Matthew's family tree shows his sixteenth-century ancestor, Lord William Howarde.

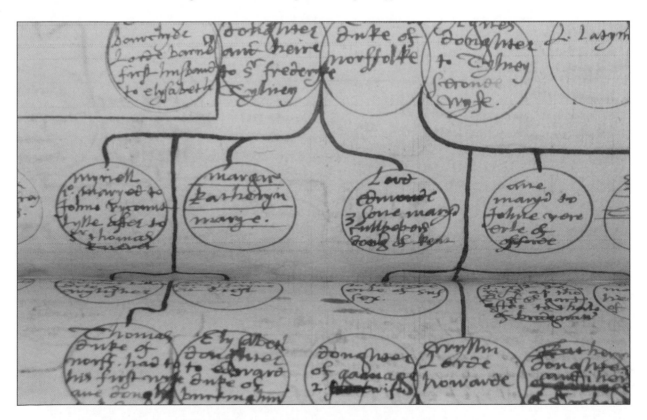

Trouble-shooting Guides

The aim of these troubleshooting guides is to provide step-by-step routes through the records, focusing on eight major themes covered in this book. Details about the information contained in the documents are contained in the main chapters about them; these are pared down flowcharts showing you what you can and can't find from the records, structured as a series of questions and points that take you to a set of records or a new archive.

Chart 1: Army Service Records, First World War

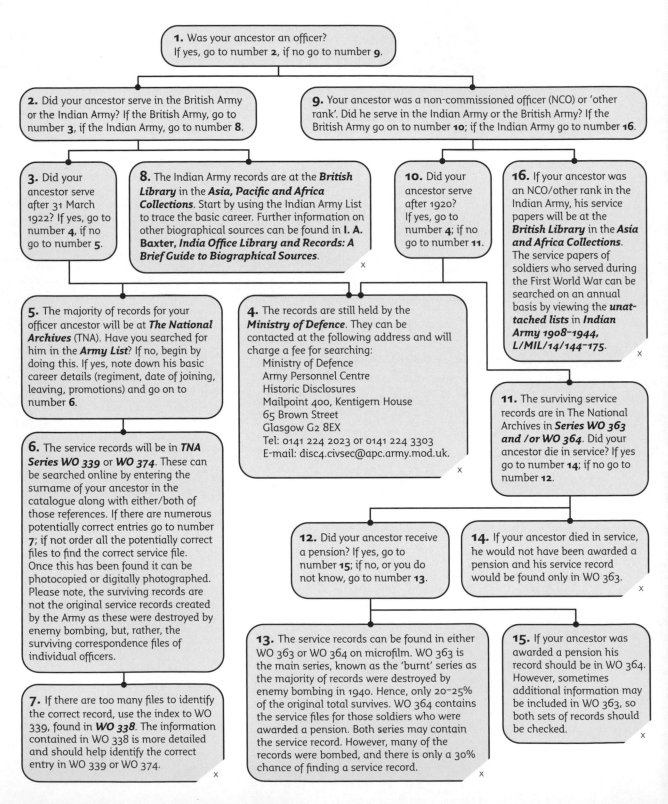

1. Was your ancestor an officer?
If yes, go to number **2**, if no go to number **9**.

2. Did your ancestor serve in the British Army or the Indian Army? If the British Army, go to number **3**, if the Indian Army, go to number **8**.

9. Your ancestor was a non-commissioned officer (NCO) or 'other rank'. Did he serve in the Indian Army or the British Army? If the British Army go on to number **10**; if the Indian Army go to number **16**.

3. Did your ancestor serve after 31 March 1922? If yes, go to number **4**, if no go to number **5**.

8. The Indian Army records are at the **British Library** in the **Asia, Pacific and Africa Collections**. Start by using the Indian Army List to trace the basic career. Further information on other biographical sources can be found in **I. A. Baxter, *India Office Library and Records: A Brief Guide to Biographical Sources***.

10. Did your ancestor serve after 1920? If yes, go to number **4**; if no go to number **11**.

16. If your ancestor was an NCO/other rank in the Indian Army, his service papers will be at the **British Library** in the **Asia and Africa Collections**. The service papers of soldiers who served during the First World War can be searched on an annual basis by viewing the **unattached lists** in **Indian Army 1908-1944, L/MIL/14/144-175**.

5. The majority of records for your officer ancestor will be at **The National Archives** (TNA). Have you searched for him in the **Army List**? If no, begin by doing this. If yes, note down his basic career details (regiment, date of joining, leaving, promotions) and go on to number **6**.

4. The records are still held by the **Ministry of Defence**. They can be contacted at the following address and will charge a fee for searching:
    Ministry of Defence
    Army Personnel Centre
    Historic Disclosures
    Mailpoint 400, Kentigern House
    65 Brown Street
    Glasgow G2 8EX
    Tel: 0141 224 2023 or 0141 224 3303
    E-mail: disc4.civsec@apc.army.mod.uk.

11. The surviving service records are in The National Archives in **Series WO 363 and /or WO 364**. Did your ancestor die in service? If yes go to number **14**; if no go to number **12**.

6. The service records will be in **TNA Series WO 339** or **WO 374**. These can be searched online by entering the surname of your ancestor in the catalogue along with either/both of those references. If there are numerous potentially correct entries go to number **7**; if not order all the potentially correct files to find the correct service file. Once this has been found it can be photocopied or digitally photographed. Please note, the surviving records are not the original service records created by the Army as these were destroyed by enemy bombing, but, rather, the surviving correspondence files of individual officers.

12. Did your ancestor receive a pension? If yes, go to number **15**; if no, or you do not know, go to number **13**.

14. If your ancestor died in service, he would not have been awarded a pension and his service record would be found only in WO 363.

7. If there are too many files to identify the correct record, use the index to WO 339, found in **WO 338**. The information contained in WO 338 is more detailed and should help identify the correct entry in WO 339 or WO 374.

13. The service records can be found in either WO 363 or WO 364 on microfilm. WO 363 is the main series, known as the 'burnt' series as the majority of records were destroyed by enemy bombing in 1940. Hence, only 20-25% of the original total survives. WO 364 contains the service files for those soldiers who were awarded a pension. Both series may contain the service record. However, many of the records were bombed, and there is only a 30% chance of finding a service record.

15. If your ancestor was awarded a pension his record should be in WO 364. However, sometimes additional information may be included in WO 363, so both sets of records should be checked.

Chart 2: Army Service Medals, First World War

1. Did your ancestor receive any medal for his service? All personnel that fought abroad were entitled to a campaign medal. If no, go to number **2**; if yes, or you do not know, go to number **3**.

3. It is possible to check to see if your ancestor received a medal by searching **online** on **The National Archives website** in **Documents Online** at www.nationalarchives.gov.uk/documentsonline. The National Archives have placed all the medal index cards for every medal issued during the First World War online, searchable by name. Begin by searching for a card by entering the name and any other details known. If a card is not found go back to number **2**; if a card is found, go to number **4**.

2. There is no further research to be done if your ancestor did not receive a medal.

×

4. If the medal awarded was a 1914 Star, 1914/1915 Star, a British War and Victory Medal, a Territorial Force Medal or the Silver War Badge, then they were campaign medals. Further research can be done by going to number 5. If the card is an index for an award of a Distinguished Conduct Medal (DCM), Military Medal (MM), Meritorious Service Medal (MSM) and Mention in Despatches, they were gallantry medals. Further research can be done by going to number **6**.

5. The Medal Index Card for campaign medals has numerical references to the **Medal Rolls** found in **The National Archives Series WO 329**.

×

6. If your ancestor was awarded a gallantry medal then the citation of this award can be followed up in the **London Gazette**. The card should give the date of when the citation was made in the paper. This reference can be searched online at www.gazettes-online.co.uk. It may also be possible to find out the reasons behind the award of the medal in local or national newspapers.

×

× denotes end of the line with your research

Chart 3: Early Army Service Records

1. Did your ancestor serve before the formation of a regular standing army in 1660? If yes go on to number **2**; if no, go to number **6**.

2. If your ancestor served in the feudal period (up to 1485) go to number **3**. If your ancestor served during the Tudor and early Stuart periods (1485–1642) go to number **4**. If your ancestor served during the Civil War and Interregnum (1642–60), go to number **5**.

3. Any records for the medieval and modern period will be in The National Archives (TNA), scattered through many Chancery and Exchequer rolls such as *TNA Series E 101*. Further guidance on these records can be obtained by referring to TNA's *online research guide* on *Medieval and Early Modern Soldiers*.

x

4. Any surviving records for soldiers during the Tudor and Stuart periods are held in either TNA or local repositories. A comprehensive guide to these records can by found in *Tudor and Stuart Muster Rolls – A Directory of Holdings in the British Isles by J. Gibson and A. Dell* (Federation of Family History Societies, 1991) detailing which records are held locally and which are at TNA.

x

5. Records for soldiers during the Civil War period or Interregnum survive mainly for officers and mostly in TNA. Start by searching the relevant *Calendar of State Papers*. A published list of officers serving for both sides in 1642 has been compiled by Edward Peacock in *Army Lists of the Roundheads and Cavaliers*. Details for those fighting on the Royalist side can be found in **W. H. Black (ed.), *Docquets of Letters Patent 1642–6*** (1837) and **P. R. Newman, 'The Royalist Officer Corps, 1642–1660', in the *Historical Journal*, 26 (1983)**. If your ancestor served in the New Model Army refer to **R. R. Temple, 'The Original Officer List of the New Model Army', *Bulletin of the Institute of Historical Research*,** or *The New Model Army*, a history written by I. Gentles.

x

8. Once you have the basic outline of your ancestor's army career it is possible to start searching through the original service records. These are held in two main series at TNA, *WO 25* and *WO 76*. There is a partial card index to these records in the reading rooms at TNA. It may also be worthwhile checking whether your ancestor received a pension. Pension records are also at TNA, in the Pay Master General *Series PMG*. Further guidance can be obtained by referring to the TNA research guide on Officers' Records, 1660 to 1913. Once you have finished searching through the service records, turn to number **9** for further options.

9. It may well be possible to find further details about your ancestor through a variety of alternative sources:
- *Regimental Museums*. Once you know which regiment your ancestor was in, contact the appropriate museum to see if they have any records of your ancestor in particular or general information relating to the campaigns he would have been involved in.
- *Imperial War Museum*. This museum holds private papers donated by soldiers serving at the time, including personal journals.
- *The Times* newspaper and the *London Gazette*. Both publications would carry details of commissions, promotions and medal awards of individual officers. Additionally, *The Times* would also carry reports of any big campaigns officers fought in.
- *Dictionary of National Biography*. The DNB would often document the careers of prominent officers and could give useful summaries and personal details of the life of an officer.

x

6. If your ancestor served after 1660 and was an officer, go to number **7**. If he served as an ordinary soldier ('other rank'), go to number **11**.

7. The earliest records for officers being commissioned are within the *Calendars of State Papers, Domestic* starting in the late 1600s. However, the main sets of records begin in the early eighteenth century. Begin by searching the *Army Lists*. From 1702 to 1764 they are available in manuscript form in *TNA Series WO 64*. From 1754 they are available annually and in published format and can be found at TNA, the British Library, the National Army Museum and the Imperial War Museum. They are very useful for tracing an officer's career, from granting of first commission, to promotion and eventual retirement. If you find him in the Army List, go to number **8**. If not, go to number **10**.

10. Check to see if your ancestor was perhaps not an officer, but an ordinary soldier. For details on how to do this go to number **11**. Your ancestor may have served in the Indian Army. Turn to number **8 on Chart 1** for details of how to search these records. Your ancestor may have rejoined the Army after the onset of the First World War. If so his records would be with the First World War records see **Chart 1**.

12. Service records are held in *TNA Series WO 97* and are actually discharge papers, created when a soldier was discharged from the Army and awarded a pension. Hence service records do not survive for every soldier, but rather for soldiers awarded a pension (although after 1883 most service papers do survive). If your ancestor served as a soldier between 1760 and 1854 go to number **13**. If your ancestor served between 1854 and 1882, go to number **14**. If your ancestor served between 1883 and 1913, go to number **15**.

13. From 1760 to 1854 it is possible to search by individual name online in *TNA's catalogue*. If you do not find your ancestor in these records he was not awarded a pension and it is unlikely that you will find any further information. However, if you do know the regiment he served in, go to number **16**.

14. From 1855 to 1882 it is necessary to know your ancestor's regiment in order to make a search feasible. If you do not have this information it is unlikely that you will find his service record. If you do know the regiment and do not find him, it is likely he was not awarded a pension. However, turn to number **16** for further options.

15. From 1883 to 1913 the series is arranged in simple alphabetical order and most individuals will be included. If you do not find your ancestor in this series he may have gone on to serve during the First World War. For details on how to check these records refer to the First World War Army flowchart, **Chart 1**.

11. The majority of Army personnel were ordinary soldiers, 'other ranks', and not officers. The main series of service records begin in 1760. If your ancestor served prior to that then go to number **16**. If he served between 1760 and 1913 then go on to number **12**.

16. If you know approximately when and which regiment he served in, go to number **17**. If you do not have this information it is unlikely you will be able to retrieve his service record.

17. If you do know approximately when and with which regiment your ancestor served, it is worthwhile consulting the *Muster Rolls* and *Pay Lists* in *TNA Series WO 11*, *WO 12*, *WO 13* and *WO 16*. The earliest Muster Rolls start in approximately 1732 and are useful for locating soldiers who served prior to 1760. The series finish towards the end of the nineteenth century and are arranged by individual battalion of each regiment. Each regiment compiled on a monthly or quarterly basis the names of each and every serving officer and soldier in that month or quarter and where they were stationed at that particular time. Additionally, they list when a soldier first enlisted and when he was discharged and are therefore a useful alternative in the absence of a service record for individual soldiers.

Chart 4: Naval Service Records

1. Was your ancestor an officer or rating of the Royal Navy? If he was an officer, see number **2**; if he was a rating, go to number **6**.

2. The modern origins of the Royal Navy lie in the late seventeenth century, during the reign of Charles II. If your ancestor was in the Royal Navy before 1660 go on to number **3**. If he was recruited after 1660 go to number **4**.

3. Any surviving documentation for individuals employed in the Navy prior to 1660 will be in general *Chancery*, *State Paper* and *Exchequer* series in The National Archives (TNA). If your ancestor was involved in a well-known naval battle then published histories of such conflicts may be worthwhile investigating, which can be found at libraries and archives.

4. Start by referring to the many published sources available for Royal Naval officers post-1660, such as:
- *The Commissioned Sea Officers of the Royal Navy, 1660–1815* by Syrett and DiNardo
  The book is now available to search online as part of the databases on the Ancestry website.
- *Naval Biographical Dictionary* by William R. O'Bryne (1849)
  This book has also been digitized and can be searched online as part of the Ancestry website. It lists all officers alive at the time of publication.
- *Biographia Navalis* by John Charnock (1797)
  A survey of all naval officers serving between 1660 and 1797, from the rank of captain to admiral.
- *Lives of the British Admirals* by Dr John Campbell
  A list of all admirals serving up to 1817.
- *Royal Naval Biography: or, Memoirs of the services of all the flag-officers, superannuated rear-admirals, retired-captains, post-captains, and commanders, whose names appeared on the Admiralty list of sea officers at the commencement of the present year, or who have since been promoted; illustrated by a series of historical and explanatory notes ... With copious addenda* by John Marshall (1823–25)
- *The Navy List*
- *The New Navy List*
  This is particularly useful as along with all the information provided in the official lists it gives biographies of officers.
- Other relevant sources such as *The Times* and *The Dictionary of National Biography*.
After this, go on to number **5**.

5. The majority of original service records are found in TNA in the following series:
- **ADM 196** (1756–1966). Most records are from 1840 to 1920. It can be accessed by using the partial card index and the integral indexes within the series. The records cover all commissioned officers entering the Navy until May 1917 and warrant officers entering until 1931. If your ancestor was recruited after that period refer to number **11**.
- **ADM 29** (1802–1919). The series contains service records for some warrant officers who were seeking pensions or medals.
- **ADM 9** (1817–48) and **ADM 11** (1741–1903). The series contains various surveys conducted on commissioned and warrant officers during the nineteenth century.
- Naval Office pay registers in **ADM 18**, **ADM 22–25** and **PMG 15** (1668 to 1920), which contain service records for officers on full or half pay.
- Succession books in **ADM 6**, **ADM 7**, **ADM 11** and **ADM 106** (1673 to 1849), which state who held which position onboard a particular ship at any given time and are indexed by ship and name.
- Passing certificates (organized by rank of officer).
  ○ Lieutenants' certificates have been indexed in *Royal Naval Lieutenants: Passing certificates 1691–1902*. The original documents are in **ADM 107** (1691–1832), **ADM 6** (1744–1819) and **ADM 13** (1854–1902).
  ○ Masters' certificates are in **ADM 106** (1660–1830) and **ADM 13** (for the second half of the nineteenth century).
  ○ Gunners' certificates are in **ADM 6** (1731–1812, not complete) and **ADM 13** (1856–67).
  ○ Pursers' certificates are in **ADM 6** (1813–20) and **ADM 13** (1851–89).
  ○ Boatswains' responsibility was for the sails of the ship and also summoning other seamen to their duties. Their certificates are held in **ADM 6** (1810–13) and **ADM 13** (1851–87).
  ○ Surgeons' passing certificates are held in **ADM 106** for the eighteenth century.
- Confidential reports are in **ADM 196** (1884 to 1943) and give frank accounts of the suitability of officers seeking promotion by their commanding officers.
You may also wish to check if there are any surviving pension records for your ancestor. Turn to number **9** for guidance on this.

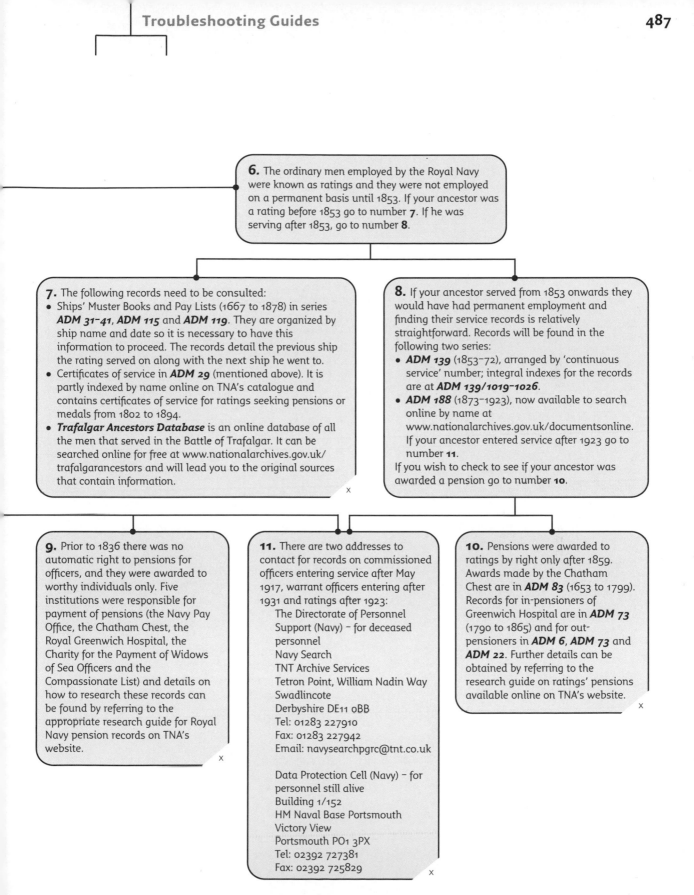

6. The ordinary men employed by the Royal Navy were known as ratings and they were not employed on a permanent basis until 1853. If your ancestor was a rating before 1853 go to number **7**. If he was serving after 1853, go to number **8**.

7. The following records need to be consulted:
- Ships' Muster Books and Pay Lists (1667 to 1878) in series **ADM 31-41**, **ADM 115** and **ADM 119**. They are organized by ship name and date so it is necessary to have this information to proceed. The records detail the previous ship the rating served on along with the next ship he went to.
- Certificates of service in **ADM 29** (mentioned above). It is partly indexed by name online on TNA's catalogue and contains certificates of service for ratings seeking pensions or medals from 1802 to 1894.
- **Trafalgar Ancestors Database** is an online database of all the men that served in the Battle of Trafalgar. It can be searched online for free at www.nationalarchives.gov.uk/trafalgarancestors and will lead you to the original sources that contain information.

8. If your ancestor served from 1853 onwards they would have had permanent employment and finding their service records is relatively straightforward. Records will be found in the following two series:
- **ADM 139** (1853-72), arranged by 'continuous service' number; integral indexes for the records are at **ADM 139/1019-1026**.
- **ADM 188** (1873-1923), now available to search online by name at www.nationalarchives.gov.uk/documentsonline. If your ancestor entered service after 1923 go to number **11**.

If you wish to check to see if your ancestor was awarded a pension go to number **10**.

9. Prior to 1836 there was no automatic right to pensions for officers, and they were awarded to worthy individuals only. Five institutions were responsible for payment of pensions (the Navy Pay Office, the Chatham Chest, the Royal Greenwich Hospital, the Charity for the Payment of Widows of Sea Officers and the Compassionate List) and details on how to research these records can be found by referring to the appropriate research guide for Royal Navy pension records on TNA's website.

11. There are two addresses to contact for records on commissioned officers entering service after May 1917, warrant officers entering after 1931 and ratings after 1923:

> The Directorate of Personnel Support (Navy) – for deceased personnel
> Navy Search
> TNT Archive Services
> Tetron Point, William Nadin Way
> Swadlincote
> Derbyshire DE11 0BB
> Tel: 01283 227910
> Fax: 01283 227942
> Email: navysearchpgrc@tnt.co.uk
>
> Data Protection Cell (Navy) – for personnel still alive
> Building 1/152
> HM Naval Base Portsmouth
> Victory View
> Portsmouth PO1 3PX
> Tel: 02392 727381
> Fax: 02392 725829

10. Pensions were awarded to ratings by right only after 1859. Awards made by the Chatham Chest are in **ADM 83** (1653 to 1799). Records for in-pensioners of Greenwich Hospital are in **ADM 73** (1790 to 1865) and for out-pensioners in **ADM 6**, **ADM 73** and **ADM 22**. Further details can be obtained by referring to the research guide on ratings' pensions available online on TNA's website.

Chart 5: Merchant Seamen Service Records

1. Systematic service records were not kept until 1835. Was your ancestor in the Merchant Navy prior to 1835? If yes, go to number **2**; if no, go on to number **3**.

2. Prior to 1835 it can be tricky to find appropriate records for your ancestor (especially if he was not a master of a ship). The following may contain some information:

- *Crew Lists*
  They were only kept for maritime disputes prior to 1747 and can be found in The National Archives (TNA) *Series HCA* (High Court of Admiralty) and *DEL* (Records of the High Court of Delegates). After 1747 the records can be found in *Series BT 98* (although many do not survive prior to 1800), but they are arranged by relevant port and are without a name index.
- *The Corporation of Trinity House petitions*
  This organization was responsible for providing charitable support to mariners and their families. In order to receive such support the mariners would have to submit petitions. These petitions survive from 1787 to 1854 at the Guildhall Library with an online index on *www.origins.net*.
- Apprenticeship records
  ○ *Local archives* may have records for pauper boys forced to serve on ships as apprentices (this practice started after 1704).
  ○ TNA *Series IR 1* has records of taxes placed on the apprenticeship system from 1710 to 1811, with indexes for masters (1710–62) and boys (1710–64).
  ○ TNA *Series HCA 30/897* contains apprenticeship records for those indentured to fishermen in the South East from 1639 to 1664.
  ○ *BT 167* has a register of apprentices for the port of Colchester for 1704 to 1757 and 1804 to 1844.
  ○ *BT 150* has an index for apprentices between 1824 and 1953 although only a limited number of actual indenture records survive.
- *The Marine Society*
  This society, founded in 1756, also provided charitable support to merchant mariners. Their records are at The *National Maritime Museum*, including registers of boys who were sent to sea from the Society from 1756 to 1958.

x

3. If your ancestor was a master or mate go to number **4**; if he was an ordinary seaman go to number **5**.

4. Records for masters or mates begin in 1845. There are a variety of sources that may be consulted depending on when he served:

- The Alphabetical Register of Masters are in *BT 115* (1845–54)
- Certificates of competency and service are from 1845–1969 and are held in the following TNA series depending on the date and would also be published in the *London Gazette*:
  ○ *BT 143/1* and *BT 6/218–219* has certificates from 1845 to 1850.
  ○ *BT 122–126* and *BT 128* contain certificates of competency and service from 1845 to 1925. These records are indexed by *BT 127*.
  ○ *BT 139*, *140* and *142* has engineer certificates from 1862 to 1921. The indexes can be found in *BT 141*.
  ○ *BT 129* and *130* contain certificates for skippers and mates of fishing vessels from 1882 to 1921 with the index in *BT 138*.
  ○ Certificates for cooks were introduced in 1908 and indexes to them are in *BT 319* for 1913 to 1956. The National Maritime Museum holds the actual registers for 1915 to 1958.
  ○ *BT 352* contains a combined index and register to all the different types of certificates issued from 1910 to 1969.
  ○ Lloyd's Captains' Register was begun in 1869 and contained an alphabetical list of masters from that date. The series continued until 1947 and is useful for supplementing information after 1913.

x

5. Records of seamen began in 1835 and form five main registers or series of records lasting till 1972 (with gaps). If your ancestor joined after 1972 turn to number **7**. The five series are as follows:

- The First Register of Seamen's Services (1835–36) is in **BT 120**, organized alphabetically.
- The Second Register of Seamen's Services (1835–44) is in **BT 112**. Part 1 (1835 to February 1840) is organized numerically and indexed by **BT 119/1**. Part 2 (December 1841 to 1844) is arranged alphabetically. There appears to be a gap from March 1840 to November 1841.

From 1845 to 1853 there is no register available but rather a ticketing system was used for seamen when leaving the country. The actual tickets were retained by the seamen and have not survived. However, there is an alphabetical index to seamen in **BT 114**. The ticket numbers refer to the ticket registers in series **BT 113**.

- The Third Register of Seamen's Services (1853–57) is in **BT 116**, listed alphabetically.

Between 1858 and 1913 there was no system of registering seamen. It is only possible to search for individuals using crew lists. If your ancestor served in this period, go to number **6**.

- The Fourth Register of Seamen's Services (1913–41) is in the following series, although the cards for 1913–18 were destroyed. The original cards are in **Southampton City Archive**, although TNA has microfilm copies in the following series:
  - **BT 351** is an index to all seamen awarded the Mercantile Marine Award for participating during the First World War
  - **BT 350** (known as the 'CR 10' cards) is an index created in 1918 to the cards from 1918 to 1921
  - **BT 349** (known as the 'CR 1' cards) is the main set of register cards, arranged alphabetically by surname. It is essential to note down the discharge number as the other series are organized by this number
  - **BT 348** (known as the 'CR 2' cards) contains further career details and is organized by discharge number
  - **BT 364** is a combined index of the CR 1, CR 2 and CR 10 cards and is also arranged by discharge number. Most cards are for seamen who served after 1941 and, therefore, it is possible to find the number by using the series for the Fifth Register of Seamen, **BT 382** (see below)
- The Fifth Register of Seamen's Services (1941–72) has two types of records:
  - 'Seamen's pouches'. These pouches contain the CR 1 and CR 2 cards from previous service and any subsequent paperwork from later service. Only 50% of the pouches survive, and can be found in **BT 372** (the majority of pouches), **BT 390** (for seamen working on Royal Navy ships during the Second World War) and **BT 391** (men working on special operations during the Second World War). They can all be searched online by name on TNA's catalogue, although only BT 372 has been catalogued fully.
  - 'Docket Books' in series **BT 382**. These are arranged alphabetically by surname in eight sub-series, although the majority of records will be in the first two series.

6. Crew lists and agreements can be used to fill in the gap between the Third and Fourth Registers of Seamen's Services and also survive for later periods. However, they can only be used if you know the name of the ship as there are no name indexes. The lists can be found in the following archives:

- *The National Archives:* They have a random 10% sample from 1861 to 1938 and then from 1951 to 1972, found in **BT 99**, **BT 100**, **BT 144** and **BT 165**. The crew lists during the Second World War are held separately, in **BT 381**, **BT 380** and **BT 387**. After 1972 only a 2% sample has been retained. Until 1994 the remainder have been destroyed apart from the lists retained by the National Maritime Museum; see below. All lists after 1994 are still with the Registry of Shipping and Seamen (RSS) in Cardiff.
- *The National Maritime Museum:* The Museum holds the remaining 90% of the above crew lists for the years 1861 and 1862. It also has the remainder for every year ending in 5 from 1865 onwards (apart from 1945).
- *Local Archives and County Record Offices:* Certain local repositories may also store crew lists for the years 1863 to 1913 for their local ports. The website www.crewlist.org.uk can provide further details regarding which lists such archives may store.
- *The University of Newfoundland:* The Maritime History Group in the University of Newfoundland, Canada, also retains significant holdings relating to crew lists and agreements. The Group has the remaining crew lists not found elsewhere for the years 1863 to 1938 and then from 1951 to 1976. The Group has also produced three indexes by ship number for their earlier collection.
  1. An index for its crew lists from 1863 to 1913
  2. An index for its crew lists from 1913 to 1938
  3. An index of crew lists deposited with local repositories and archives

All three indexes can be found at the Guildhall Library and TNA also has a copy of the first index.

7. The Sixth Register of Seamen's Services from 1973 is currently held by the **Registry of Shipping and Seamen** in Cardiff and is only available to either the individual or, in cases of decease, to the next of kin. The RSS can be contacted at the following address:

Registry of Shipping and Seamen
Maritime and Coastguard Agency
Anchor Court
Keen Road
Cardiff CF24 5JW

Chart 6: Royal Marine Service Records

1. Was your ancestor an officer of the Royal Marines or an 'other rank'? If an officer, go to number **2**; if an 'other rank', go to number **6**.

2. Officers' service records do not survive prior to 1793. If your ancestor served before 1793 go to number **3**. If he served after 1793 then go to number **4**. If he was appointed after 1925 go to number **5**.

3. The following sources may contain details:
- The **Army List** contains details of officers on full or half pay from 1740.
- Separate Marine Officer Lists were started in 1755 and The National Archives (TNA) has copies of these in series **ADM 118/230-236** (1757-1860) and **ADM 192** (1760-1886).
- Records for appointments and commissions of officers are in **ADM 6/405** (1703 to 1713) and **ADM 6/406** (1755 to 1814).

4. The main sets of service records begin after 1793 for officers enlisting prior to 1925. If your ancestor enlisted after that date go to number 5. Royal Marine officers are included in the **Navy Lists** from 1797 onwards, so start by looking at these. The main body of service records is in **ADM 196** in pieces ADM 196/58-65, 83 and 97-116. Further guidance can be obtained by referring to The National Archives research guide on Royal Marine Officers. Additional sources may also be relevant, such as obituaries (in **The Times** or the **Globe and Laurel**, the magazine of the Royal Marines), Admiralty Leave Books (in **ADM 6/200-206** for 1804 to 1846), Half Pay Lists (in **ADM 6/410-414**) and records of commissions and appointments (see number **3**). Once you have searched all the appropriate records in TNA, go to number **8**.

5. If your ancestor was appointed after 1925 his records will still be with the Royal Marines. They can be contacted at:
    DPS(N)2
    Building 1/152
    Victory View
    PP36
    HM Naval Base Portsmouth
    PO1 3PX

6. If your ancestor was an 'other rank' marine, the first thing to ascertain is which division he was in, as the records are organized on this basis. The divisions were Portsmouth, Plymouth, Chatham or Woolwich. This can be found out by the following means:
- Establish the nearest division to where your ancestor lived, as most men would have been recruited from their local area.
- Check **ADM 171** for any medals that your ancestor may have been awarded, as the rolls would also provide the division and service number.
- Garth Thomas's guide, **The Records of the Royal Marines**, has a table that can be used to find the correct division (if you know your ancestor's company number and when he served).
- If you know which ship your ancestor served in, refer to the **Navy List** as up until 1947 those serving on a particular ship would have belonged to the same division as the home port of the ship.
- The attestation forms in series **ADM 157** (see below) has a partial card index that gives the appropriate division.
Once you have discovered this information, go on to number **7**.

7. The main sets of records worth consulting are as follows. If your ancestor enlisted after 1925 refer to number **5**:
- **ADM 157**: Attestation forms (1790 to 1925). The majority are arranged by date of discharge (apart from for Chatham division, which are arranged by enlistment until 1883). There is also a partial card index for this series available for consultation in TNA.
- **ADM 158**: Description books (c.1750 to 1940). They are arranged by division, then company and date of enlistment, and by first letter of surname (not strictly alphabetically).
- **ADM 159**: Service registers (1842-1936). They are arranged by service number and division. The service number can be obtained in the index found in **ADM 313**.
- Discharge Books are held in **ADM 183** (Chatham), **ADM 184** (Plymouth) and **ADM 185** (Portsmouth).
Once you have searched all the appropriate records in TNA turn to number **8**.

8. Other than in TNA you may find relevant information about your ancestor in particular or about the Royal Marines in general by visiting the **Royal Marine Museum** in Southsea (www.royalmarinesmuseum.co.uk).

Chart 7: RAF Service Records

1. Was your ancestor male or female? If male, go to number **2**; if female go to number **10**.

2. Was your ancestor an officer or an airman? If he was an officer go to number **3**; if an airman, go to number **7**.

10. Women were also employed in the air force during the First World War, in the WRAF (Women's Royal Air Force). The WRAF was formed with the RAF but disbanded by 1920. Service records only survive for airwomen, not for officers, in **AIR 80** (indexed by **AIR 78**). The service was re-established for the Second World War in 1939 as the WAAF (Women's Auxiliary Air Force). Service records for this institute are still with the RAF at the address given in number **6**.

3. Was your ancestor an officer of the Royal Naval Air Service (RNAS) or Royal Flying Corps (RFC) and left the service before March 1918? If yes, go to number **4**; if no, go to number **5**.

7. If your ancestor was in the Royal Flying Corps (RFC) or Royal Naval Air Service (RNAS) and left either service before the formation of the RAF on 1 April 1918, go to number **8**. If he was still in service on 1 April 1918, go to number **9**.

4. Officer records for RFC personnel who began their service in 1914 and left before March 1918 have been forwarded to the main series for RAF officers in The National Archives (TNA) series **AIR 76**. These officers will also be found in the relevant **Army Lists** prior to their transfer. RNAS officer records for those serving from 1914 and leaving by March 1918 are in TNA series **ADM 273**. These men can also be traced in the **Navy List** during this period. ADM 273 is organized numerically and there are indexes within the series and among the finding aids at TNA. If your ancestor stayed with these two forces after March 1918 go on to number **5**.

8. Records for RFC men who left by March 1918 will be with the ordinary Army service records in TNA **WO 363** and **WO 364** (see **Chart 1** for further details on these records). Additionally, the biographies of the first men to join the RFC between 1912 and August 1914 (service numbers 1 to 1,400) can be found in **A Contemptible Little Flying Corps** by **Webb and McInnes**. RNAS personnel who left by 1 April 1918 are in the normal ratings' service records series in **ADM 188** (see **Chart 4** for further details).

9. Service records for the first 329,000 men who made up the RAF upon its formation on 1 April 1918 are in **AIR 79** arranged by service number (with an alphabetical index in **AIR 78**). Records for those with a service number of 329,000 and higher are still with the RAF, who can be contacted at the same address as given in number **6**. Additionally, records for men with service numbers between 1 and 329,000 who served in the Second World War are also still with the RAF (contact at the same address).

5. The RAF was formed on 1 April 1918 by the merging of the RFC and RNAS. If your ancestor was an officer in either force after 1 April 1918 he would automatically have become an RAF officer. TNA contains the service records for all these men in series **AIR 76**. However, the series only contains records for individuals up to 1919. If your ancestor was still in service after that date, go on to number **6**.

6. The **Air Force Lists**, published since March 1919, will give basic details for individuals still employed in the RAF after 1919. The original service records are with still with the RAF. They can be contacted at the following address:
ACOS (Manning) 22E
Room 5
Building 248A
RAF Innsworth
Gloucester GL3 1EZ

Chart 8: Immigration

1. Did your ancestor come from Ireland?
If yes, go to number **2**; if no, go to number **3**.

2. As Ireland was part of the United Kingdom until 1922 Irish people settling in Britain were not considered immigrants/aliens and no specific records were created for their arrival. Hence, there are no immigration records for these individuals, although you may obtain some details by going to number **16**.

3. Did your ancestor arrive before or after the passing of the Aliens Act of 1793? If before, go to number **4**; if after, go to number **8**.

4. Was your ancestor naturalized or made a 'denizen'? If the answer is yes, turn to number **5**; if no, go to number **7**.

5. Begin by searching the *indexes* published by the *Huguenot Society of Great Britain*. The Society has compiled a list of immigrants (including non-Huguenots) arriving from 1509 to 1800 from the many sources that contain information about immigrants during this period. These indexes can be found at most major archives and libraries. If you find your ancestor in this index you can follow this up by viewing the original documents listed in the entry. If you do not find your ancestor, go on to number **6**.

7. There is no single documentary series for individuals arriving before 1793. The earliest records of immigrants and aliens are in various sources at TNA. Records may be found in the Chancery (**C**), Treasury (**T**), Exchequer (**E**), Customs (**CUST**) or State Paper (**SP**) series. Further guidance can be found in the appropriate immigration research guide on TNA's website. If you do not find your ancestor amongst these records then it is unlikely that any documentation survives for them in the UK. If you are aware of approximately when and where they arrived from, try to see whether any records survive in the country of origin.

x

6. There are other records for individuals becoming denizens. If your ancestor was made a denizen prior to 1509 there may be records in *Chancery Rolls* at The National Archives (TNA). If your ancestor was naturalized prior to the Naturalization Act of 1844 it would have been through a private Act of Parliament. These Acts have also been indexed by the Huguenot Society and the original acts are at the *House of Lords Record Office*. If your ancestor was naturalized from 1844 onwards, go to number **11**.

8. Monitoring and recording the presence of immigrants and aliens increased after 1793. Immigrants now had to register with the local JP and send information certificates to the Home Office. If you know where your ancestor settled in the UK check with your *local archive* for any surviving information, as the Home Office copies of these certificates do not survive. If you do not know where your ancestor settled, go to number **9**.

9. TNA has a number of records for aliens from the nineteenth century onwards created by the Home Office. Certificates were created for all aliens living in England and Scotland under the 1826 and 1836 Aliens Acts. The certificates survive from 1836 to 1852 in *HO 2* and are indexed in *HO 5/25-32*. HO 5 also has a variety of out-letters and entry books relating to aliens and naturalizations till 1909. *HO 3* contains lists of immigrants listed by masters of ships, although there is no index, from 1836 to 1869. If you do not find your ancestor in these records go to number **10**.

10. Your ancestor may have arrived at a later date. If they arrived between 1878 and 1960 (excluding 1889) check the inward passenger lists in TNA series *BT 26*. They list arrivals of individuals on vessels who set off outside Europe (however, the ships may have stopped in ports en route in Europe). As they are arranged by port of arrival it is necessary to have this information before you begin this search. If you do not find your ancestor in this series, check whether they were naturalized by turning to question **11**.

11. The Naturalization Act of 1844 simplified the process of immigrants becoming British citizens, and finding the records for individuals who were naturalized after this date is relatively straightforward. Each prospective new citizen presented a memorial to the Home Office staking his or her claim. These memorials are now in TNA series *HO 1* (up to 1871), *HO 45* (1872–78), *HO 144* (1879–1933) and *HO 405* (1934–48) and can be searched online by name in TNA's catalogue subject to closure rules after 1922 (which can be reviewed upon request). If you do not find your ancestor in this series and he or she arrived in the twentieth century, there are additional sources to search. Go to number **12**.

12. If your ancestor arrived as a result of population displacement caused by the First World War go to number **13**. If they arrived under similar circumstances or as refugees as a result of the Second World War go to number **14**. If they arrived as a Jewish immigrant, go to number **15**. If they arrived after the Second World War from the Commonwealth countries, go to number **16**.

13. In 1914 the Alien Registrations Act required all foreigners to register with the local police and registration cards were duly issued. Any surviving cards will be with the *local archive* or a very small number are in TNA series *MEPO 35*. Belgian refugees (about 2,000) arriving after the German invasion were registered with the Ministry of Health and issued with 'history cards'. They are held in TNA series *MH 8*.

14. Waves of European refugees arrived during the 1930s and 1940s. The Czechoslovak Refugee Trust records can be found in TNA series *HO 294* (although only a small number of personal files exist). Some Poles who had fought with the Allies settled in the UK after the Second World War. They were assisted by the Polish Resettlement Corps (PRC) whose records are in TNA series *WO 315*. General records for Hungarian refugees arriving after the Uprising in 1956 are in TNA series *HO 352*, although little is included on individuals.

15. Apart from the general immigration records it may be possible to find more details about your Jewish ancestors through the following records:
- Surviving *synagogue records*. Further information about the location of these records can be found by contacting the British Board of Deputies (www.bod.org.uk).
- The *NIPR* has listed all synagogues present in Britain prior to 1838 in its third volume.
- The *Jewish Chronicle* has birth, marriage, death and obituary details of members of its community and can be searched online on www.jc.com.
- The *Jewish Genealogical Society of Great Britain* also has holdings of interest. Further details are available on its website on www.jgsgb.org.uk
- The *London Metropolitan Archives* has records of the Poor Jews' Temporary Shelter, established in 1886 and involved in the assistance of Jews fleeing Nazi persecution. The Archives also has the records of the Jewish Refugees Committee.

16. If your ancestor was an immigrant from the Afro-Caribbean or South Asian community, the best place to start is by referring to the website *www.movinghere.org.uk*. It is dedicated to the history of immigration over the past 200 years and has specific sections on Caribbean, South Asian, Jewish and Irish immigrant experiences, along with advice on how to begin researching an ancestor from these communities.

Chart 9: Emigration

1. Did your ancestor leave on a voluntary or involuntary basis? If on an involuntary basis, go to number **2**; if voluntary, go to number **11**.

2. If your ancestor was transported abroad, go to number **3**. If your ancestor left as an assisted migrant go to number **10**.

3. Transportation was used as a humane system to punish convicts. Criminals would be sent to North America until the American War of Independence (1775–83) and thereafter to Australia. If your ancestor went to North America, go to number **4**. If your ancestor was sent to Australia, go on to number **5**.

10. Many emigrants went to North America and the West Indies in the seventeenth and eighteenth centuries as assisted migrants. Details of these individuals can also be found in **Coldham**'s book (mentioned in number **4**). Additional information may be located in TNA's **series SP**, **CO** and **PC** for North America. **Local archives** may have records for assistance granted to individuals mentioned in parish vestry minutes.
x

4. Approximately 150,000 individuals were transported to North America and the West Indies between 1615 and 1776. If your ancestor was one of these people, begin by referring to the book by Coldham, **The Complete Book of Emigrants in Bondage, 1614–1775** (Baltimore: Genealogical Publishing Co. Inc., 1988). This also gives details of trial and conviction, which can be followed up in the appropriate sources, along with references to contemporary sources found at The National Archives (TNA).
x

5. The first convicts arrived in Australia in 1788. A very good introduction to this system can be found online at **www.convictcentral.com**. If your ancestor was on one of the first or second fleets to arrive, further details can be obtained by referring to books by P. G. Fidlon and R. J. Ryan (eds.), **The First Fleeters** (Sydney, 1981) and R. J. Ryan (ed.), **The Second Fleet Convicts** (Sydney, 1982) respectively. Otherwise, go to number **6**.

6. There is no uniform index to every criminal transported to Australia, and relevant information is scattered in a number of documents. If you are approximately aware of when and where your ancestor was tried then you can refer directly to the original documents by going on to number **8**. If you know your ancestor was tried in Ireland, go to number **9**. Otherwise, start by referring to two books by David T. Hawkings, **Bound for Australia** (Chichester, 1987) and **Criminal Ancestors: A guide to historical criminal records in England and Wales** (Stroud, 1992). These publications have many trial transcriptions and lists of those who were transported. If you find your ancestor in these sources they will give you relevant trial details to follow up in criminal documents. If you do not find your ancestor, then go to number **7**.

7. Refer to the **index** to the **New South Wales Convicts and Ships**, which gives the names of all convicts arriving in New South Wales and Van Dieman's Land between 1788 and 1842 and is available in The National Archives library. Many censuses and musters were also taken at various points in the nineteenth century. A good number of these were published and list individuals and where they were tried. A list of these can be found on TNA's research guide for transportation. If you find your ancestor in these lists turn to number **8**.

8. You now have the key details for finding the original documents – the trial date and the name of the ship they arrived on. It is now possible to refer to the **transportation registers** in TNA series **HO 11** and **trial records** (in TNA and local archives. Please refer to Chapter 27 for further details). It may well be worth referring to petitions for clemency in TNA **series HO 17** and **HO 18**. They contain petitions from the convicts' families appealing for clemency after conviction and can give detailed personal information and the circumstances surrounding the crime.
x

9. If your ancestor was tried in Ireland records will be at The National Archives of Ireland. You can search for your ancestor online and free of charge at **www.nationalarchives.ie** (although original records do not survive prior to 1836 unless a petition was made on behalf of the convict).
x

11. Most individuals emigrated on a voluntary basis due to economic factors. If your ancestor travelled between 1890 and 1960 go to number **12**. If they went before 1890, go on to number **13**.

12. Begin by searching the outward passenger lists. They were kept by ships' masters and recorded each individual travelling from the UK or Ireland for final destinations outside of Europe and give the age, occupation and residence of each person travelling. These lists can now be searched online by passenger name on the website ***www.findmypast.com*** from 1890 to 1939. Thereafter, the originals are held at TNA and are only searchable by name of ship and port of departure.

13. If your ancestor went as a free settler to North America, go to number **14**. If your ancestor went to Australia, go to number **15**. If your ancestor went to New Zealand, go to number **16**; if to South Africa, go to number **17**; and if to other parts of the British Empire, go to number **18**.

14. Migration to North America began in the sixteenth century. Start by searching the ***State Papers Colonial, American and West Indies***, available to search by name on CD ROM. Additionally, refer to the published sources listed in Chapter 23. If your ancestor left in the early twentieth century, try searching on the online source ***www.ellisisland.org***, which lists individuals arriving at this immigration station in New York from 1892 to 1924. Most immigration records for people settling in Canada are still in Canada. Microfiche copies of the Hudson Bay Company are held in TNA ***series BH 1***. These include journals of early settlers. Otherwise refer to the collections available online on ***Canada's National Archive website***, specifically http://www.collectionscanada.ca/immigrants/ index-e.html (an online exhibition relating to migrants arriving in Canada), http://www.collectionscanada.ca/archivianet/passenger/001045-100.01-e.php (online database of passenger lists but not searchable by name), or http://www.ist.uwaterloo.ca/~marj/genealogy/thevoyage.

16. If your ancestor settled in New Zealand begin by referring to the ***history of immigration placed online by the New Zealand government*** available on http://www. teara. govt.nz/NewZealanders/NewZealandPeoples/HistoryOf Immigration/en. The earliest migrants arrived via New South Wales and it may be possible to trace them by referring to Australian sources first. Otherwise records of the New Zealand Company (active till 1858) can be viewed in TNA ***series CO 208***. The majority of records are with the National Archives of New Zealand (***www.archives.govt.nz***).

17. Many Britons also settled in South Africa from the nineteenth century onwards. Only a very limited amount of information is available through the general passenger lists mentioned in number **12** and in TNA series relating to ***Army and Navy pension awards to ex-personnel*** settling there in ***WO*** along with the ***Colonial Office Correspondence***. Refer to the South African genealogical society for guidance on searching in South Africa (***www.gensa.info***).

15. Free settlers started to arrive to Australia from the nineteenth century onwards. Begin by searching the ***Colonial Office papers for New South Wales*** held in TNA and a number of census returns in TNA series ***HO 10***. Australian archives also hold ample documentation for migrants. Migration to Australia was not controlled centrally until 1901 and therefore the National Archives of Australia (***www.naa.gov.au***) has little information prior to that date compared to the ***state archives***. Refer to these state archives to establish the exact documentation held by them to see if there would be any relevant records for your ancestor.

18. If your ancestor went to India, go to number **19**. If they settled elsewhere throughout the globe the best starting point on collections for individual countries is visiting the country pages found on ***www.cyndislist.com*** or ***www.worldgenweb.org***. Both websites contain country-specific guidance. The Society of Genealogists also has collections for a number of different countries and details of their holdings can be found in their online catalogues at ***www.sog.org.uk***.

19. India was never a mass emigration destination, although many individuals did live there whilst part of the colonial settlement. The ***British Library*** holds the archives of the British Raj, including ***parish registers*** for Britons living in the continent, along with biographical notes of individuals living there. Further information on the nature of their holdings can be found on the website at http://indiafamily.bl.uk/UI/. Assistance on researching British or Anglo-Indian families in India can be found by contacting the ***Families in British India Society*** at www.fibis.org.

SECTION FIVE

Key Resources

The aim of this section is to provide you with a toolkit of key resources to help you with your research. The topics covered in this section are ones that don't fit naturally into the main body of the text, which has been designed to introduce subject areas of family history by theme and then describe the associated records. Instead, these topics are more by way of reference, to be used as and when you need them.

Origins and Meanings of Popular Surnames

Listed below are the 100 most popular surnames, drawn from both historic sources (the census returns) and modern electoral lists. If you can't find a surname listed below, some resources are listed at the end to help you track it down.

Top 100 Surnames from 1881

1   SMITH – Old English for a 'metal worker' or 'blacksmith'

2   JONES – Son of John, from the Welsh version of John, Ioan

3   WILLIAMS – Son of William (William Germanic for 'will' or 'resolve helmet')

4   BROWN – Old English for brown-haired or skinned

5   TAYLOR – Old French for tailor, tailleor

6   DAVIES – Son of Davys or David

7   WILSON – Son of Will / William

8   EVANS – A Welsh version of the name John

9   THOMAS – Aramaic for twin, an apostle

10 ROBERTS – Germanic for 'fame bright'

11 WALKER – Old English for the occupation of fuller (someone who would step on cloth during the 'fulling' process)

12 JOHNSON – Son of John (John being Hebrew for 'Jehovah has favoured')

13 WOOD – Meaning someone living near a wood

14 WHITE – Meaning white or fair hair or complexion

15 ROBINSON – Son of Robin, a diminutive for Robert (see Roberts)

16 WRIGHT – Old English for 'carpenter' or 'joiner'

17 THOMPSON – Son of Thomas

18 CLARK – Originally referring to 'a man in religious order, cleric', during the Middle Ages became 'cleric or secretary' as writing was mostly done by clergy members

19 HALL – Old English for someone residing or working in a hall or manor house

20 HUGHES – Germanic for 'heart or mind'

21 JACKSON – Son of Jack, a diminutive of John

22 EDWARDS – Old English for 'prosperity' or 'happiness guard'

23 GREEN – Old English for residing near the village green

24 TURNER – From the occupation of turner, someone who makes items from wood, bone or metal by using a lathe

25 LEWIS – Germanic for 'loud battle', used as the translation for the Welsh name Llewelyn

26 SCOTT – One coming from Scotland

27 HILL – One living on or near a hill

28 HARRIS – Harry, Harry being the usual Middle English pronunciation of Henry

29 MARTIN – A diminutive of Martius or Mars, the Roman God of war

30 COOPER – Middle English term referring to makers of buckets, casks or tubs

31 WATSON – Son of Wat(t), a diminutive of Walter (Germanic for 'mighty army')

32 MORRIS – From the Latin Mauritius meaning 'dark' or 'Moorish'

33 HARRISON – (Son) of Harry, Harry being the usual Middle English pronunciation of Henry

34 YOUNG – Used to differentiate the younger of two men, possibly for father and son

35 DAVIS – Son of Davys or David

36 WARD – Old English weard meaning 'watching' or 'guarding'

37 KING – Working in the Royal Household, or performing as a King in a pageant, from the Old English Cyng

38 BAKER – Literally a baker

39 MITCHELL – A popular version of the name Michael, also Old English for 'big'

40 ANDERSON – Son of Andrew (Andrew meaning 'manly' in Greek, the first disciple)

41 MORGAN – Old Welsh for 'circling sea' or 'brightness'

42 MOORE – Old French Maur meaning 'the Moor', also for one living by the moor

43 JAMES – A form of the name Jacob

44 CLARKE – Originally referring to 'a man in religious order, cleric', during the Middle Ages became 'cleric or secretary' as writing was mostly done by clergy members

45 BELL – Shortened version of Isabel, shortened version of Latin Bellus or Old French Bel meaning 'beautiful', or being a bell-ringer

46 COOK – An occupational surname for cook, or seller of cooked meats

47 SHAW – Old English for 'thicket' or 'small wood', indicating someone who lived in or near such a place

48 PARKER – Old French for parquier, meaning an individual who is responsible for a park

49 ALLEN – The name of a Welsh and Breton saint, became a popular name with the Bretons who arrived after the Norman conquest

50 MILLER – An occupational surname for a miller

51 PHILLIPS – Greek for one who is 'fond of horses', an apostle

52 SIMPSON – Son of diminutive of Simon, Simon originating from the Hebrew Shimeon

53 PRICE – From Welsh 'ap' (son of) 'Rhys', may also be a metonym for the occupation of price fitting

54 ROBERTSON – Son of Robert (Robert Germanic for 'fame bright')

55 CAMPBELL – Scottish Gaelic for caimbeul, meaning 'wry or crooked mouth'

56 RICHARDSON – Son of Richard, Richard being Germanic for 'powerful brave', the name being brought by the Normans

57 MARSHALL – From the old French mareschal, meaning 'tender of horses (mares)', especially caring for their medical needs

58 MCDONALD – Son of Donald (Donald Scottish Gaelic for 'world mighty')

59 GRIFFITHS – From the Old Welsh Griph-iud, where 'iud' indicates lord or chief

60 CARTER – From Middle English cart(e), meaning someone who originated from Scandinavia

61 LEE – From the locality of a lea

62 BENNETT – A diminutive of Benedict (Latin Benedictus, meaning 'blessed one')

63 STEWART – Occupational surname for steward, an official responsible for being the keeper of the household

64 WILKINSON – Son of Wilkin (Wilkin a diminutive of William with the Flemish suffix -kin)

65 RICHARDS – Son of Richard, Richard being Germanic for 'powerful brave', the name being brought by the Normans

66 BAILEY – Old French for bailiff, an official of the crown

67 COX – Many possibilities: a diminutive of Cook, or from the Cornish coch (meaning 'red'), a nickname for young apprentices or servants (from the strutting barnyard animal) and therefore attached to Christian names such as Wilcock, Hancock

68 ELLIS – A version of the Middle English name Elias, a diminutive of the Hebrew Elijah, meaning 'Yahweh is God'

69 GRAY – Old English for grey-haired or possibly pale-faced

70 ADAMS – Hebrew for the colour 'red'

71 THOMSON – Son of Thomas (Aramaic for twin, the apostle)

72 COLLINS – Diminutive for Col-in whereby Col is a nickname for Nicholas. The name in Ireland is derived from the Irish O Cullane meaning 'descendant of Whelp'

73 CHAPMAN – Old English for 'merchant' or 'trader'

74 FOSTER – Possibly from the Middle English foster for foster-parent or nurse, may also be diminutive of forester (forster)

75 MASON – Occupational surname for stonemason, from Norman Machen

76 WEBB – From Old English webbe or webba, meaning 'weaver'

77 BARKER – Originally Old French bercher, meaning 'shepherd', later Middle English for tanner (stripping the bark from wood to use in the tanning process)

78 HUNT – From the occupation of a hunter

79 MURRAY – From the county of Moray in Scotland, a 'seaboard settlement'

80 ROGERS – Son of Roger (Roger Germanic for 'fame spear')

81 MILLS – Either a diminutive of the surname Miles (ambiguous origins, possibly from Latin for 'soldier' or Germanic milo, meaning 'merciful'), or residing near a mill, or son of Mill

82 POWELL – From Welsh apHowell (son of Howell)

83 RUSSELL – From Old French rous-el, being a diminutive of rous ('red')

84 GIBSON – Son of Gibb, a shortened form of Gilbert (Old German for 'pledge' or 'hostage bright')

85 HOLMES – From Old Norse holmr, living near a flatland by a fen or land with streams around it

86 KNIGHT – Old English for a soldier or feudal tenant required to be a mounted soldier

87 OWEN – Connected to Ewan (Ewan is derived from the Greek Eugene or Eugenics, meaning 'well born')

88 JENKINS – Diminutive of John, the kin part indicating son

89 BARNES – From residing or working near a barn, or the geographical location of Barnes in South West London

90 GRAHAM – Old English for 'gravelly homestead' or 'homestead of Granta'

91 LLOYD – From the Welsh Llwyd, meaning 'grey'

92 PEARSON – Son of Piers (an Old French version of Peter, Peter being a Greek name meaning 'stone' or 'rock')

93 FISHER – From the occupation of a fisherman

94 FLETCHER – From the occupation of a fletcher (an arrow-maker)

95 HENDERSON – Son of Henry (from the Old German Haimric, Henric meaning 'home rule')

96 PALMER – From the Old French palmer, paumer meaning 'pilgrim' (one who returned from the Holy Land with a palm branch)

97 ROSS – A diminutive of Rose, having various meanings: Scots Gaelic for 'cape', Irish Gaelic for 'wood', Cornish and Welsh for 'moor', Germanic origin meaning 'fame kind'

98 KELLY – Anglicized version of Irish surname Ceallaigh, meaning 'descendant of Ceallach (war)'

99 DAWSON – Son of Daw, a diminutive of David (a Hebrew name originally meaning 'darling' and later 'friend')

100 DIXON – A diminutive of Dick which is in turn a diminutive of Richard (Richard being Germanic for 'powerful brave', the name being brought by the Normans)

Top 100 Surnames from 1998

1 SMITH – Old English for a 'metal worker' or 'blacksmith'

2 JONES – Son of John from the Welsh version of John, Ioan

3 WILLIAMS – Son of William (William Germanic for 'will' or 'resolve helmet')

4 BROWN – Old English for brown-haired or skinned

5 TAYLOR – Old French for tailor, tailleor

6 DAVIES – Son of Davys or David

7 WILSON – Son of Will / William

8 EVANS – A Welsh version of the name John

9 THOMAS – Aramaic for twin, an apostle

10 JOHNSON – Son of John (John being Hebrew for 'Jehovah has favoured')

11 ROBERTS – Germanic for 'fame bright'

12 WALKER – Old English for the occupation of fuller (someone who would step on cloth during the 'fulling' process)

13 WRIGHT – Old English for 'carpenter' or 'joiner'

14 ROBINSON – Son of Robin, a diminutive for Robert (see Roberts)

15 THOMPSON – Son of Thomas

16 WHITE – Meaning white or fair hair or complexion

17 HUGHES – Germanic for 'heart or mind'

18 EDWARDS – Old English for 'prosperity' or 'happiness guard'

19 GREEN – Old English for residing near the village green

20 HALL – Old English for someone residing or working in a hall or manor house

21 WOOD – Meaning someone living near a wood

22 HARRIS – Harry, Harry being the usual Middle English pronunciation of Henry

23 LEWIS – Germanic for 'loud battle', used as the translation for the Welsh name Llewelyn

24 MARTIN – A diminutive of Martius or Mars, the Roman God of war

25 JACKSON – Son of Jack, a diminutive of John

26 CLARKE – Originally referring to 'a man in religious order, cleric', during the Middle Ages became 'cleric or secretary' as writing was mostly done by clergy members

27 CLARK – Originally referring to 'a man in religious order, cleric', during the Middle Ages became 'cleric or secretary' as writing was mostly done by clergy members

28 TURNER – From the occupation turner, someone who makes items from wood, bone or metal by using a lathe

29 HILL – One living on or near a hill

30 SCOTT – One coming from Scotland

31 COOPER – Middle English term referring to makers of buckets, casks or tubs

32 MORRIS – From the Latin Mauritius, meaning 'dark' or 'Moorish'

33 WARD – Old English weard, meaning 'watching' or 'guarding'

34 MOORE – Old French Maur, meaning 'the Moor', also for one living by the moor

35 KING – Working in the Royal Household, or performing as a King in a pageant, from the Old English Cyng

36 WATSON – Son of Wat(t), a diminutive of Walter (Germanic for 'mighty army')

37 BAKER – Literally a baker

38 HARRISON – (Son) of Harry, Harry being the usual Middle English pronunciation of Henry

39 MORGAN – Old Welsh for 'circling sea' or 'brightness'

40 PATEL – An Indian name originating mainly in the province of Gujarat. A caste title meaning 'chief' or 'landlord'

41 YOUNG – Used to differentiate the younger of two men, possibly father and son

42 ALLEN – The name of a Welsh and Breton saint, became popular with the Bretons who arrived after the Norman conquest

43 MITCHELL – A popular version of the name Michael, also Old English for 'big'

44 JAMES – A form of the name Jacob

45 ANDERSON – Son of Andrew (Andrew meaning 'manly' in Greek, the first disciple)

46 PHILLIPS – Greek for one who is 'fond of horses', an apostle

47 LEE – From the locality of a lea

48 BELL – Shortened version of Isabel, shortened version of Latin Bellus or Old French Bel, meaning 'beautiful', or being a bell-ringer

49 PARKER – Old French for parquier, meaning an individual who is responsible for a park

50 DAVIS – Son of Davys or David

51 BENNETT – A diminutive of Benedict (Latin Benedictus meaning 'blessed one')

52 MILLER – An occupational surname for miller

53 COOK – An occupational surname for cook, or seller of cooked meats

54 PRICE – From Welsh 'ap' (son of) 'Rhys', may also be a metonym for the occupation of price fitting

55 CAMPBELL – Scottish Gaelic for caimbeul, meaning 'wry or crooked mouth'

56 SHAW – Old English for 'thicket' or 'small wood' indicating someone who lived in or near such a place

57 GRIFFITHS – From the Old Welsh Griph-iud, where 'iud' indicates lord or chief

58 KELLY – Anglicized version of Irish surname Ceallaigh, meaning 'descendant of Ceallach (war)'

59 RICHARDSON – Son of Richard, Richard being Germanic for 'powerful brave', the name being brought by the Normans

60 SIMPSON – Son of diminutive of Simon, Simon originating from the Hebrew Shimeon

61 CARTER – From Middle English cart(e), meaning someone who originated from Scandinavia

62 COLLINS – Diminutive for Colin whereby Col is a nickname for Nicholas. The name in Ireland is derived from the Irish O Cullane, meaning 'descendant of Whelp'

63 MARSHALL – From the Old French mareschal, meaning 'tender of horses (mares)' especially caring for their medical needs

64 BAILEY – Old French for bailiff, an official of the crown

65 GRAY – Old English for grey-haired or possibly pale-faced

66 STEWART – Occupational surname for steward, an official responsible for being the keeper of the household

67 COX – Many possibilities: a diminutive of Cook, or from the Cornish coch (meaning 'red'), a nickname for young apprentices or servants (from the strutting barnyard animal) and therefore attached to Christian names such as Wilcock, Hancock

68 MURPHY – From the Irish O Murchadha, meaning 'descendant of Murchadh', a sea-warrior

69 ADAMS – Hebrew for the colour 'red'

70 MURRAY – From the county of Moray in Scotland, a 'seaboard settlement'

71 RICHARDS – Son of Richard, Richard being Germanic for 'powerful brave', the name being brought by the Normans

72 ELLIS – A version of the Middle English name Elias, a diminutive of the Hebrew Elijah, meaning 'Yahweh is God'

73 ROBERTSON – Son of Robert, Robert Germanic for 'fame bright'

74 WILKINSON – Son of Wilkin (Wilkin a diminutive of William with the Flemish suffix -kin)

75 FOSTER – Possibly from the Middle English foster for foster-parent or nurse; may also be diminutive of forester (forster)

76 GRAHAM – Old English for 'gravelly homestead' or 'homestead of Granta'

77 CHAPMAN – Old English for 'merchant' or 'trader'

78 MASON – Occupational surname for stonemason, from Norman Machen

79 RUSSELL – From Old French rous-el, being a diminutive of rous ('red')

80 POWELL – From Welsh apHowell (son of Howell)

81 WEBB – From Old English webbe or webba, meaning 'weaver'

82 ROGERS – Son of Roger (Roger Germanic for 'fame spear')

83 HUNT – From the occupation of a hunter

84 MILLS – Either a diminutive of the surname Miles (ambiguous origins, possibly from Latin for 'soldier' or Germanic milo meaning 'merciful'), or for residing near a mill, or son of Mill

85 HOLMES – From Old Norse holmr, living near a flatland by a fen or land with streams around it

86 OWEN – Connected to Ewan (Ewan is derived from the Greek Eugene or Eugenics, meaning 'well born')

87 PALMER – From the Old French palmer, paumer meaning 'pilgrim' (one who returned from the Holy Land with a palm branch)

88 MATTHEWS – Hebrew Matthias, meaning 'Gift of God'

89 GIBSON – Son of Gibb, a shortened form of Gilbert (Old German for 'pledge' or 'hostage bright')

90 FISHER – From the occupation of a fisherman

91 THOMSON – Son of Thomas (Aramaic for twin, the apostle)

92 BARNES – From residing or working near a barn, or the geographical location of Barnes in South West London

93 KNIGHT – Old English for a soldier or feudal tenant required to be a mounted soldier

94 LLOYD – From the Welsh Llwyd, meaning 'grey'

95 HARVEY – A Breton name meaning 'worthy of battle', introduced after 1066

96 BARKER – Originally Old French bercher, meaning 'shepherd', later Middle English for tanner (stripping the bark from wood to use it in the tanning process)

97 BUTLER – Old French for bouteillier, meaning 'butler or servant responsible for the wine-cellar'

98 JENKINS – Diminutive of John, the kin part indicating 'son'

99 REID Scottish version of the surname Read (meaning 'red')

100 STEVENS – Son of Stephen, a Greek name meaning 'crown' or 'garland'

If you didn't find a particular surname listed, there are a couple of websites which may be of help tracking down its meaning: www.ramsdale.org/surname.html and www.nameseekers.co.uk.

Definitions of Historic Occupations

On many occasions, you may discover an occupation listed on a civil registration certificate or census return that leaves you scratching your head as to what the job actually entailed. Here are some of the more obscure lines of work that kept our ancestors busy – if not always clean, satisfied, happy or safe. They have been drawn

Occupation Meanings

Accipitrary – An individual who catches birds of prey

Agister – An individual working as an official in a Royal Forest

Amanuensis – An individual taking diction or copying from manuscripts professionally

Armiger – An esquire, an individual who was entitled to bear heraldic arms

Accoucher / Accoucheuse – An individual assisting a woman giving birth, a midwife

Balister – An individual working as a crossbow man

Blemmere – An individual working as a plumber

Boniface – An innkeeper or landlord of an inn

Bunter – A female individual who would collect rags or bones

Biddy – A female servant, usually of Irish origin

Calciner – An individual who produced quicklime by burning bones

Carnifex – An individual working as a butcher

Cashmarie – An fishmonger selling fish inland

Couranteer – An individual working as a journalist or newspaper writer

Camerist – A chamber woman or lady's maid

Decretist – An individual with expertise in decretals (the nature of decrees)

Dexter – An individual working as a dyer

Drover – An individual who would drive herds of cattle

Duffer – An individual selling inferior-quality goods, often pretending they were more valuable then they were

Departer – An individual who would separate and refine metals

Earer – An individual working as a ploughman

Eremite – A recluse or hermit

Erite – A heretic

Estafette – An individual working as a mounted courier

Eyer – An individual who made the eyes in needles

Farandman – A traveller or stranger, usually working as a merchant

Farrier – An individual who shoed horses and worked as a horse doctor

Fiscere – A name for a fisherman

Fower – An individual working as a street cleaner

Ganneker – An individual working as an alehouse keeper

Gelder – An individual who castrated animals

Graffer – An individual working as a notary

Gummer – An individual who would enlarge the spaces between the teeth of a saw

Hayward – An individual responsible for guarding the fences or enclosures of the parish and ensuring cattle would not break through

Headborough – An individual working as the parish constable

Hellier / Hillier – An individual working as a slater or tiler

Huckster – An individual retailing small goods in petty shops, or a pedlar

Intendant – An individual in charge of or directing a public or government business

Intelligencer – An individual working as a spy or secret agent

Jagger – An individual working as a fish peddler or carrier

Jerquer – An official of HM Customs who examined ships and ensured that the duty had been paid

Jongleur – A roaming ministrel or entertainer

Justiciar – The head political and judicial officer of the Crown

Keeker – An official responsible for inspecting or overseeing a colliery

Keeler / Keelman – An individual working on a barge known as a keel

Kempster – An individual working as a comber of wool

Knacker – An individual working as a harness maker or saddler

Knoller – An individual working as a bell toller

Lapidary – An artificer who worked in cutting or engraving precious or semi-precious stones

Lattener – An individual working with or in the production of latten (a metal alloy)

predominantly from historic census returns, and if an occupation is not listed below, there are more definitions of historic occupations at www.genuki.org.uk and www.rmhh.co.uk/occup.html.

Leavelooker – A municipal officer inspecting food selling in markets

Leech – A physician or doctor

Lumper – An individual working as a labourer responsible for unloading cargoes of timber

Manciple – A steward or individual responsible for supplying provisions

Mango – A dealer in slaves

Mealman – An individual dealing in meal or flour

Mudlark – An individual who scavenged in the tidal river banks or sewers

Naperer – An individual responsible for the table linen in a royal household

Nimgimmer – A doctor or physician

Orrery maker – An individual making an orrery (a mechanical model displaying the movement of the earth and the moon)

Ostiary – An individual working as a doorkeeper of a church or monastery

Panter – An individual responsible for the pantry

Paperer – An individual employed to pack newly made needles into papers to bind them

Pargeter – An individual working as an ornamental plasterer

Pelterer – An individual working with animal skins

Pettifogger – A lawyer working in petty or small cases

Phrenologist – An individual assessing someone's character by examining the shape of their cranium

Poller – A hair cutter or barber

Quister – An individual working as a bleacher

Raker – A street cleaner

Ratoner – An individual working as a rat catcher

Rubbisher – An individual working in a quarry and separating the small stones

Sandesman – An envoy, ambassador or messenger

Scavenger – A street cleaner who was employed by the local parish

Schrimpschonger – An artisan who would carve bone or ivory

Seneschal – A high ranking steward for royalty or senior aristocracy

Skepper – A maker of skeps (baskets or hampers)

Spallier / Spalliard – A labourer working in tin-mining usually performing the more menial jobs

Sutler – An individual who sold provisions to soldiers and would reside in garrison towns

Tawer – A manufacturer of white leather

Thirdborough – A petty constable of a township

Tiger – An informal name for a boy working as a groom or pageboy

Tippler – An innkeeper or seller of alcohol

Tipstaff – A sherriff's official or bailiff or constable

Tonsor – Latin word for barber

Topman – A type of seaman

Topsman – The leading drover in charge of cattle

Trusser – An individual responsible for tying and bundling hay

Ulnagar / Alnagar – An official responsible for inspecting the quality of woollen goods

Vaginarius – A sheath or scabbard maker

Venator – A hunter

Verderer – An official of the Royal Forest

Villein – An occupier of land owned by the Lord of the Manor who paid dues to the Lord to be able to occupy it

Vulcan – A blacksmith or ironworker

Wabster / Wobster – A weaver

Wainwright – An individual building wagons

Wantcatcher – An individual catching wants (moles)

Wetter – A person employed in the printing process whose role was to dampen the paper prior to printing

Whacker – Another term for a drover

Willeyer – An individual employed to feed a willey machine (a revolving machine used in the textile industry)

Xylographer – An individual operaring the xylograph machine (a wood engraving machine)

Yeoman – A farmer in possession of his own piece of land

Zitherist – A player of the musical instrument, the zither

Genetic Genealogy

One of the major growth areas in family history research techniques is the use of DNA testing to prove genetic connectivity. The tests themselves have been available on the market for a number of years, and now the cost has fallen dramatically whilst the sophistication of the range of tests available has expanded. In addition, the power of the Internet to store and share data has, inevitably, opened up the possibility of sharing test results with other individuals, and a range of online communities has grown up. Furthermore, the creation of control groups using DNA samples from around the world has also made it possible for people to trace their ethnic roots. The collision of science and history is one of the more fascinating spheres of genealogy, and promises to be an area with the greatest potential growth in years to come.

DNA Testing

There are various tests available. The most popular are either tests for the Y-chromosome covering the paternal line, which can allow either paternity testing, linkage to a particular haplogroup (a distinct section of the population that shares common Y-chromosome features) or, when results are compared, the number of generations removed from a common ancestor; or mitochondrial DNA testing of the maternal lineage. This allows similar assessments to be made about connectivity to a halpogroup.

Further tests can be used to determine the ethnic origin of an individual, and even to begin mapping human population movements across the globe, provided sufficient numbers of DNA tests are carried out against sample populations. These tests have been around for a number of years, and have frequently been used in these various ways on television.

There are many companies, both UK and US, that offer DNA tests to individuals, and the range of test options is explained on their websites, depending on whether you want to prove a link between two living relatives (a paternity test, for example) or distant ancestors; an ethnic profile; or connectivity to a haplogroup. Prices vary from company to company, and some of the main ones are EthnoAncestry and Oxford Ancestors in the UK, and Sorenson Genomics in the US (whose test kits, prepared by their subsidiary company Identigene, are used by some of the newly emerging DNA social networking websites run by Ancestry and GeneTree, described in more detail below). Testing

CASE EXAMPLE

DNA testing

DNA technology was used to examine athlete **Colin Jackson's** ethnic background for Who Do You Think You Are?, and was of enormous help when focusing on the genealogical paper trail. We already knew that his family hailed from the Caribbean, but the proportion of sub-Sahara African DNA in his sample was quite low, being only 55 per cent. It was quite a surprise to discover that 38 per cent of Colin's DNA was of European extraction, which was traced to his Scottish great-grandfather who had a child with his Jamaican housemaid – and even more of a shock to discover that there was 7 per cent native Caribbean blood in his veins.

is quick and easy to do, and usually does not need a blood sample. Most require a simple swab sample to be taken from the inside of the mouth, sealed and sent back to the laboratory for analysis. You will then be sent your results in strict confidence; it is then up to you what you choose to do with them.

Social Network Sites

Just as people were keen to upload and share family trees and pedigrees via sites such as Genes Reunited, Familysearch, Ancestry and My Heritage, new social networking sites are emerging where you can upload your DNA profile, and the growing interest in the subject looks set to revolutionize the way we consider our genetic roots and blood ties with the past.

Amongst the current market leaders are two main platforms, GeneTree and DNA Ancestry (an offshoot of the main Ancestry site). Not only do they encourage you to create a personal profile and upload your DNA test results, but also they offer comparative datasets of genetic samples from around the world – allowing you to plot where your ethnic origins are, and how far back in time your ancestors moved around. Furthermore, they have both created the capacity to upload your family trees, providing researched documentary and historical context to support the scientific results. The key feature of both sites is to draw upon the power of social networking – sites such as Facebook and YouTube have shown how popular online interactivity is these days – and encourage 'self-connectivity'; individuals who have taken a DNA test are asked to submit their results, place them in their haplogroup, and allow them to be connected with other users via 'common' genetic ancestors, as well as showing where their original genetic make-up places them over time via technology mapping population movements. However, as with all social network sites, their success depends on numbers of users submitting information in this manner, in this case in terms of both DNA results and supporting family trees.

At present, each of these main sites has its own particular advantages. DNA Ancestry taps into Ancestry's existing worldwide network of research and uploaded family trees, for those who wish to submit their DNA to the site. However, the site has not acquired a DNA sample database against which to compare uploaded results, making ethnic connectivity much harder to achieve until critical mass is gained. By way of contrast, GeneTree has collected several hundred

KEY RESOURCES

Test kits are available from a wide range of companies online. Here are the web addresses of the ones mentioned above:

EthnoAncestry
www.ethnoancestry.com

Identigene
www.dnatesting.com

Oxford Ancestors
www.oxfordancestors.com

Genetic genealogy social networking sites are proliferating online. Here are the ones mentioned above:

GeneTree
www.genetree.com

DNA Ancestry
www.ancestry.co.uk/dna

thousand actual DNA samples from ethnic groups covering 170 countries worldwide, providing a far wider and more detailed DNA test base against which to compare results. This places them as the UK market leader, offering mapping services, personal privacy controls and self-management resources.

Other sites and services are appearing, however, so the advice is to shop around, find the site that most suits your needs – or whose interface and upload facilities you are most comfortable with – and sign up. Of course, the sensible thing to do would be to connect to as many as possible, to maximize your chances of finding a genetic match. Most of the sites are free, and contain privacy policies that you should examine carefully before putting personal – and genetic – information online.

National Archive Profiles
The National Archives (TNA)

Location:	Ruskin Avenue, Kew, Surrey, England TW4 4DU
Opening times:	Monday, Friday 0900–1700
	Tuesday 0900–1900
	Wednesday 1000–1700
	Thursday 0900–1900
	Saturday 0930–1700
	Closed Sundays and public holidays
	Annual stocktaking closure
	(see website for details)
Website:	www.nationalarchives.gov.uk
Contact email:	Record enquiry contact forms available online
	Departmental email contacts available online
Contact number:	Main switchboard +44 (0)20 8876 3444
Transport notes:	Free car park available
	Buses: nearest bus route R68
	Other local routes: 65, 237, 267 391
	Rail: nearest overground station Richmond
	Rail: nearest underground station Kew Gardens
Entry requirements:	Reader's ticket, valid for 3 years
	Application online or onsite
	Valid proof of ID required onsite to complete registration
	Passport/national ID card
	Driving licence

	Debit/credit card
	Photograph taken onsite, incorporated onto card
Facilities:	Disabled access, but contact archive in advance of visit (see website for details)
	Café
	Free Internet café
	Shop
	Exhibitions
	Online research guides to collections

Key Collections

The National Archives holds the records of central government for England, Wales and the United Kingdom in its various guises over the past millennium. TNA was formed in 2003 after the Public Records Office (PRO) and Historic Manuscripts Commission (HMC) were merged, and its collection of original documents dates back as far as the Domesday Book. While its main purpose is to preserve documents of national importance, this does encompass a vast amount of material relating to individuals from all over the United Kingdom, from Army service records to criminal records. Since the closure of the Family Records Centre (FRC) in March 2008, TNA has provided free access to English and Welsh civil registration indexes and census returns via its computer terminals, from which you can also access its subscriptions to numerous other genealogical databases.

National Archives of Scotland (NAS)

Location:	HM General Register House
	2 Princes Street, Edinburgh, Scotland EH1 3YY
Opening times:	Monday – Friday 0900–1645
	Closed weekends and public holidays
	See website for details
	No annual stocktaking
Website:	www.nas.gov.uk
Contact email:	enquiries@nas.gov.uk
Contact number:	Main switchboard +44 (0)131 535 1314
	Departmental contact numbers on website
Transport notes:	No car parking facilities
	Buses: numerous routes in city centre
	Rail: Edinburgh Waverley station within walking distance

Entry requirements:	Reader's ticket, valid for 3 years
	Personal application
	Photographic ID required, and proof of address
Facilities:	Two main sites: General Register House and West Register House
	No onsite café on either site but city centre food outlets nearby
	Disabled access, but contact archive in advance of visit (see website for details)
	Online research guides to collections

Key Collections

The National Archives of Scotland (NAS) is the central archive for Scottish governmental records dating back to the twelfth century as well as the repository for some Scottish business, estate, family and church records. However, some documents relating to Scottish history after the union of Scotland and England in 1707 have been deposited at The National Archives in Kew, such as military service records, some ship passenger lists and records of immigration. If you have Scottish heritage to investigate you may find the National Library of Scotland (NLS) useful, as this is where legal deposits, maps, newspaper collections and some manuscripts are held; this is also in Edinburgh. The General Register Office of Scotland is next door to NAS, and there is a specialist family history service that links the heraldic collections of the Lord Lyon.

National Library of Wales (NLW)

Location:	Aberystwyth, Ceredigion, Wales SY23 3BU
Opening times:	Monday – Friday 0930–1800
	Saturday 0930–1700
	Closed Sundays and public holidays
	Sometimes closed in severe weather (see website for details)
Website:	www.llgc.org.uk
Contact email:	Record enquiry contact forms available online
	Departmental email contacts available online
Contact number:	Main switchboard +44 (0)1970 532800
Transport notes:	Free car park
	Buses: station in the centre of Aberystwyth, 15 minute walk
	Rail: Aberystwyth train station next to bus station

Entry requirements:	Reader's ticket, valid for 5 years
	Registration online or onsite
	Two forms of ID required, one showing address
	Photograph taken onsite, incorporated onto card
Facilities:	Onsite café
	Disabled access, but contact archive in advance of visit (see website for details)
	Exhibitions
	Limited online research guides to collections
	Library shop

Key Collections

Although Wales does not have a national institution like Scotland or Ireland, the National Library of Wales acts as an equivalent centre, combining a national collection of books with original manuscripts, maps and copies of relevant family history material. In particular, it holds a large collection of Welsh parish registers on microfilm, along with copies of the national GRO indexes and census returns relating to Wales. It is the repository for diocesan will registers for Wales, along with Welsh assize material prior to 1831.

While TNA holds most of the records regarding the administration of Wales, the National Library of Wales has collections of importance relating to Welsh history including newspapers, maps, books, manuscripts, pictures, photographs and electronic resources. Therefore, if you are researching Welsh ancestry you may find it useful to visit both The National Archives in Kew and the National Library of Wales, depending on your ancestors' occupations.

National Archives of Ireland (NAI)

Location:	Bishops Street, Dublin 8, Ireland
Opening times:	Monday – Friday 1000–1700
	Closed weekends and public holidays
	Annual media closure
	(see website for details)
Website:	www.nationalarchives.ie
Contact email:	mail@nationalarchives.ie
Contact number:	Main switchboard +353 (0)1 407 2300
Transport notes:	No parking facilities but car park nearby
	Buses: numerous routes in city centre
	Rail: Pearse Station nearest station

Entry requirements:	Reader's ticket, valid for up to 3 years
	Registration onsite (form can be downloaded online)
	Photographic ID required
	Passport/national ID card
	Driving licence
	International student card
	Social security card
	Travel pass
	Employment ID
Facilities:	Limited disabled access, contact archive in advance of visit
	Limited online research guides to collections

Key Collections

The National Archives of Ireland hold records from government departments that detail the history of the modern Irish State. Established in 1988, it assumed the roles of the State Paper Office (created in 1702) and the Public Record Office of Ireland (formed in 1867). It is a great source for family historians researching Irish ancestors, as it is a repository for some of the few genealogical records that survived the fire at the Public Registry Office in Dublin in 1922. As a result of this fire, the majority of the National Archives' material now dates from the nineteenth and twentieth centuries.

Public Record Office of Northern Ireland (PRONI)

Location:	66 Balmoral Avenue
	Belfast, Northern Ireland BT9 6NY
Opening times:	Monday – Wednesday, Friday 0900–1645
	Thursday 1000–2045
	Closed weekends and public holidays
	Annual stocktaking (see website for details)
Website:	www.proni.gov.uk
Contact email:	proni@dcalni.gov.uk
	Record enquiry contact forms available online
Contact number:	Main switchboard +44 (0)28 9025 5905
Transport notes:	Free car park, limited space
	Disabled parking
	Buses: Metro Route 8 or 9
	Rail: Balmoral Halt nearest station

Entry ***requirements:***	Reader's ticket
	Registration onsite
	Photographic ID required
	Passport
	Driving licence
	Student photocard
	Photograph taken onsite, incorporated onto card
Facilities:	Disabled access, but contact archive in advance of visit (see website for details)
	Onsite café
	Online research guides to collections
	Exhibitions

Key Collections

The Public Record Office of Northern Ireland (PRONI) contains government departmental records, court, public bodies' and local authority records for Northern Ireland as well as some records deposited by private individuals and companies. It acts as a County Record Office for the whole of Northern Ireland and, as such, unlike any other national archive elsewhere in the United Kingdom, it has an immense collection of private, local and ecclesiastical records.

Useful Website Addresses

British Library: www.bl.uk

British Library Newspaper Library: www.bl.uk/collections/newspapers

Family Search/Family History Centres: www.familysearch.org.uk

Federation of Family History Societies: www.ffhs.org.uk

General Register Office (GRO): www.gro.gov.uk

General Register Office for Scotland (GROS): www.gro-scotland.gov.uk

General Register Office of Ireland (GROI): www.groireland.ie

General Register Office, Northern Ireland (GRO Northern Ireland): www.groni.gov.uk

Institute of Heraldic and Genealogical Studies (IHGS): www.ihgs.org.uk

Jewish Genealogical Society: www.jgsgb.org.uk

Principal Registry of the Family Division: www.hmcourts-service.gov.uk

Society of Genealogists: www.sog.org.uk

The Wellcome Library: www.library.wellcome.ac.uk

Acknowledgements

Book
Instrumental in preparing the text for this book were Laura Berry and Sara Khan, who have been actively involved in the genealogical research for *Who Do You Think You Are?* Both are trained historians; they have undertaken research work in archives up and down the UK and Ireland, and have distilled this knowledge into numerous chapters for this volume.

Television Series
Wall to Wall would like to thank the production team who weave the genealogical research with historical context and personal stories to create the incredible stories that make up the *Who Do You Think You Are?* series and to the celebrities who give so much of their time and trust to enable that to happen.

Picture Credits

Alamy/TNT magazine p170(t), Alamy/Michael Jenner p180; Sally Cole pp8, 345, 361; Ancestry.co.uk, part of The Generations Network pp78, 105; College of Arms p56; Corbis pp95, 254, 269, 448, Corbis/Bettmann pp153, 203, 292, Corbis/Hulton-Deutsch Collection pp173, 201, 249, 257, 287, 349, 369, 386, 397, Corbis/Underwood & Underwood pp245, 325, Corbis/Stefano Bianchetti p276; Crew List Index Project (CLIP) p224; Getty Images/Hulton Archive pp12, 178, 191, 209, Getty Images/Topical Press Agency pp297, 341; Mary Evans pp290, 307, 315, 343, 429, 436, 459, 480, Mary Evans/John Chaffin of Taunton p3, Mary Evans/William Hamilton's Indenture of Apprenticeship, 1839 p358, Mary Evans/Barnaby's p413; Federation of Family History Societies p69; Sarah Newbery p22; Photolibrary.com ppxii, 6, 496; Science & Society/Science Museum Pictorial p295, Science & Society/NMeM Daily Herald Archive p367, Science & Society/Sayle/Daily Herald p391; The Bridgeman Art Library/Giraudon/Musée des Beaux-Arts, Angers, France p115; The National Archives pp36, 41, 99, 101, 139, 143(l), 143(r), 170(b), 202, 183, 185, 186, 202, 218, 229, 238, 378, 379, 416; The National Archives of Scotland/www.nas.gov.uk pp38, 91; Topfoto pp72, 97, 122, 150, 216, 221, 235, 251, 328, 355, 435, 451, Topfoto/National Archives p452, Topfoto/World History Archive p381, Topfoto/Image Works p408, Topfoto/Balean p426, Topfoto/Public Record Office/HIP p443; Trustees of Genuki/www.genuki.org.uk p63; Wellcome Library London p402.

Images on pages 37, 101(t), 107, 116, 145, 169, 194, 215, 231, 261, 281, 298, 301, 313, 331, 342, 362, 376, 394, 405, 440, 444, 453, 478 and 479 reproduced with kind permission of Wall to Wall Media.

Picture research Sally Cole/Perseverance Works Ltd.

Index